MOTOR MANUALS

A SERIES FOR ALL MOTOR OWNERS AND USERS

VOLUME I

AUTOMOBILE ENGINES

IN THEORY, DESIGN, CONSTRUCTION,
OPERATION, TESTING & MAINTENANCE

BY

ARTHUR W. JUDGE

Associate of the Royal College of Science, London; Diplomate of Imperial College of Science and Technology (Petrol Engine Research); Whitworth Scholar; Associate Member of the Institution of Mechanical Engineers; Associate Fellow of The Royal Aeronautical Society; Member, Society of Automotive Engineers (U.S.A.)

Fifth and Revised Edition
Reprinted

First Edition,   1925
Second Edition, 1931
Third Edition,   1937
Reprinted,   1940
Fourth Edition, 1942

Co., Ltd.,

PREFACE TO FIFTH EDITION

The necessity of another reprinting of this book has

ation in
)f a new
ients in
10d has
ie publi-
lt times
in much

1 on the
ver out-
: design
hamber
r maxi-
valves,
, crank-
intings,
, crank-
ts.
ind des-
type is
w types
referred
turbine
various
ogether
hat will

DGE.

CONTENTS

CHAPTER I

THE COMBUSTION PROCESS IN THEORY AND PRACTICE

Introduction.—Modern mechanical transport on land, sea and in the air has been made possible only by the development of the steam and internal combustion types of engine. The use of steam engines is now confined to the heavier class of transport and includes steamships, locomotives and steam road vehicles, tractors and rollers, although motor-cars propelled by steam engines have attained a considerable degree of success in the past, and there are still one or two satisfactory makes of steam car in service in America. The Doble steam motor-car, externally indistinguishable from a petrol engine type, silent in operation, extremely flexible and economical in fuel, represents the modern development of this type of engine.

Although the present volume is confined to that extensive class of engine designated the internal combustion engine, yet a few remarks on the steam type may not be out of place here, if only to keep before the reader the past successes and the potential possibilities, even to-day, of this engine. Invented by James Watt towards the end of the seventeenth century for stationary engine purposes, the steam engine was apparently first applied to road locomotion (or automobile) purposes by a French engineer, Cugnot, in 1769, when a four passenger vehicle attained a speed of 2½ m.p.h. for a period of about 15 minutes—the time corresponding to the capacity of the small boiler. Cugnot, a year or two later, built, with the help of the French Government, a steam tractor for artillery haulage purposes; it was designed to haul a load of 4½ tons, but was not used owing to the French Revolution.

A*

9

Adverse popular opinion, the prejudice of stage-coach interests and legislation (the tolls on steam cars were ultimately raised to £2 against the three shillings for stage-coaches) practically killed the steam auto-mobile until in 1896 conditions changed. Prior to this no less than 54 Bills were introduced into Parliament on the subject of road vehicle taxation, and the mechanically propelled vehicle was only allowed to proceed along the road under the guidance of a man, who preceded it, carrying a red flag; its maximum speed, therefore, could not exceed about 4 m.p.h. The passing of the Locomotives on Highways Act in 1896 removed these restrictions, and real progress in automobile development has continued from this date.

In 1902 a Serpollet steam car attained a speed of 75 m.p.h. at Nice, thus beating the record of 65.79 m.p.h. previously made by Jenatzy on an electrically-driven car in 1899. Serpollet's record was subse-quently beaten several times by petrol cars, but in 1906 the latter's records were swept off the board by those set up by a Stanley steam car, which attained a speed of 127.5 m.p.h. on the Ormonde Beach sands.   In 1909 a huge Benz petrol-engine racer raised this figure to 140.9 m.p.h., and from this period the steam car has lapsed into obscurity as a speed record breaker, although it is but fair to add that the subsequent success of the steam car in hill-climbing competitions was so pronounced that ultimately it was debarred by the Automobile Clubs from competing.   Here we must leave it and turn to its confrère, the petrol-type engine, but in passing it should be added that there is little doubt that if all of the design, skill, research work and energy which has been devoted to the latter had been similarly applied during the past 15 to 20 years to the former, a different state of affairs might now exist.

Types of Combustion Engine.—The steam engine may be termed an "external combustion" engine, for the fuel (coal or heavy oil in this case) is burnt in a separate boiler, and the steam generated at a pressure is led into the cylinder of the steam engine.

In the case of the "internal combustion" engine, which comprises all types of explosion engine—petrol, heavy oil, gas and Diesel types—the fuel is burnt, or exploded within the engine itself.

Internal Combustion Engines.—An examination of the history and development of this type reveals the interesting fact that it is possible, although not always practical, to employ a very wide range of solid, liquid and gaseous fuels, in conjunction with air, or oxygen, as the explosive medium. Thus the earliest authentic record credits Huyghens, in 1680, with the invention of an engine in which gunpowder was exploded in order to drive the air out of a cylindrical vessel, closed at one end and fitted with a piston in its band. The result of forcing the air out, and of the cooling down of the combustion products, caused a partial vacuum within the cylinder, so that the piston was forced into it by the external atmospheric pressure. This inward travel of the piston was utilized to raise weights and to work mechanisms. Coal dust is another fuel which is used for internal combustion engines and much experimental work has been in progress with this type of engine in recent years.

Hydrogen gas was employed by Cecil in 1820 as the exploding agency, whilst Lenoir—to whom we owe a great debt for his very marked progress—in 1860 constructed several hundreds of engines using coal gas. These engines belonged to the "non-compression" class, and were very wasteful in fuel—using about seven times as much as a modern gas engine of the same power. It is interesting to note that electrical ignition was employed on these engines

In 1876 Otto invented his famous "Otto-cycle" engine, in which the charge of gas and air to be exploded was compressed before it was ignited. This pre-compression of the charge is the fundamental secret of success of the modern automobile and other engines; by compressing the charge a very consider-able increase in the explosive effort (and therefore the power output of the engine) and a marked reduction in the quantity of fuel used, result. Gaseous fuels applicable to internal combustion engines include hydro-gen (H), carbon monoxide (CO), acetylene (C_2H_2), coal-gas (a mixture of various combustible gases such as hydrogen, marsh-gas, carbon monoxide and hydrocarbons), water-gas (carbon monoxide and hydrogen mixture), methane, butane, blast-furnace, producer and other gaseous mixtures. Liquid fuels include the heavier tarry products obtained after the fractional distillation of coal and of petroleum; these fuels are employed in heavy-oil compression-ignition and Diesel type engines. The lighter fuels include petrols, paraffins, benzole, alcohol and similar fuels of a spirituous nature.

Generally speaking, any fuel which contains either hydrogen (H) or carbon (C) is capable of combustion in oxygen; this fact, well known to chemists, accounts for the wide range of fuels available for internal combustion engines as we have seen.

The Process of Combustion in Engines.—It is as well, at the outset, to have a clear idea of the process of combustion; this constitutes the basic principle of all internal combustion engine operation.

Combustion is simply the name given to the process of chemical combination of a combustible, such as any of the fuels previously mentioned, and oxygen; in this process, heat is evolved, and this heat is utilized in expanding the gaseous products of combustion; the expansive effort of these gases is utilized in engines to give the " working " or " power " stroke.

Confining our remarks to the case of liquid fuels, these as we have seen are organic substances composed of the chemical elements, carbon and hydrogen (often with the addition of oxygen). These elements are usually defined by the capital letters C H and O respectively, and the composition of the fuel is expressed by a chemical formula including these.

Thus, the hydrocarbon fuel *hexane*, which is the principal constituent of pure petrol has the formula C_6H_{14}. This formula expresses the fact that one molecule of hexane contains 6 atoms of carbon and 14 of hydrogen. Since the atom of carbon weighs 12 times that of hydrogen,[*] it follows that the carbon and the hydrogen weigh 6 × 12 (or 72) and 14 × 1 (or 14) respectively. Expressed as percentages, these weights are 83.6 and 16.4 respectively, so that one pound, say,

* The oxygen atom is 16 times, and the nitrogen atom 14 times, as heavy as that of hydrogen.

of petrol, of this type, contains 0.836 lb. carbon and 0.164 lb. hydrogen.

The carbon, in combustion with oxygen, forms carbonic acid gas, denoted by the formula CO_2; the hydrogen forms water or H_2O. The combustion products are therefore carbonic acid gas and water (or steam) in the case of petrol and air combustion. The full statement, in chemical language, of the process is as follows : —

$$C_6H_{14} + \left(6 + \frac{14}{4}\right) O_2 = 6CO_2 + \left(\frac{14}{2}\right) H_2O$$

$$\text{or } C_6H_{14} + \frac{19}{2} O_2 = 6CO_2 + 7H_2O$$

Interpreting this relation in ordinary language, 1 cubic foot of petrol vapour will require 9½ cubic feet of oxygen to burn it completely, and the resulting combustion products (or exhaust gases) consist of 6 cubic feet of carbonic acid gas and 7 cubic feet of water vapour.

It can be shown from this result that 1 lb. of petrol will require 15.2 lb. of air* to explode it completely, a fact which is of importance to carburettor designers and users of automobile engines.

Benzole requires less air (about 13.3 lb.) and alcohol also requires less air (about 9.0 lb.), per lb. of fuel, for complete combustion.

Detonation.—If pure oxygen is used in a petrol engine instead of air (which is a mixture consisting of 76.8 per cent. nitrogen and 23.2 per cent. oxygen, by weight), the explosion will be almost instantaneous and very violent; exceedingly high pressures will result, and unless the engine is specially designed to withstand these, disastrous results may occur.

Again, if acetylene gas be used in a petrol engine, the explosion is also most violent; many engine cylinders have been wrecked by inexperienced individuals experimenting with acetylene.

In both cases this extremely rapid and violent explosive effect is termed *detonation*.

* 1 lb. of air contains 0 768 lb. nitrogen and 0.232 lb. oxygen.

Fortunately, the atmospheric air contains the neutral or inert gas, nitrogen, in the proportion of nearly $\frac{4}{5}$. This gas will not burn with oxygen, and its presence damps down this detonation tendency. It is possible, in the case of petrol engines, to obtain detonation by using too high a compression with a given fuel. Thus if the compression pressure is much over 90 lb. per sq. in. when paraffin is used, detonation occurs. For petrol of the modern high octane commercial type, benzole and pure alcohol, the limiting compression pressures are approximately 170, 180 and 200 (min.) lb. per sq. in., respectively. An engine with a badly-designed combustion chamber or one which normally runs " hot," will begin to detonate at lower compressions than these.

Although many theoretical explanations have been advanced to account for the phenomenon of detonation, it is only more recently that experimental investigations have given a more satisfactory indication of the causes. Thus it has been established that detonation occurs when the rate of inflammation or burning of the mixture exceeds a certain critical value; the latter depends upon several factors including the nature of the fuel used, compression ratio, combustion chamber design, ignition timing, etc.

It is now believed that the earlier stages of the burning of the charge after the spark occurs are normal in character, but that towards the end of the combustion process, namely, at about the last 25 per cent. of the flame travel in the combustion chamber, a sudden burning effect of the remaining part of the charge occurs, which gives rise to a very rapid rate of pressure rise which produces impact effects on the combustion chamber walls, usually termed detonation or knock.

That this explanation is substantially correct is indicated by the fact that it is possible to design combustion chambers based upon the results of its application which enable higher compression ratios to be employed with a given grade of fuel without detonation, than are possible in ordinary engines.

Pre-Ignition.—Another cause of engine knocking, which is quite distinct from detonation effects, since it *always occurs before the ignition spark*, whereas the latter always occur afterwards, is that known as pre-ignition. It is usually due to the inefficiency of the engine cooling arrangements, so that all of the surplus heat cannot be disposed of, with the result that the temperature within the combustion chamber increases progressively until some projecting part, such as the electrodes of the sparking plug or a protruding piece of carbon deposit, becomes so hot that it ignites the charge before the spark occurs at the sparking plug. The piston is actually ascending when this happens and it therefore receives a retarding force. Generally, once pre-ignition commences, its effects become more and more enhanced with the result that the engine loses power and eventually stops.

The most common causes of pre-ignition are excessive carbon-deposit and running with the ignition too far retarded, thus causing a gradual increase in the cylinder temperature which the engine cooling system will not cope with. Another cause is that of sparking plug electrodes protruding beyond the inside surface of the combustion chamber.

Fuels.—Before passing on to a consideration of the mode in which this process of combustion of a fuel with air is applied and utilized to its best advantage in modern engines, it will be necessary to say a few words concerning the principal properties of suitable fuels.

The best fuels are those which contain the highest proportions of hydrogen in their composition, since this latter element has the greatest heating value. By *Heating* or *Calorific Value* is meant the heat generated during combustion of the fuel with air or oxygen. It is usually defined as the total number of heat units (in this country, British Thermal Units*) evolved during the combustion of a given weight of the element or fuel.

In the following table is given the principal properties of various fuels. Hydrogen has heating value of 62,000 B.T.U.'s.

---

* The British Thermal Unit is the quantity of heat required to raise the temperature of 1 pound of water through 1 degree Fahrenheit — 778 foot pounds in work units.

TABLE No. 1.
Properties of Fuels.

Name.	Chemical Symbol.	Heating Value. B.T.U.'s per lb. (Lower.)*	Density. Wt. per Gallon at 15° C.	Maximum Compression Pressure which can be used.	Fuel Consumption† in Modern Engine of Efficient Design, with 5.1 Compression Ratio.
			lbs.	lbs. per sq. in.	lbs. per I.H.P. hour.
Light Petrol (Aromatic, F r e e, or Aviation)	C_6H_{14}	19,100	7·18	105·5	0·415
Medium Petrol ...	Mixture	19,000	7·27	118·0	0·421
Heavy Petrol ...	Mixture	18,790	7·67	140·5	0·425
Paraffin (Illuminating Oil)	Mixture	19,000	8·13	86·0	0·523
Benzene ...	C_6H_6	17,300	8·84	180	0·458
Alcohol (Ethyl) ...	C_2H_6O	11,470	7·98	> 204	0·663
Methylated Spirits ...	Mixture	10,200	8·21	163·5	0·740
Ether...	—	—	7·35	47·5	—
Carbon	C	14,540	—	—	—
Hydrogen	H	52,500	Gaseous	—	—
Carbon Monoxide ...	CO	4,329	,,	—	—
Ethylene	C_2H_4	21,350	,,	—	—
M e t h a n e (Marsh Gas)	CH_4	23,510	,,	—	—

* This is the total heat of combustion, less the latent heat of the steam formed by the combination of hydrogen in the fuel to water vapour.

† Values obtained from Ricardo's variable compression engine tests.

From the motorist's point of view, the best fuel at a given price per gallon is the one giving the greatest density with a good heating value combined with a high octane rating. For a compression pressure below 90 lb. sq. in., paraffin would be the best from this viewpoint.

Petrol.—This is a light highly volatile spirit obtained from petroleum by distillation. Its weight varies from 7 to 7¾ lb. per gallon. It contains about 84 and 16 per cent., respectively, of carbon and hydrogen. About 15 parts, by weight, of air to 1 of petrol are required for complete combustion, but modern engines will work on mixtures of from about 9 to 1 (rich in petrol) up to about 20 to 1 (weak in petrol). Inflammable vapour is given off at ordinary atmospheric temperatures. Petrol attacks most varnishes, and dissolves rubber, oils and fats.

Highest Useful Compression Ratio.—Most commercial petrols are mixtures of other fuels belonging principally to certain groups or series, known as Paraffins (these include pentane, hexane and heptane), Naphthenes, Olefines and Aromatics.

Each of these series confers certain properties upon the petrol. Of these, the more important is that of the Aromatic Series (including benzine, toluene and xylene) which enables the working compression ratio of the engine in which it is used to be raised.

For each series of fuels there is a limiting compression ratio, known as the *Highest Useful Compression Ratio* (H.U.C.R.) above which the engine will not work satisfactorily owing to detonation effects.

For aromatic petrols the H.U.C.R. ranges from about 6 to 7, whilst for petrols containing paraffin and naphthene series fuels the value seldom exceeds 5·5.

The petrols containing aromatic fuels usually have a density of ·74 to ·76, and will enable compression pressures up to 150 lb. per sq. in. to be used, whilst with pure aromatic fuels compression pressures of at least 180 lb. per sq. in. are possible.

Values of the highest useful compression pressure are given in the last column but one of Table No. 1.

Benzole.*—The commercial benzole is a mixture of benzene (C_6H_6) with other spirits such as toluene and xylene. The 90 per cent. grade used in automobile engines boils at 80° C. to 100° C., and weighs about 8·7 to 8·9 lb. per gallon at 15° C. At higher temperatures it weighs less. It has a heating value of 16,700 to 17,500 B.T.U.'s per pound (lower value). Its principal advantage when used in petrol engines is that it has a higher detonation pressure, which enables the engine to maintain its power better on hills or under loads, when it otherwise would "knock" and overheat with petrol. A higher compression pressure

* A mixture of equal parts of benzole and petrol, known as "50/50 Mixture," possesses advantages over pure petrol in the matter of fuel economy and freedom from "knocking."

can be used, with a greater power output, and there is a better fuel economy than with petrol. In the case of most motor-cars the mileage per gallon is about 15 to 20 per cent. better than with petrol. Owing to the fact that benzole is heavier by about 15 per cent. than petrol, it becomes necessary to weight the carburettor float, otherwise the jet level will be lowered. A smaller jet can be used with benzole. Benzole requires about $13\frac{1}{2}$ lbs. of air per lb. for complete combustion, which is rather less than for petrol; it has a mixture range of about 11 to 1 to 19 to 1, and with the richer mixtures is apt to deposit much carbon and to gum up the valves. Benzole attacks most varnishes strongly, so that varnished cork floats (which have been used in certain American cars) should not be used, and care should be taken not to spill any of the spirit on the coachwork of automobiles.

Paraffin.—This fuel is sometimes used in low compression petrol engines, but for the best results an efficient vaporizing and mixture heating device is required. The heat of the exhaust gases may be used for this purpose. Paraffin is heavier than petrol (8·10 to 8·2 lb. per gallon). When used in petrol engines, it is generally necessary to have a petrol supply for starting purposes. Its heating value is about 19,000 to 19,500 B.T.U.'s per lb., and inflammable vapour is given off at about 80° Fahr. (26° C.).

Alcohol.—This spirit is obtained from vegetable and organic sources, such as sugar- starch- and cellulose-containing products. Sugar-beet, mangolds, potatoes, artichokes, grain, molasses, and mahua flowers are amongst the more important alcohol sources.

Alcohol has the chemical formula C_2H_6O, and in the pure state has a density of 7·95 lb. per gallon, and heating value of 12,900 B.T.U.'s per lb. (higher). For industrial and fuel purposes it is necessary to render alcohol unfit for human consumption. This is accomplished by adding a denaturant to make it unpalatable. The denaturants used include crude wood spirit, petroleum naphtha, pyridin, and benzene. Methylated spirits usually consist of ethyl alcohol (80 per cent.), wood spirit (10 per cent.), petroleum naphtha (0·5 per cent.), and water (9·5 per cent.).

When benzole is used as a denaturant it increases the heating value; thus a 50 per cent. alcohol-benzole mixture would have a heating value of about 15,300 B.T.U.'s per lb.

Alcohol can be used in petrol engines, although to obtain the best results, considerably higher compressions are necessary, namely from 160 to 200 lb. per sq. in. Less air is required for combustion, namely about 9 lb. to 1 lb. of pure alcohol, so that a larger jet should be employed. It is also necessary to heat the mixture in the carburettor or induction pipe, as alcohol-vapour and air are not combustible below about 20° C.

Alcohol blended fuels are now available commercially, a typical one being Cleveland Discol. Many such fuels are a mixture of petrol, benzole and alcohol.

Anti-Detonation Fuels.—Fuels, such as petrols and paraffins, are known to detonate, when the compression pressures used exceed a certain limiting value. By adding a small quantity of another special chemical or fuel it is possible to raise the compression pressure considerably without experiencing detonation. Thus if about 1 to 2 per cent. of aniline be added to petrol, very much higher compressions can be used. Ethyl iodide, xylidine, and tetra-ethyl lead are also effective in this respect; these liquids are termed " stabilizers," and their solutions with petrol are known as anti-knock fuels.

Ethyl petrol as sold commercially contains one part tetra ethyl lead in 1,300 parts of petrol. It enables an appreciably higher compression to be used.

Octane Value of Fuels.—Considerable research has been undertaken with the object of improving petrol engine fuels, to enable higher compressions to be used without detonation occurring.

New blends of fuels made by special refining methods and synthesis have been produced, whilst the performances of engines using ordinary fuels have been improved by adding stabilizers, or " anti-knock " constituents to them.

In order to compare these fuels from the point of view of their anti-knock qualities it is now the custom to use a standard fuel, having two constituents, viz.,

Iso-octane and *Heptane*; both are hydrocarbon fuels. The former is a fuel having a higher anti-knock value than all ordinary fuels used in modern petrol engines. The latter, on the other hand, is a very poor fuel of bad " knocking " properties.

For this reason iso-octane is given the arbitary rating of 100, and heptane, 0. Any other fuel will have an *Octane Value*, intermediate between these two fuels.

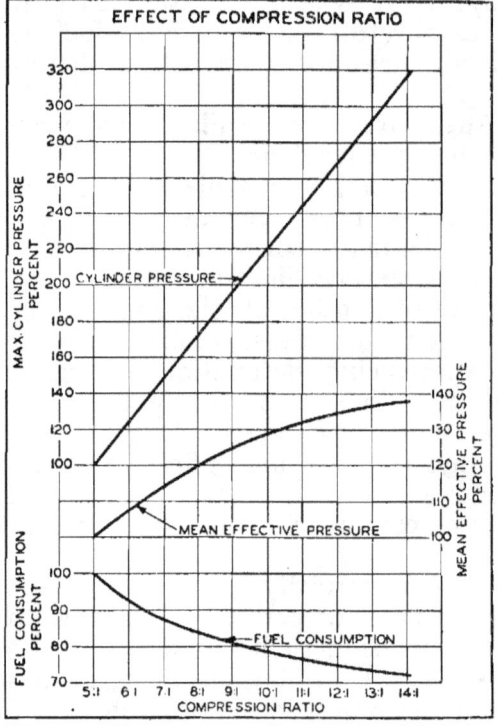

Fig. 1.—Performance Curves for Modern Petrol Engines.

The octane value of any fuel represent the percentage of iso-octane in the m i x t u r e of heptane and iso-octane which has similar knocking properties—when tested in a standard engine—to that of the fuel.

The usual octane values of commercial m o t o r petrols r a n g e from 60 for the cheaper petrols to 75/77 for the anti-knock petrols, such as Ethyl petrol. Modern aircraft engines operate with fuels of octane values from 80 upwards; for starting or take-off purposes fuels of 100 octane rating and even higher values are used.

Advantages of High Octane Fuels.—If the octane value of a fuel is increased the compression ratio can be raised and for each octane value there is a corresponding compression ratio that can be employed before detonation begins. In this connection the following relationship between the octane value and the highest

useful compression ratio has been determined experimentally : —

Octane Value ...	50	60	70	80	90	100
Highest Useful Com-pression Ratio...	5:1	5·3:1	5·7:1	6·2:1	6·8:1	7·7:1 to 8·0:1

As the compression ratio is increased the power output
is raised, since the engine develops a higher brake
mean effective pressure.

Another result of using a higher octane fuel, with
its corresponding highest useful compression ratio is
that, owing to the better thermal efficiency, the fuel
consumption per B.H.P. per hour is reduced.

These results are illustrated graphically in Fig. 1,
which shows the results obtained from a modern design
of petrol engine, using fuels of various octane values
which enabled the compression ratios shown in the
graph to be employed without detonation.

It will be noted that the maximum cylinder pressures
increase more rapidly than the mean pressures and since
the weight of the engine is largely dependent upon the
maximum allowable cylinder pressure, a limit is set,
in practice, to the highest compression ratio that can
usefully be employed.

It is considered to be more satisfactory to use high
octane fuels of 90 and above in supercharged engines
in order to obtain maximum power output without
unduly high cylinder pressures and this is done in most
aircraft engines.

Obtaining Power from Fuels.—We have seen that
every fuel, when exploded or burnt with oxygen
or air, is capable of generating a very large amount
of heat in proportion to its bulk.   Now it is the
primary object of internal combustion engines to
utilize this heat to the best advantage, and for this
purpose it is necessary to explode the fuel-air
mixture in a closed vessel, or cylinder, fitted with a
close-fitting plug, or piston, capable of sliding backwards and forwards as shown diagrammatically in
Fig. 2.   Let us suppose that the piston shown is a heavy
one, and at first is near to the bottom of the cylinder.
If, now, a mixture of petrol vapour and air in the
proportions of 1 part to 15 parts be introduced

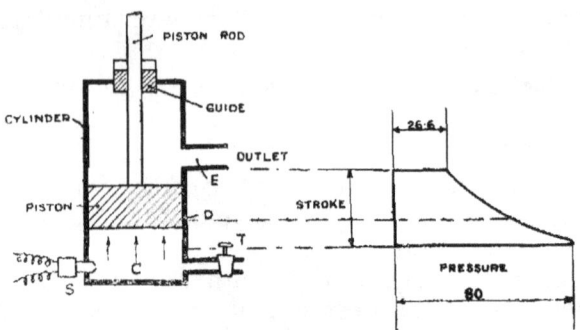

Fig. 2—A Simple Type of Engine.

through the tap T into the chamber C, the tap turned off, and the mixture exploded by causing a high voltage spark to pass between the ends of two wires, or electrodes S, a relatively large amount of heat will be developed. This heat raises the temperature of the gases formed during combustion, namely carbonic acid gas and water vapour, and causes them to expand. This expansion will tend to raise the weight of the piston, if it is not too heavy, forcing it upwards, until its lower edge uncovers the outlet orifice E, when the gases escape into the atmosphere. The piston will then fall, and in doing so will compress the remaining exhaust gases in the cylinder.

If the piston has a diameter of, say, 3 inches, and the mixture is at atmospheric pressure before explosion, the instantaneous pressure generated when the spark passes will be about 80 lb. per sq. in., which is equivalent to a total force (= pressure × area of piston) of 570 lb. As the piston moves upwards the pressure diminishes until, when the orifice E is uncovered, it falls to atmospheric value (14·7 lb. per sq. in.).

Let us now consider what the pressure, within the cylinder, will be when the piston has moved upwards to any given position. There is a physical law, known as Boyle's law, which states that the pressure of a gas at a given temperature varies inversely as its volume; if the volume is doubled, the pressure will be halved, if trebled, it will be reduced to one-third, and so on.

If we can keep our cylinder warm, it is easy to estimate what the pressure will be at any piston position. If the volume of the cylinder under the piston is, say, 1 cubic foot in the position shown in Fig. 1 and 3 cu.ft. when E is about to be uncovered, then since as we have seen the explosion pressure is 80 lb. per sq. in., this value will drop at E to one-third, that is 26·6 lb. sq. in. In an intermediate position, such as at D, where the volume is 2 cu. ft., the pressure will be 80 × $\frac{1}{2}$ = 40 lbs. per sq. in. The manner in which the pressure varies throughout the piston's stroke is shown graphically in the R.H. diagram, the lengths of the horizontal lines representing the pressures.

The area of such a pressure-stroke diagram represents the work accomplished by the explosive mixture: it is from such diagrams, known as *Indicator Diagrams*, that the power developed can be ascertained. It is only necessary to measure the average pressure length of the diagram, and to multiply by the length of stroke in order to obtain the mechanical work done during the piston's stroke. An example will serve to make this clear.

Example.—If the piston in Fig. 2 has a diameter of 3 inches, and an effective stroke of 3 inches, estimate the average work done per explosion, assuming the mixture before explosion to occupy 3 inches depth of the cylinder. If the piston makes 1,000 such strokes every minute estimate the horse-power.

$$\text{The piston area } = \pi\frac{(3)^2}{4} = 7\cdot06 \text{ sq. in.}$$

The initial pressure is 80 lb. per sq. in. and the final value 40 lb. sq. in., as we have stated before.

The average pressure measured from a diagram similar to that shown in Fig 2 is about 58 lb. sq. in.

Hence the work done per stroke

= Total force on Piston × Stroke.
= 58 × 7·06 × $\frac{3}{12}$ ft. lb.
= 102·4 ft. lb.

The effort exerted during the explosion and expansion of the piston would be sufficient to raise a weight of 102·4 lb. through a distance of a foot.

Now power = rate of doing work, i.e., work done per unit time, and 1 horse-power = work done at the rate of 33,000 ft. lb. per minute.

In the present example the work done per minute = 1,000 × 102·5 ft. lb.

Hence the equivalent horse power $= \dfrac{1000 \times 102\cdot5}{33,000} = 3\cdot1$

In passing it should be mentioned that in practice the hot gases would cool down as the piston moved upwards, due to the piston and cylinder metal extracting some of their heat, so that the expansive force would be less than we have estimated.

Application to Engines.—We have now shown that the heat energy of combustion of a fuel with air is capable of being employed for doing useful work, although our example was a somewhat crude one.

Otto and Langen, in their famous gas engine of 1866, utilized this energy of combustion in the following

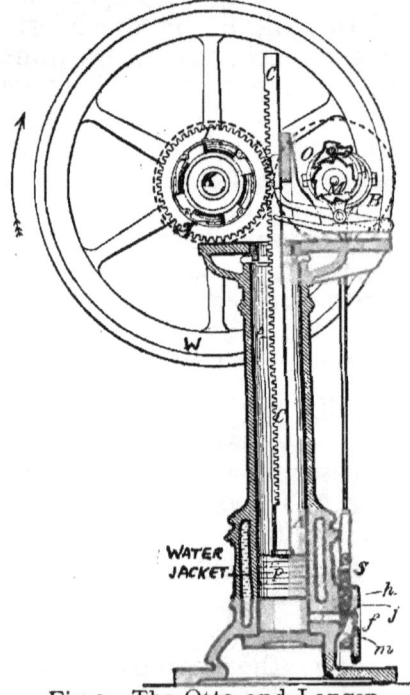

Fig 3—The Otto and Langen
Engine of 1886.

manner. They employed a long vertical cylinder, as depicted in Fig. 3, fitted with a piston P, the rod of which consisted of a toothed rack C, engaging with a gear wheel I attached through a ratchet device to a flywheel W on the shaft K. A mixture of gas and air was admitted to the cylinder through the lower R.H. passage i, by means of a slide valve S worked by an eccentric O and rod. This was simply a device to admit the gas and air mixture at the proper time. The mixture in the cylinder was ignited by

means of a flame f in the cover of the slide
valve. The explosion forced the piston upwards, and
during its passage the rack C engaged with the
toothed wheel T, but the latter, in virtue of its free-
wheel device, did not turn the fly-wheel. The power
stroke in this case was the downward, or return one,
due to the weight of the piston and rack, together
with the atmospheric pressure on top of the piston
(there being a vacuum underneath, due to the cooled
exhaust gases). The ratchets in the free-wheel engaged
during the down stroke, so that the piston rack turned
the fly-wheel in the direction shown. The explosion
was very violent and there was a good deal of noise and
rattling of the gear, but nevertheless many of these
engines were installed in this country and in France
and worked quite satisfactorily.

Similar to Lenoir's gas engine of 1860, this type
belonged to the non-compression class, in which the
explosive charge was ignited at atmospheric pressure.
All modern engines work on the *compression* principle,
the charge being compressed to several atmospheres
pressure before ignition. The result is a very considerable increase in power, reduction in size of engine,
more rapid working, and much better fuel economy.

The Otto Principle.—The majority of modern engines
work on a principle of operation first enunciated by
M. Beau de Rochas (in 1862) and applied by Otto, in
which compression of the charge was the important
feature. In order to understand this principle it will
be necessary to outline briefly the mechanism of a
simple type of engine in order to familiarize the
reader with the common automobile engineering terms
employed, and to show how the method is carried out.
Fig. 4 represents in outline the mechanism of the
simple one-cylinder type internal combustion engine.
It consists of a cylinder, having a closed end, in which
a hollow piston can slide inwardly and outwardly,
by a fixed amount, termed the *Stroke*. In order to
define, or limit the stroke of the piston, and also to
convert the reciprocating movement of the latter into
the more convenient form of rotary motion, the
piston is provided with a hinge-joint and pin; the latter
is known as the *Gudgeon, Piston or Wrist Pin*. A rod,

having pin bearings at each end, known as the *Connecting Rod*, connects the piston with a cranked arm, or *Crank*, attached to a shaft (*Crankshaft*) capable of rotating in fixed bearings A fly-wheel is usually attached to the crankshaft, in the case of one and two cylinder engines, for a specific reason to be mentioned later. It follows that if the crank is rotated in the direction shown, it will cause the piston to reciprocate in the cylinder, the joints at the ends of the connecting-rod obviating any constraint.   Similarly if the piston is reciprocated, it will cause the crankshaft to turn.

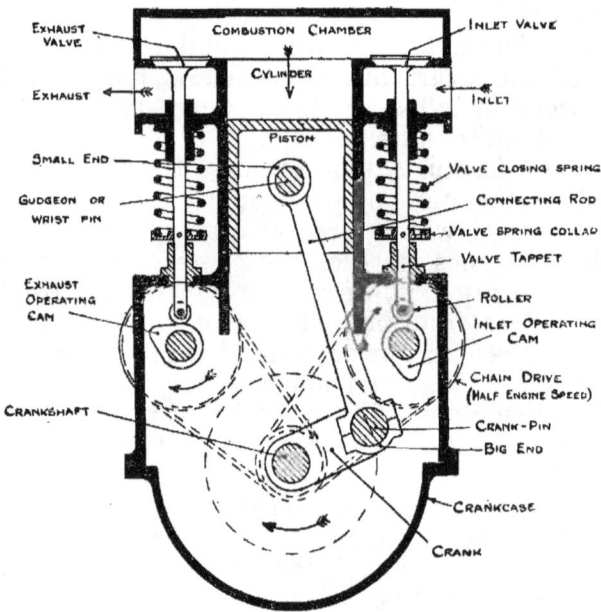

Fig. 4.—Illustrating the components of the petrol type engine and their names.

In the Otto method of working, the piston receives one power impulse stroke once every two revolutions of the crankshaft, that is, once every four strokes. For this reason the Otto cycle is known as the *Four Stroke* or *Four Cycle* principle.   During the other three cycles the momentum of the fly-wheel has to carry the connecting-rod and piston through their movements.   There is thus one useful and three idle strokes per cycle.   Having made this clear, we can now consider a complete cycle in more detail, and follow each successive stroke, commencing with the one termed the *Suction Stroke*, and illustrated in Fig 5 (*a*). In the head, or *Combustion Chamber* of the cylinder, two orifices or ports are arranged; each orifice can be completely sealed from the inside by suitable discs, termed Valves. One valve, known as the *Inlet Valve*, regulates the admission of the explosive mixture, or charge, whilst the other, or

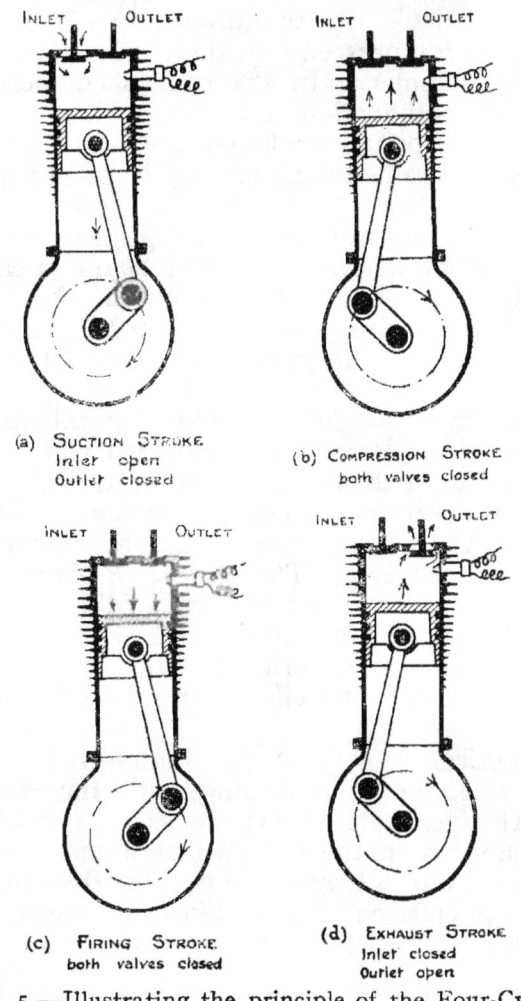

(a) SUCTION STROKE.
Inlet open
Outlet closed

(b) COMPRESSION STROKE
both valves closed

(c) FIRING STROKE.
both valves closed

(d) EXHAUST STROKE
Inlet closed
Outlet open

Fig. 5.—Illustrating the principle of the Four-Cycle Engine.

Exhaust or *Outlet Valve,* controls the exit of the burnt or exhaust gases. These valves are operated by simple mechanical devices.

In Fig. 5 (*a*) the piston is moving downwards, and the inlet valve is open. The suction created by the piston draws the explosive charge into the cylinder. When the piston reaches its lowest position (or *Outer Dead Centre*), the inlet valve is arranged to shut down on its seating, thus sealing the cylinder contents. On the upward stroke, Fig. 5 (*b*), the charge is compressed into the combustion chamber end of the cylinder. In the case of modern petrol engines the combustion chamber volume is about one-sixth of the piston stroke volume, and the compression pressure attained is about 145 lb. per sq. in. prior to ignition.

Just at the moment when the piston reaches the upper end of its stroke, an electric spark is caused to pass inside the combustion chamber at the *Sparking Plug.** The resulting explosion forces the piston downwards, the burnt gases expanding at the same time. This is the *Firing Stroke,* Fig 5 (*c*). Just before the end of the firing stroke, the exhaust valve, which until this moment has remained closed, opens, and allows the exhaust gases to flow out into the atmosphere or into a silencing device. The final upward stroke of the piston forces the exhaust gases out of the cylinder. This is the *Exhaust Stroke,* Fig. 5 (*d*). At the end of this stroke the exhaust valve closes, leaving the combustion head full of burnt gases at about atmospheric pressure. These gases have no real detrimental effect, but dilute the incoming charge.

The Practical Aspect of the Four-Stroke Cycle.— Hitherto we have been dealing with the theoretical aspect of the Otto cycle; it is proposed to consider what actually happens in the cylinder of a modern engine during the cycle of operations. In this respect it must be emphasized that unlike the steam engine with its slower speeds of 150 to 300 r.p.m., and its

* Indicated by the coiled wires and spark-gap shown on the upper right hand side of cylinder. [Fig. 5 (*d*).]

large cylinders and pistons, the petrol type engine has comparatively diminutive cylinders (from 2 to 5 inches diameter in the case of automobile engines), so that its individual impulses are relatively small. Owing, however, to the rapidity of the combustion and expansion of the gases, the piston moves very quickly when working normally, and to facilitate these movements the crankshaft revolutions are high, namely, from 1,000 to about 5,000 r.p.m. The petrol type automobile engine gives only a fraction of its full power at low speeds, so that for slow road speeds and for hill climbing gear-boxes or variable gears (which enable the torque at the road wheels to be kept high) become essential.

Commencing with the suction stroke, the entering gases are comparatively cool and at a pressure slightly below atmospheric. They enter the cylinder with a very high velocity; in normal cases the speed of the fresh charge through the inlet pipe and valve port varies from 120 to 240 feet per second, that is, from about 82 to 164 miles per hour.

The high velocity charge swirls around the combustion chamber in a state of agitation which persists not only during the suction, but also the compression stroke. Owing to this *Turbulence*, as it is termed, the combustion of the fuel particles is extremely rapid, for their movements spread the flame much quicker than would otherwise be the case. The degree of turbulence depends partly upon the design or shape of the combustion chamber, and also upon the position of the valves, and upon the engine speed. In certain cases the shape of the combustion chamber is specially designed so as to promote turbulence. But for this turbulence effect, petrol engines would only work at low speeds.

At the end of the compression stroke, the pressure of the charge at full throttle opening varies from 120 to 150 lb. per sq. in. in most automobile engines. This pressure varies with the *Compression Ratio*; this is the ratio of the combustion chamber, plus the piston swept-volume to the combustion chamber volume. In Fig. 6 it is the ratio of clearance volume (G) + stroke volume, to the clearance volume (G).

This ratio in modern automobile engines ranges from about 5·5 : 1 to 7·0 : 1 for petrol, and 7 : 1 to 9 or 10 : 1 for alcohol and anti-knock fuel mixtures. Immediately after the spark passes at the plug and the mixture ignites, the pressure rises rapidly to about 3 to 4 times the compression pressure value, after which it falls rapidly as the piston descends, until at the moment the exhaust valve opens its value is usually from 30 to 50 lb per sq. in., dropping quickly to a pound or two above atmospheric pressure as the exhaust valve opens fully.

During the exhaust stroke, the burnt gases being ejected are at a pressure of a few pounds per square inch above atmospheric value, and there is thus a certain load on the piston, acting against it, or opposing its motion. With large exhaust valves, opening for the longest possible period, and an efficient silencer this loss of power due to back pressure of the exhaust can be minimized.

The following table will serve to show the approximate relation between the compression ratio, compression pressure, explosion and average pressure (over the stroke) in the modern engine working upon commercial petrol as a fuel.

TABLE No. 2.
Showing the Pressures Occurring in Petrol Engines.

Compression Ratio.	Compression Pressure. lb. sq. in.	Maximum Explosion Pressure. lb. sq. in.	Average Pressure during Stroke. lb. sq .in.
3·5	66	230	62
4·0	80	275	76
4·5	95	335	90
5·0	110	390	105
5·5	128	450	121
6·0	147	520	140
6·5	166	580	157
7·0	184	650	175
7·5	202	720	192

NOTE.—These are " gauge " pressures, that is pressures above atmospheric.

Engine Temperatures.—The temperature of the charge just before ignition depends chiefly upon the compression ratio; its value ranges from about 390° C. for a ratio of 3·5 up to 480° C. for one of 7·5.

The temperature during explosion is very high for a short period, and ranges from 1500° C. up to over 2000°.* Owing to these very high temperatures it becomes necessary to provide means for cooling the cylinder and piston, otherwise these would soon become red hot. The two methods adopted in the case of automobile engines are the water-jacketed cylinder, and the air-cooled radiating fins ones; there was also an engine which employed the lubricating oil to cool the cylinder walls and barrel.

The inlet valve is kept relatively cool by the petrol vapour which flows past it; its temperature seldom exceeds 250° C., whereas that of the exhaust valve, which is exposed to the hot exhaust gases, varies from 600° C. to 700° C. The top of the piston also gets fairly hot, since the heat cannot readily get away; it varies from about 300° C. to 350° C. in average engines. It is owing to this high "crown" temperature that the lubricating oil splashed up on the under side usually becomes carbonized.

Graphical Representation of Cylinder Pressures.—To those accustomed to reading graphs, the cylinder pressure — or indicator — diagram will be found useful in interpreting the Otto, or four-stroke cycle of operations, upon which most automobile engines work. Referring to Fig. 6, the piston and cylinder are shown in outline below, and the pressure diagram above. The base AB of the pressure diagram corresponds to the piston stroke. During the suction stroke AgB the

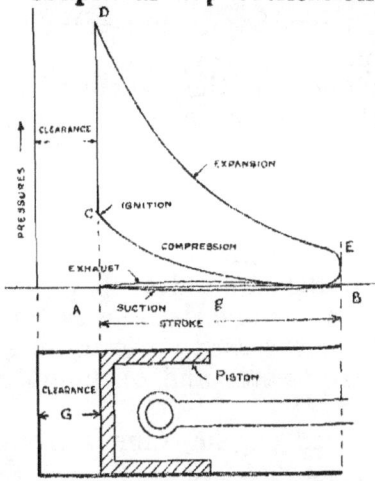

Fig 6.—Graphical method of illustrating Cylinder Pressures.

* Iron melts at 1530° C., and platinum at 1750° C. The temperatures of explosion only occur for a small fraction of a second, so are ineffective in melting the metal

pressure in the cylinder G is slightly below atmospheric, and during the next inward or compression stroke it rises along the line B C to the compression pressure represented by the height AC corresponding to the inner dead centre piston position. The pressure at any intermediate position is represented by the height of the pressure ordinate at that place.

At C the spark occurs, and the pressure rises suddenly to the value AD. As the piston is forced outwards, the pressure falls along the expansion line DE, until at E the exhaust valve opens, and the pressure drops quickly to a value slightly above atmospheric during the exhaust stroke BA (upper line).

Horse Power.—If the pressure scale is expressed in lb. per sq. in., and the stroke AB in feet, the area of the diagram BCDE will represent the work done per cycle, that is per two revolutions of the engine. The average height of this diagram gives the indicated mean effective pressure, or, as it is termed, the I.M.E.P.

If p_m = the I.M.E.P. in lb. sq. in.

l = length of stroke in feet

A = piston area in sq. in. = 0.7851 (diameter)^2

and N = engine speed in r.p.m.

then the horse power developed in the cylinder, or *Indicated Horse Power* (I.H.P.) is given by the relation

$$\text{I.H.P.} = \frac{P_m.l.A.N.}{66,000}$$

The horse power available at the crankshaft, known as the *Brake Horse Power*, or B.H.P., is less than this owing to the power loss due to friction of the piston and the bearings, and to the power used up in driving the valve mechanism, magneto, water and oil pumps, etc.

The B.H.P. in a well-designed engine, ranges from 80 to 90 per cent. of the I.H.P.

For taxation purposes, the Treasury employ a horse power formula which takes into account the cylinder bore and the number of cylinders only. The formula is as follows:

$$\text{Rated H.P.} = \frac{d^2 \cdot n}{2 \cdot 5}$$

i.e., the square of the bore (d), in inches, multiplied by the number of cylinders (n), divided by $2\frac{1}{2}$.

Thus a four-cylinder engine of 3-inch bore will have a rated H.P. for taxation purposes of $\frac{3 \times 3 \times 4}{2 \cdot 5} = 14 \cdot 4$

Actually such an engine will develop about 50 B.H.P., at its maximum speed, of, say, 3,500 r.p.m.

It may be considered that the logical way to obtain the greatest benefit from engines taxed by the Treasury rating method would be to employ engines of relatively long strokes in relation to their bores, and thus to obtain a greater cylinder volume and therefore a larger quantity of mixture in each cylinder, for the same diameter of cylinder.

Whilst it is true that more power would be developed for each working stroke of the engine, this beneficial result is offset by the fact that the longer the stroke of the engine in relation to its bore the slower must be the maximum engine speed, since there is a limiting velocity of the piston in traversing the cylinder.

Actually, it is an advantage to employ a rather longer stroke than one equal to the cylinder bore, since the hot gases resulting from the combustion of the charge will thus expand more and give a correspondingly greater amount of useful work on the piston before the exhaust valve opens and the pressure is released.

In practice a compromise has been made between the two alternatives of a short-stroke high speed and long-stroke low speed engine by adopting a stroke-bore ratio, for modern engines, lying between $1 \cdot 2$ and $1 \cdot 5$. Most of the American car engines (1941) had a stroke-bore ratio of $1 \cdot 2$ to $1 \cdot 3$.

The Two-Cycle Engine.—There is another cycle of operations, upon which the petrol engine can work, in which only two strokes, or one revolution of the crank-shaft, are necessary for a complete cycle.

This method, first invented by Sir Dugald Clerk in 1880, consists in partly compressing the mixture in a separate cylinder, or in the crank-case of the engine itself, and admitting this mixture into the cylinder

B

when the piston is at the outer end of its stroke. It is compressed fully on the inward stroke of the piston, exploded in the usual way and expanded until the piston, when nearing the outer end of its expansion stroke, uncovers an orifice, or port, through which the burnt gases escape. In most motor-cycle two-stroke engines, the piston next uncovers a second port leading to the crank-case, or separate compression cylinder, so that the mixture is admitted to the cylinder. In such cases the piston takes the place of the inlet and exhaust valve for these operations.

Referring to Fig. 7 which illustrates the Enfield two-stroke engine, the cycle of operations is as follows: Diagram 1 shows the piston at the bottom of its expansion stroke, with the remaining exhaust gases flowing out of the exhaust port B (Diagram 3) on the left, and the fresh, slightly compressed mixture flowing into the cylinder from the crank-case through what is termed the *Transfer Port* D. On the upward stroke of the piston (Diagram 2) the fresh charge is compressed; at the same time the suction created below the piston, due to its upward travel, draws in fresh mixture from the petrol and air mixing device, known as the *Carburettor*, and to which we shall refer later, through the opening C (Diagram 3). In this latter diagram the piston has reached the top of its compression stroke, and is just about to be exploded by the electric spark at the plug E. Diagram 4 shows the piston moving downwards on the ensuing expansion stroke, and uncovering the exhaust port B, the transfer port D being just about to uncover when a fresh cycle of operations occurs.

This common type of engine is known as the *Three Port* one, from the fact of there being the three ports B, C, and D. There are no valves in this type, the piston performing the duties of the valves in the four-stroke type, so that the absence of valves and valve gearing enables such engines to be made more economically than four-stroke ones. The inwardly opening valve shown in Diagram 3 at A is merely a motor-cycle control device. A Bowden lever on the handle-bar enables this valve to be opened, thus releasing the compression and preventing the engine

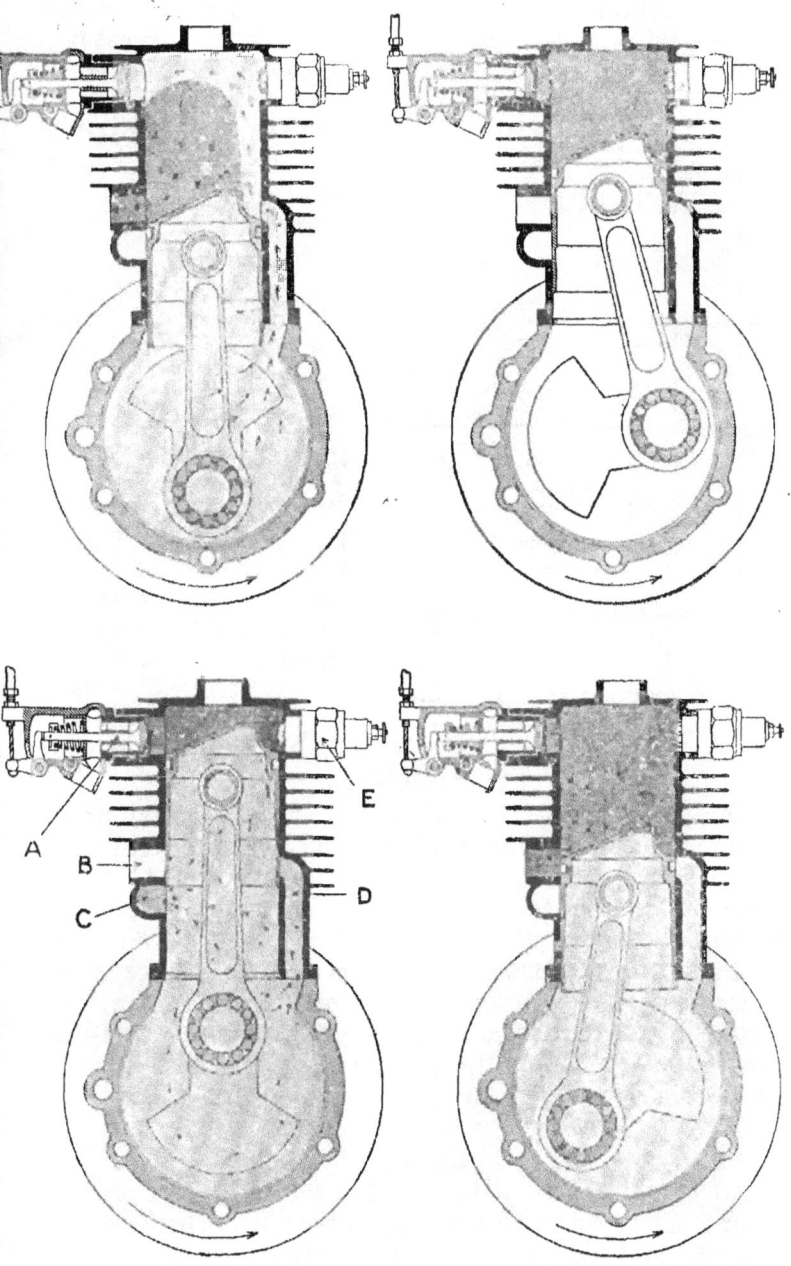

Fig. 7.—Illustrating the operation of the Two-Cycle Engine.

from working. This *Compression Release Valve* enables the rider to stop his engine when necessary; it has nothing to do with the two-stroke cycle of operations.

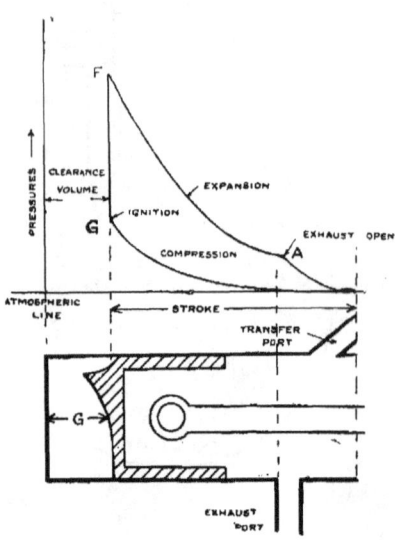

Fig. 8.—Graphical method for Two-Cycle Engine Pressures.

It is not proposed to consider the mechanical details, or the different types of two-stroke engine, here, but to deal only with the theoretical principles. The manner in which the pressures in a two-stroke engine vary is illustrated in Fig. 8. This diagram of pressures is somewhat similar to that shown in Fig. 6, except for the fact that there are no suction and exhaust lines (these are the two "strokes" missing in the present case), and that the "toe" of the diagram is different.

The pressures developed in the case of a two-stroke engine are appreciably lower than in the four-stroke one of the same dimensions, and at the same speed. This is due to the fact that it is difficult to charge the cylinder during the short inlet port opening interval, and also that the engine tends to run much hotter, since it has twice as many explosions in the same time as the four-stroke engine. Moreover, since both exhaust and inlet ports are open together, a part of the fresh charge blows right through into the exhaust, and is lost; this results in a smaller effective charge for compression and firing. In order to minimize this charge loss, it is usual to shape the piston as shown in Fig. 7, namely, with a baffle on the crown, so as to deflect the incoming charge, and to obstruct its passage to the exhaust. The results of scientific tests made on three-port engines, fitted with deflector pistons, indicate a loss of charge through the exhaust port, however, varying from 15 to 25 per cent., in spite of the deflector. The net result of the disadvantages mentioned are that the ordinary two-stroke engine gives only about 50 to 60, instead of 100 per cent. more power for a given cylinder capacity, than the four-stroke engine.

Horse Power of Petrol Engines.—The automobile engineer is more concerned with the actual horsepower delivered at the crankshaft (and available for propulsion purposes) than with the power developed in the cylinder itself. It has, therefore, become customary to deal with the shaft, or brake horse-power (B.H.P.) in making comparisons of engines.

As shown in Chapter IX, the B.H.P. can be measured accurately with a power brake or dynamometer, and in this respect the following relation is of interest when studying the possible limits of performance of four-cycle petrol engines:—

$$B.H.P. = \frac{B.M.E.P \times \text{Displacement} \times R.P.M.}{1,008,460}$$

Where B.M.E.P. = Brake mean effective pressure in lbs. per sq. inch.

Displacement = Total cylinder volume swept by pistons. (cu. inch.)
= Piston area × piston stroke × number of cylinders.

R.P.M. = Revolutions per minute of engine.

In order to obtain the *best performance* from an engine of given dimensions (or displacement), one can either increase the B.M.E.P., or the R.P.M., or both.

Car Engine Performances.—As car engines are taxed upon the basis of cylinder diameter the aim of designers has been to increase, progressively, the horse-power per given cylinder capacity. Since, however, most car engines have about the same ratio of bore-to-stroke, this is equivalent to increasing the horse-power per unit of piston area, i.e., the H.P. per sq. in.

The methods adopted to obtain the increased horsepower are by (1) *Increasing the compression ratio.* (2) *Increasing the average and maximum engine speeds.* (3) *Improving carburation* and (4) *Improving the charge or volumetric efficiency.* It has been possible to increase the compression ratio by using fuels of higher octane value so that at the higher compression there is no tendency of the engine to detonate; further, by the use of carefully-designed combustion chambers higher compression ratios can be used than were hitherto possible for the same grade of fuel. Yet another factor has been the more universal adoption of aluminium alloy pistons and cylinder heads (in some instances the cylinders themselves are of aluminium alloy). This has enabled some of the surplus heat of combustion to be got rid of more readily by conduction so that excessive temperatures do not occur.

The usual compression ratios employed in more recent engines range from $6 \cdot 0 : 1$ to $7 \cdot 5 : 1$; in this connection the smaller bore engines can employ a higher compression ratio than the larger ones, owing to their better heat dissipation. In the case of American car engines, the compression ratios of all the various makes, according to published statistics, lie between the two extremes of $6 \cdot 1 : 1$ and $7 \cdot 4 : 1$, with an average value of $6 \cdot 5 : 1$.

In regard to *engine speeds*, the maximum values have increased appreciably in recent years, until in 1941, the usual maximum speeds for the smaller engines have been from 3,800 to 4,500 R.P.M. and for the larger engines, namely, those of 12 H.P. (rating) and above, from about 3,400 to 4,000 R.P.M.

In the case of racing car engines the maximum engine speeds are considerably higher, ranging from about 4,500 to 8,000 R.P.M. The highest engine speeds are those employed in the smaller sizes of engine, since most engines, irrespective of cylinder diameter, run at about the same mean piston speeds. It can be shown that if the latter speed is the same for all engines the larger bore ones must run more slowly than the smaller bore ones.

Useful Methods of Expressing Horse Power.—
It is of interest to note that in the case of a well-designed car engine the usual output expressed in terms of the area of the piston lies between $1 \cdot 4$ and $2 \cdot 0$ H.P. per sq. in. of piston area (maximum output values).

Another useful method of expressing the outputs of modern engines is in terms of cylinder capacity, that is, as the horse-power per cubic inch or per litre of cylinder volume. The latter quantity is the displacement or product of the piston area by the stroke, by the number of cylinders.

An analysis of recent car engine maximum output figures indicates that these lie between about 0·45 and 0·55 H.P. per cu. in.

The corresponding values in terms of H.P. per litre are obtained by multiplying the previous values by 61, since a litre is equal to 61 cu. in. It is thus seen that the ordinary car maximum outputs lie between about 27 and 34 H.P. per litre. Certain makes of car engine, however, give appreciably higher values, whilst small supercharged racing type motor car and motor cycle engines have given, on the test bed, values between 150 and 200 H.P. per litre. Modern aircraft engines of the air-cooled radial type give from 40 to 50 H.P. per litre, whilst water-cooled ones give from 50 to 65 H.P. per litre.

Charge or Volumetric Efficiency.—During the suction stroke of a four-cycle engine the piston, in moving from its top to bottom dead centre, should theoretically draw into the cylinder a volume of mixture at atmospheric pressure equal to its area multiplied by its stroke, i.e., its working volume. In practice the quantity of mixture drawn into the cylinder is always appreciably less than this theoretical amount, so that it is usual to term the ratio of the actual to the theoretical amounts the charge or volumetric efficiency.

$$\text{Thus Charge or Volumetric Efficiency} = \frac{\text{Actual Charge Weight}}{\text{Theoretical Charge Weight}} \times 100 \text{ (per cent.)}$$

The greater this efficiency the more power will be developed from a given size of engine. The usual value of the volumetric efficiency in the case of a well-designed engine is from 80 to 85 per cent. at full throttle opening. The principal reasons for the inability of the engine to obtain a full weight of charge during the suction stroke are as follows : —(1)Throttling effects due to presence of carburettor restrictions to mixture flow, presence of bends in inlet manifold, changes of section of manifold and inlet port, etc. (2) Heating of the incoming charge by the hot inlet port, cylinder combustion chamber, walls, piston, hot exhaust valve, remaining hot exhaust gases in clearance space, from previous exhaust stroke, etc.

In the case of supercharged engines, whilst it is of equal importance to minimise the results due to (1) and (2), the fact that the mixture is forced into the cylinder under pressure, instead of relying upon engine suction, ensures that a greater quantity of mixture will enter the cylinder during the suction stroke.

Horse Power and Air Consumption.—It has been shown by Ricardo that there is a definite relation between the weight of air entering the cylinder per minute and the indicated H.P. developed and that this relationship is, within fairly small limits, the same for all hydrocarbon fuels. This statement, however, does not mean that all engines of similar dimensions will give the same power output, since other factors—such as the thermal and volumetric efficiencies, mixture strength and mechanical efficiency—must also be taken into account.

It is, however, useful to note that for engines having the same mechanical efficiency and operating upon rich mixtures, using either petrols, benzole or alcohol, the indicated horse-power can be estimated, approximately, from the following relation : —

$$I.H.P. = k \times E \times Lbs. \text{ of air consumed per hour.}$$

Where $k =$ a constant which, for various hydrocarbon liquid fuels varies between the limits of $1 \cdot 94$ and $2 \cdot 00$, and $E =$ thermal efficiency of the engine. The deduction that can be made from this relationship is that for the *maximum power the engine should consume the greatest amount of air per unit time.*

It is also of interest to remember that the results of tests by Ricardo show that, irrespective of the fuel employed in the engine, the amount of energy liberated for each pound weight of air consumed is approximately 1,300 B.T.U's, for rich mixtures.

Estimating the Mean Effective Pressure.—For engine design purposes it is useful to be able to estimate, approximately, the mean effective pressure that will be developed. This can be done as follows : —

Assuming that the engine is working on petrol having a calorific value of 18,290 B.T.U.'s per lb. and with the correct air-petrol proportions of 15 : 1 by weight, it can readily be shown that the heat energy of the mixture works out at 45 ft. lb. per cu. in.

If the volumetric and also the thermal efficiencies were at their ideal values, namely, 100 per cent., the theoretical indicated M.E.P. would be given by the following relation : —

$$\text{Theoretical I.M.E.P} = 12 \times 45$$
$$= 540 \text{ lbs. per sq. in.}$$

In practice the indicated thermal efficiency of a well-designed petrol engine would be of the order of 30 to 35 per cent., say, 32·5 per cent. for a compression ratio of about 6 : 1. The volumetric efficiency would be about 80 per cent. for such an engine.

The actual I.M.E.P. is obtained by multiplying the theoretical I.M.E.P. by the actual thermal efficiency and also by the volumetric efficiency. Taking the values previously mentioned we have : —

$$\text{Actual I.M.E.P.} = 540 \times 3.25 \times \cdot 80.$$
$$= 140 \text{ lbs. per sq. in.}$$

The corresponding brake M.E.P. is then obtained from the last value by multiplying it by the mechanical efficiency of the engine, which is about 85 to 90 per cent., say, 87·5. So that we have : —

$$\text{B.M.E.P.} = 140 \times \cdot 875$$
$$= 122.5 \text{ lbs. per sq. in.}$$

From the latter value, the probable horse-power output of an engine of a given bore, stroke, speed and number of cylinders can be estimated by the use of the formula given on page ----.

The Efficiency of Petrol Engines.—It is interesting to consider how the high speed petrol engine compares with other types of prime mover in the matter of power output and fuel consumption, and to see what proportion of the fuel is usefully employed. The best method of comparison, perhaps, is that in which the power output for a given supply of combustion heat is considered. The method, which gives true results for any type of engine, whether working with steam, heavy oils, gas or petrol, involves a quantity known as the *Thermal Efficiency;* actually all engines are compared on the basis of their thermal efficiency. The thermal efficiency of any type of engine

B*

unit is simply the ratio of the useful work or power obtained to the heat supplied from the fuel. The useful work is defined by the Indicated Horse Power and the heat supplied, by the product of the quantity of fuel used per given time and its calorific value.

Thus if w lb. of fuel of calorific value c B.T.U.'s per lb. be used in time t minutes, and if the Indicated Horse Power be denoted by I.H.P., then we have:—

$$\text{Total Heat supplied per minute} = \frac{w\,c}{t} \text{ B.T.U.'s}$$

or, expressed in terms of work units, $778\,\frac{w\,c}{t}$ foot pounds (since 1 B.T.U. is equivalent to 778 ft. lbs. of work).

Now the useful work obtained per minute
$$= \text{I.H.P.} \times 33{,}000 \text{ ft. lbs.}$$

So that we have

$$\text{Thermal Efficiency (I.T.E.)} = \frac{\text{Useful Work obtained per minute}}{\text{Heat supplied per minute}}$$

$$= \frac{\text{I.H.P.} \times 33{,}000 \times t}{778\,w\,c} \times 100 \text{ per cent.}$$

This reduces to

$$\text{I.T.E.} = 4233\,\frac{\text{I.H.P.} \times t}{w\,c} \text{ per cent.}$$

Example.—An engine, with a mechanical efficiency of 85 per cent., uses 0·55 lb. of petrol per B.H.P. per hour. What are its indicated and brake thermal efficiencies?

$$\text{Since Mechanical Efficiency} = \frac{\text{B.H.P.}}{\text{I.H.P.}}$$

$$\text{We have I.H.P.} = \frac{\text{B.H.P.}}{0\cdot85} = 1\cdot176 \text{ B.H.P.}$$

$$\text{Fuel used per B.H.P. per min.} = \frac{0\cdot55}{60} \text{ lbs.}$$

This same quantity enables 1·176 H.P. to be developed.

$$\text{Hence I.T.E.} = 4233\,\frac{1\cdot176 \times 60}{0\cdot55 \times 18{,}000} \text{ per cent.} = 30\cdot2 \text{ per cent.}$$

The calorific value of petrol is taken at 18,000 B.T.U.'s per lb]
The Brake T.E. = 85 per cent. of this value.

$$= \frac{85}{100} \times 30\cdot2 = 25\cdot6 \text{ per cent.}$$

The indicated thermal efficiency values for well-designed petrol engines range from about 25 to 33 per cent., and the brake values from 20 to 27 per cent. Certain types of high compression engines, which run upon alcohol or anti-knock fuels, give higher values for the I.T.E., namely from 30 to 35 per cent. The Diesel engine, which utilizes compressed air at 500 to 600 lb. per square inch, into which a heavy oil fuel is sprayed, and ignites spontaneously at nearly constant pressure, yields I.T.E. values of from 35 to 45 per cent.

Steam engines are notoriously poor in thermal efficiency values; the best reciprocating types seldom exceed about 12 per cent., but with high pressures and a high degree of superheating modern marine steam turbines may attain efficiencies up to at least 30 per cent., thus making them practically as good as Diesel engines from the efficiency viewpoint. The following table indicates the thermal efficiencies of the principal types of prime mover:—

TABLE No. 3

Comparison of Brake Thermal Efficiencies of Various Types of Engines.

Type of Engine.	Brake Thermal Efficiency per cent.
Steam Locomotives and Road Vehicles ...	4 to 6
Electric Traction	8 to 12
Marine Steam Engines (Reciprocating) ...	9 to 15
Marine Steam Engines (Superheated) ...	16 to 20
Marine Steam Turbine	20 to 30
Gas Engines	22 to 26
Petrol Engines (Aircraft and Automobile) ...	20 to 27
Semi-Diesel Engines	27 to 30
Diesel Engines	30 to 36
The Still Petrol-Steam Engine	32 to 35

Interpreting these results, if a gallon of petrol is to be used as fuel, the two latter types of engine will utilize, as B.H.P., nearly one-third of this fuel, use-fully, the ordinary petrol engine about one-quarter and the steam engine from one-eighth to one-twelfth. The superiority of the internal combustion over the steam engine will therefore be evident from these results.

Efficiency of Automobile Engines.—The thermal efficiency of automobile engines varies within certain limits, due to several factors; these and their influences are enumerated, briefly, below :—

The *Thermal Efficiency*—

(1) Is generally less at part loads, and a maximum at full loads, so that the fuel consumption at part loads is heavier. Thus at quarter to half full load, the fuel consumption is from 50 to 30 per cent. greater than when the engine runs on open throttle.

(2) Is higher as the compression ratio is increased. Detonation considerations fix the upper limit of this ratio.

(3) Is higher for weaker mixtures of air and petrol. The usual ratio for maximum efficiency is about 17 to 18 parts of air to 1 part, by weight, of petrol; the ratio for complete combustion is about 14.5 to 15.0.

(4) Is dependent upon the shape of the combustion chamber. The best efficiencies are usually given by those engines having the smallest surface of combustion chamber for a given volume. The ideal shape is spherical, and the overhead valve, "pocket-less" engine, with inclined valves and dished piston, is the nearest approach to this ideal. The T-headed engine is about the worst.

(5) Increases with engine speed up to a certain limiting value, after which it decreases. The maximum efficiencies are usually obtained at speeds of from 70 per cent. to full load speed values.

(6) Depends upon the type of fuel used. For the same compressions and speeds, the " higher compression " fuels, such as alcohol, benzole and high octane petrols, give the better efficiencies.

How the Fuel's Heat is Utilized in Petrol Engines.—

We have seen that about 25 to 30 per cent. of the heat of the fuel is actually converted into useful work, or power, and the question now arises as to what becomes of the remaining 75 to 70 per cent. The latter heat is used up, or absorbed by the cylinder cooling agency, and also by the exhaust gases. The expanding gases are at a very high temperature compared with the metal of the cylinder walls, so that heat is conducted through this metal, to be carried away by the cylinder cooling water, or in the case of air-cooled cylinders, by the cooling air. A part of the heat is also lost by radiation from the heated metal surfaces. Comparatively large volumes of exhaust gases at from 500° to 700°C. (at the exhaust valve) are discharged constantly from an engine; these carry away a considerable amount of the fuel's heat.

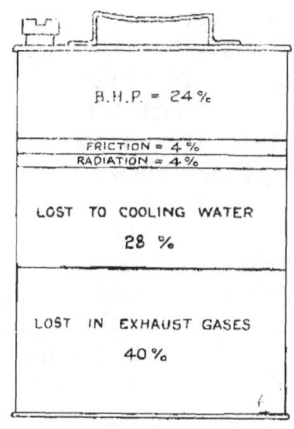

Fig. 9.—Illustrating how the Fuel's Heat is utilized in the Engine.

The results of scientific tests show that the heat of combustion of the fuel is utilized in an average petrol engine in the following way:

TABLE No. 4.

Showing how the Fuel is Utilized in a Petrol Engine

Heat available as useful B.H.P . . .	24 per cent.
Heat absorbed, due to Engine Friction .	4 ,,
Heat carried off by the Exhaust Gases . .	40 ,,
Heat carried away by the Cooling Water .	28 ,,
Heat lost by Radiation	4 ,,

100

Referring to Fig. 9 which illustrates a two gallon can of petrol, the manner in which the fuel utilized is shown by the different volumes indicated.

At first sight it would appear distinctly disappointing to motor users that three-quarters out of every gallon of petrol is wasted, so far as power is concerned.

Unfortunately it is not possible in the case of ordinary automobile engines to recover more than a small percentage of this waste energy, although with the modern tendency to employ higher compression ratios the percentage of useful energy has been increased above the valve shown in Table 4. Similarly, the adoption of the compression-ignition engine on certain classes of motor vehicle has resulted in a still greater reduction in the waste energy; in this connection it should be added that with the high compression ratios employed in this type of engine, namely, from 14 : 1 to 18 : 1, the thermal efficiencies have improved considerably, so that it is the rule to obtain indicated thermal efficiencies of 45 to 50 per cent. with normal mixtures of fuel and air and still higher efficiencies with mixtures rather weaker in fuel.

Assuming that the mechanical efficiency is 85 per cent., the corresponding brake thermal efficiencies would be 38·2 and 42·5 per cent., respectively. The corresponding percentages of heat energy of the fuel converted into useful work as B.H.P. would therefore be the same, namely, 38·2 and 42·5, respectively. Thus, by using the compression-ignition principle instead of the Otto cycle one the amount of heat wasted can be reduced by from 10 to 15 per cent. Another method of reducing the waste energy is to utilise the hot high speed exhaust gases to drive a gas turbine of the Rateau type, coupled to a centrifugal pump for supercharging the mixture supplied to the engine. This method has been applied successfully to aircraft engines and has resulted in a marked increase in power output, so that engines thus equipped can maintain their ground H.P. rating up to much higher altitudes than gear-driven supercharged type engines.

Supercharging.—The power obtainable from a given engine is determined by the amount of mixture which fills the cylinders at the beginning of the compression stroke. If we could double the quantity of air, we could obtain twice the quantity of mixture, and much higher cylinder mean pressures and powers would be realised. If we previously compress the mixture to one-half its volume, before compression in the cylinder, twice the charge weight can be obtained.

The power obtained from this doubled charge will not, however, be twice as much, due to combustion considerations into which we cannot enter here, and to mechanical power losses in connection with the pre-compression of the charge.

Supercharging is the name given to the process of increasing the charge weight in the cylinder so as to obtain more power from a given cylinder capacity.

The subject of supercharging is dealt with more fully in Chapter IV.

CHAPTER II

Development of the Petrol Engine.—Although there is some doubt as to the original inventor of the petrol engine, it appears fairly certain that the first road car driven by the explosion of benzene vapour was constructed by a German named Siegfried Marcus, of Mecklenburg, in 1864 to 1868, according to a photograph taken at about that time. In 1873, Marcus's improved vehicle was exhibited at the Vienna Exhibition; this vehicle appears to have been constructed from a hand-cart, the two rear wheels of which were replaced by the fly-wheels of the benzene engine used for driving it.

There was little progress to record between 1862 and 1875, but several inventors appear to have been working on the subject of road-car design. In 1889, a vehicle was actually built by Butler, an Englishman, and was able to travel along the roads at 4 to 12 m.p.h. This three-wheeled vehicle was driven by means of a small two-cylinder engine, of 2¼ ins. bore, driving direct on to the hub of the rear wheel, through an epicyclic gear with a reduction ratio of about 6 to 1. The first ignition system was a magneto, with a low tension make-and-break sparking device in the cylinders. Ultimately, this was abandoned in favour of a primary battery and induction coil, with a wipe contact. A float feed carburettor was used; this appears to be the first instance in which this type was employed.

On the Continent, Gottlieb Daimler, in 1886, constructed a benzene engine-driven vehicle, which was fairly successful. This was a motor-cycle and employed carburetted air in an engine working on the Otto, or four-stroke, principle. Another pioneer who was working independently and on somewhat parallel lines, was Benz, who in 1885 produced a satisfactory motor-tricycle using electric ignition. Daimler, and later Levassor, in France, used the hot tube ignition system. These early engines were virtually modified horizontal gas-engines, running at relatively low speeds, namely, from 400 to 700 r.p.m.; they were also very heavy and somewhat erratic in their running. It was only when Count de Dion and M. Bouton, two French engineers, realized that the future development of the automobile engine was along the lines of reduced weight, and increased power output, that any real advance was made. They constructed the famous De Dion Bouton engine running at 1,500 r.p.m.—a considerable speed increase over the engines of their rivals. Later events proved conclusively the soundness of their foresight and judgment. To Messrs. Panhard and Levassor also belongs much credit for constructing one of the first successful motor-cars, in 1901, from which date progress was, relatively speaking, very rapid.

It will be seen from this somewhat brief survey that the automobile engine has been developed from the early gas engines of Lenoir and Otto, by lightening the construction, improving the valve and cylinder design, employing the vapours of light spirits such as benzene and petrol, by utilizing electric in place of hot-tube ignition and by increasing the engine speed.

The Complete Engine.—If the principle of the four-stroke type of engine, explained in the preceding chapter has been properly understood, the reader will experience little difficulty in following the application of this principle to modern petrol engine practice. Although the sectional views of engines at first sight may appear to be complicated and a little bewildering, he will bear in mind the fact that these engines work on a similar system to that shown in Fig. 4, he should not find it difficult to understand engine designs. For this reason it is proposed to take our first step from the outline drawing of Fig. 4 to the rather more detailed one shown in Fig. 10. This latter illustration shows all of the essentials of a modern air-cooled petrol engine, but devoid of details which might at this stage confuse the reader.

Keeping the Cylinder Cool.—It has been stated in the previous chapter that the heat of explosion is very great, and that there is a considerable proportion of this heat conducted through the cylinder walls whilst the

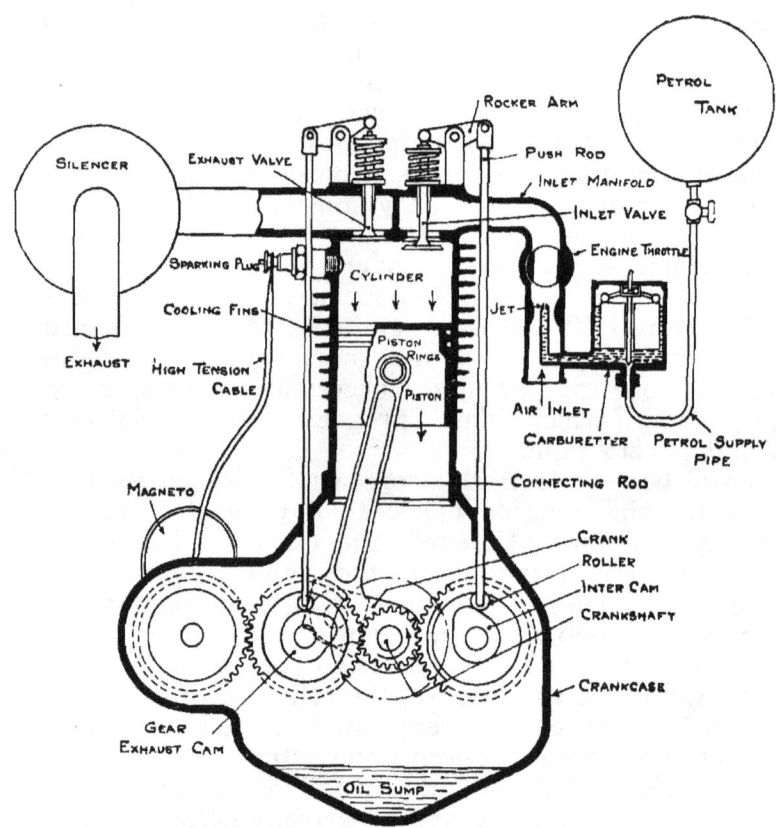

Fig. 10.—Showing the Components of the
Air-Cooled Petrol Engine.

engine is working, namely about 30 per cent. of the combustion heat. In the case of an engine of $3\frac{1}{2}$ inches cylinder diameter and 4 inches stroke (about the dimensions of a motor-cycle engine) running at 2,000 r.p.m., no less than 2,500 B.T.U.'s of heat are given to the metal of cylinder and piston every minute. This quantity of heat is sufficient to raise $1\frac{1}{2}$ gallons of water from freezing to boiling-point every minute.

If the cylinder of the engine was merely a plain metal cylinder, it would be raised to a red heat after a very short period of running if it did not seize up before; indeed one remembers several cases of early engines, in which the cooling and cylinder lubrication systems were inefficient, running for a time at a dull red heat! Not only is this bad for the wearing surfaces, but the incoming charge is liable to explode whilst it is still entering the cylinder, and thus to fire back into the carburettor. The cylinder walls and piston must therefore be kept sufficiently cool whilst the engine is working. The simplest method, and one which is employed upon certain aircraft, motor-cycle and small

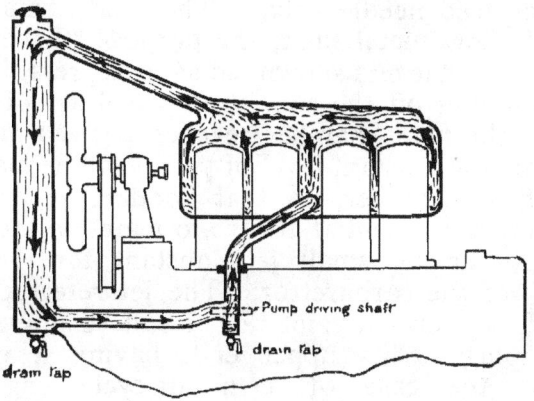

Fig. 11.—Illustrating the principle of Water-Cooling of the Cylinders of a Four-Cylinder Engine.

stationary engines, consists in providing a number of heat-radiating fins on the hottest parts of the cylinder, as shown in Figs. 4 and 17. Most air-cooled cylinders of this type are made of cast-iron, with the radiating fins cast integrally with the barrels, but aluminium alloys and also copper fins have been used; both of these metals are very much better heat conductors than cast-iron.

Fig. 11 illustrates the principle of the method of water-cooling, and shows how the heat of the cooling water is dissipated or carried away by means of a *Radiator* (shown on extreme left). The latter consists of a large number of thin metal tubes, of great surface area, through which the hot water passes, and is cooled by radiation from the large tube surface, and also a current of air blowing past the outside surfaces of the tubes.

We shall return again to the subject of cooling in a later chapter.

Supplying the Mixture to the Cylinder.—Petrol is contained in the tank or reservoir shown, and is fed to the petrol-air proportioning and mixing device known as the *Carburettor*, along the petrol supply pipe; a tap is usually provided near the tank to shut off the supply when required. Referring to Figs. 10 and 12, the petrol enters the base of the carburettor *Float Chamber*, past a tapered needle valve. The float chamber contains a hollow metal float, the purpose of which is to lift the pivoted levers shown, so as to depress the needle valve and shut off the supply of petrol as soon as the petrol in the float chamber reaches a given height. In this manner a constant level of petrol can be maintained in the float chamber, so that flooding is prevented; moreover, it is essential always to maintain the level of the petrol in the small jet constant for the proper working of the carburettor. The jet referred to consists of a vertical pipe of small diameter fitted with a plug at its upper end, having a very fine hole; in the case of a motor-cycle engine this hole is only about $\frac{1}{30}$ to $\frac{1}{50}$ inch diameter. The open end of the jet is arranged in the centre of a constricted tube, throat, or *Venturi*, the purpose of which is to increase the speed of the air which enters at the lower end, so that it will flow very quickly past the jet, and in so doing will cause a reduction of pressure below that of the atmosphere. Since the pressure on the surface of the petrol in the float chamber is atmospheric, whilst at the jet it is less than this, it follows that the petrol will flow out of the fine jet as a spray and will be carried along with the air which is sucked in during the suction stroke of the piston.

The dimensions of the air supply pipe and of the jet are carefully proportioned so as to give about the correct ratio of air to petrol, namely about 15 parts of air to 1 of petrol, by weight, but means are provided for varying the proportions between the limits of about 8 (rich) and 20 (weak) to suit certain running conditions.   It is usual in modern carburettors to provide automatic means for maintaining the mixture strength uniform over the whole speed range.

The *Throttle* shown between the carburettor and inlet valve is simply a valve provided for the purpose of governing the engine speed and power; it acts by regulating the quantity of mixture admitted to the engine. For slow running and light loads it is nearly closed:

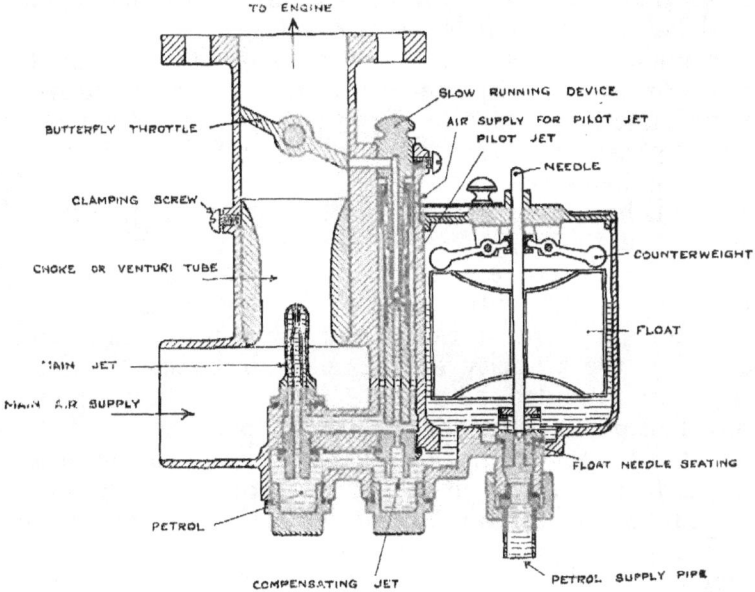

Fig. 12.—A typical Automobile Engine Carburettor.

for high speeds and full power it is opened wide.   The throttle shown in Fig. 10 consists of a cylindrical piece, capable of rotating in the outer cylindrical casing, and is provided with a cylindrical hole of the same diameter as that of the inlet pipe.   It will be seen that if this inner piece is rotated in its casing it will gradually restrict the area through which the mixture can pass to the engine.

Fig 12 illustrates a typical carburettor in detail.   In this case there are two petrol jets, namely, the *Main Jet* on the left, which corresponds to the one previously described, and the *Pilot Jet*, on the right. The purpose of the latter is chiefly to supply the very small quantity of mixture required to start from the cold or keep the engine running slowly, when there is no load on it, as when an automobile engine is running idle. There is also a compensator jet to keep the mixture strength approximately constant when the engine is working at various speeds. It is known that the simple type of carburettor illustrated in Fig. 10 can only give the correct mixture proportions at one definite speed. Above this speed the mixture obtained is richer in petrol, whilst below it is weaker.

All modern carburettors are provided with additional means, or devices, to render the mixture proportions approximately constant over the working speed range; the compensator device referred to in Fig. 12 comes under this heading.

It will be noticed that another type of throttle, namely, the *Butterfly Valve*, is employed in this case; it is shown closed, and it opens by rotating with the pin shown in the centre. The air inlet is on the lower left hand side, whilst the choke, or venturi tube for increasing the velocity of the air, surrounds the jet. Plugs are provided below the jets to facilitate their removal for cleaning or changing purposes. It is usual to fit a petrol strainer, or filter, between the tank and the float chamber needle valve in order to trap any solid particles of matter which might otherwise block up the fine petrol jet. It will also be observed that the upper end of the needle valve is accessible; by raising the needle, by hand, the level of the petrol can be raised. This procedure, which is known as " flooding " or " tickling " the carburettor, enables a rich mixture to be provided for starting the engine more readily.*

Modern Carburettor Improvements.—The later types of carburettor are mostly of the " down draught " pattern, in which the air for combustion is drawn vertically downwards through the carburettor, the latter being placed near the top of the engine. This arrangement enables the heavier mixture to assist by its weight in the induction process, whilst the

* A full account of Carburettors and Fuel Systems is given in Vol. II of this series.

number of passages and bends to the combustion chambers is minimised; it also renders the carburettor more accessible.

Carburettor main air inlets are fitted with *air cleaning devices* and, since in the past there was a certain amount of noise made by the passage of the air through the carburettor unit, many modern carburettors are also fitted with *air silencers* of the gauze or perforated plate pattern. In some cases *flame arresters*, of copper gauze, are fitted over the air inlets to stop back-fire flames.

The *starting of car engines* from the cold has been improved by the fitting of automatic devices for enriching the mixture at the time of starting. One typical starting device has a *thermostat* which, when cold, operates an air valve known as a *choke* for reducing the main air supply; as soon as the engine warms up the air-valve is opened by the device.

The mixing chamber above the jet is usually heated by exhaust gases; there is often a throttle valve for shutting off the gases when the engine's throttle is fully opened.

Acceleration pumps are also fitted for the purpose of ensuring an adequate supply of petrol for acceleration, when the accelerator is suddenly pressed down.

Finally, *a mixture control* is usually fitted for the purpose of altering the mixture proportions to suit climatic conditions, i.e., extreme heat and cold atmospheres.

Supplying Petrol to the Carburettor.—The petrol supply is carried in a tank at the rear of the car, although earlier models were fitted with a dashboard or bonnet tank to allow the petrol to feed the carburettor by gravity.

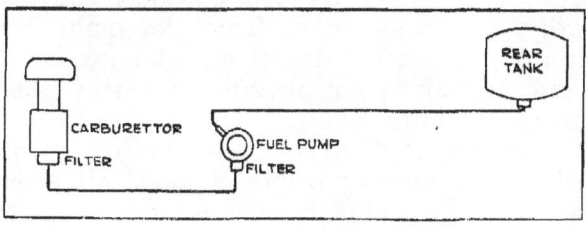

Fig. 13.—Typical Fuel Pump Arrangement.

The favoured method a few years ago was to employ a system known as the *vacuum feed*, whereby the reduced pressure in the inlet pipe of the engine was used to draw the petrol from the rear tank into a small auxiliary tank placed under the bonnet above the engine. This tank had a float mechanism to shut off the suction from engine and at the same time to open

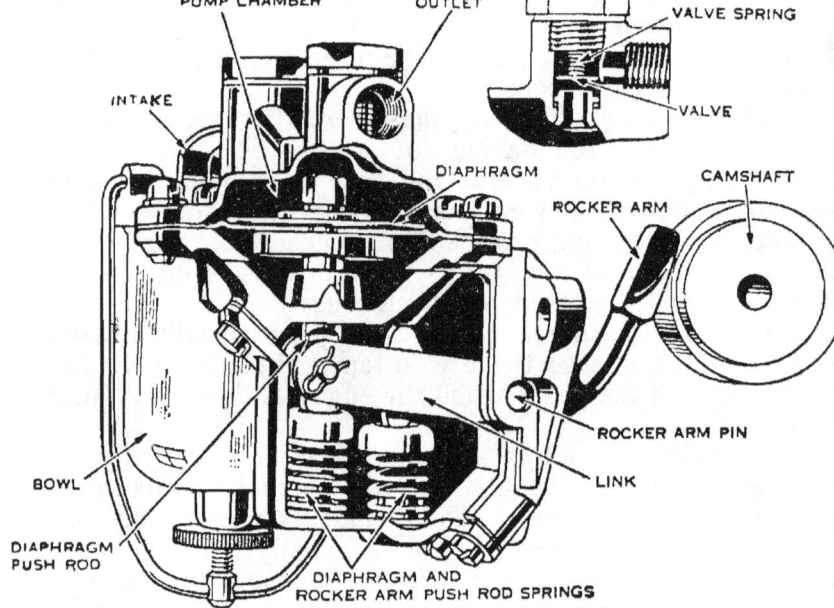

Fig. 14.—The A.C. Mechanically Operated Petrol Pump.

a valve allowing the outside air to be placed in communication with the petrol in the tank. The petrol then flowed by gravity to the float chamber of the carburettor.

The modern method is to employ a petrol pump, operated mechanically off the engine's camshaft, or electrically, to draw petrol from the main tank and force it into the carburettor; most pumps have automatic release valves to obviate excessive supply of petrol to the carburettor.

Fig. 14 illustrates a typical cam-operated petrol pump in which a flexible diaphragm clamped all around its periphery is reciprocated by means of a lever attached to a rocker arm, the other end of which engages with a cam on the engine's camshaft, the diaphragm is pulled down by the rocker arm, but is returned by the push rod springs shown. This to-and-fro action of the diaphragm draws petrol in from the main tank, through a non-return valve and then pumps it out to the carburettor through another non-return valve. If the outlet pressure exceeds a certain value a release valve on the outlet side opens and allows the petrol to flow to the suction side again. The pump illustrated has a glass bowl and a gauze-filter; the former collects any sediment present in the petrol.

Air Cleaners.—The cleaning of the air taken through the carburettor has a marked effect upon the useful life of the pistons and cylinders, so that it is now usual to fit cleaners to the air-intakes of the carburettors. These cleaners operate either upon the direct filtering principle, using felt or hair elements, to stop the dust particles, or upon the centrifugal one, whereby the air is rotated at high speed before it enters the carburettor, thus throwing any solid matter outwards into a dust collecting chamber.

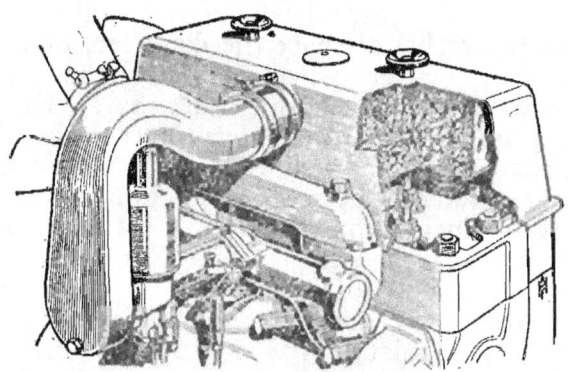

Fig. 15.—Morris Air Cleaner and Heater, on Cylinder Head.

Fig. 15 shows a method of air cleaning which has been used on Morris engines. It consists of a large domed cover for the detachable cylinder head, having a horizontal partition converting the upper part of this cover into a chamber for the retention of the air-cleaning medium, which consists of a quantity of curled white horse hair introduced into the cleaning chamber through the large aperture to which the carburettor air pipe is attached. This large aperture renders the removal of soiled hair for cleaning purposes or for the introduction of a fresh supply a very simple matter. The wire guard in the mouth of the air cleaning chamber must be replaced before putting back the car-burettor air pipe.

The air cleaner requires very little attention, but once a season the horse hair should be withdrawn from the cover, carefully washed out in petrol, and replaced. If the hair is particularly dirty it should be replaced with fresh and perfectly clean curled white horse hair.

Mechanical Interpretation of the Otto Cycle— Referring to Fig 10, the mixture from the carburettor is admitted once every two revolutions of the crank-shaft to the combustion chamber and cylinder, through the *Inlet Valve Port*. This port is opened at the appropriate moment by the depression of the *Inlet Valve*, as shown in Fig. 10. The manner in which this is carried out is as follows. On the crankshaft a gear-wheel is keyed or fixed, so that it rotates with the shaft. This gear-wheel meshes with another (shown on the right) of twice the diameter, or number of teeth: the latter wheel therefore makes one revolution to every two of the crank-shaft. Keyed to this wheel is a *Cam* (the *Inlet Cam*) which is so designed that it is cylindrical over about three-quarters of its surface, and has a curved projection on the other part. Thus during one complete rotation of the larger gear-wheel, the cam lifts the *Push-Rod* shown, for about one-quarter of its revolution, and allows it to remain stationary during the other three-quarters (i.e., when the inlet valve is closed during the compression, firing and exhaust strokes). During the inlet stroke the cam raises the push-rod, tilts the *Rocker Arm* about its central pivot or pin, and opens the inlet valve. When the roller of the push-rod returns to the plain portion of the inlet cam, the inlet valve spring closes the valve, and keeps is closed. The valve, it will be observed, opens inwards, and closes on to a conical seating; the pressures of compression and explosion therefore force the valve on to its seating and tend to render the joint quite gas-tight.

The *Exhaust Valve* is operated by its own cam, push-rod and rocker in exactly the same way, at the correct moment and for the appropriate period. It will be observed that the exhaust and inlet cams are not in the same positions relatively to the crank-shaft; this is because these cams have to open the valves at different moments, in order that the Otto cycle of operations may be carried out.

The sequence of valve opening periods can also be shown graphically in the cases of both the two and

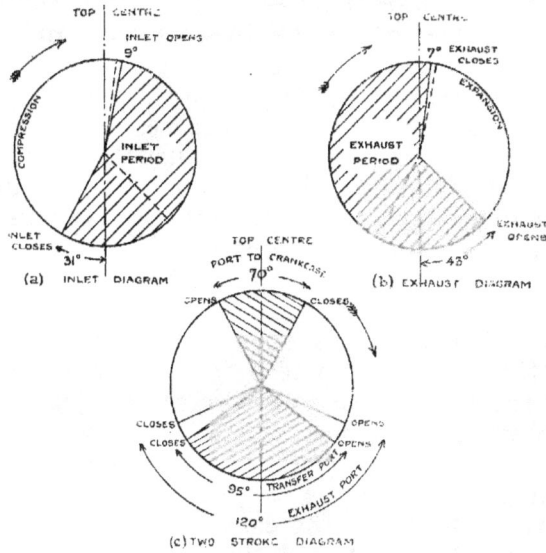

Fig. 16.—Valve Timing Diagrams.

four-stroke engines by denoting the corresponding positions of the main crank arm at which these operations occur. Referring to Fig. 16 (left), which represents the *Valve Timing Diagram* in the case of a typical four-cylinder Car Engine, the inlet valve opens when the crank is just 9° past the top centre position (i.e., a little past the piston's top position when it has descended a little). The inlet valve closes 31° after bottom centre; this extra 31° enables the incoming charge, in virtue of its momentum, to continue to flow into the cylinder in spite of the fact that the piston has begun to move up. From 9° past the top centre position both valves are closed on the firing stroke. The exhaust valve opens at 43° before bottom centre, and remains open until 7° past the top centre on the next stroke. The reason for the early opening and late closing of the exhaust valve is to obtain the longest possible time, in order to get rid of the hot gases. The inlet valve is thus open for 202°, and the exhaust for 230°. The engine in question is a high speed one, and this valve timing diagram is appropriate. For a slower type of engine the following is a good valve timing arrangement : —

(1) Inlet valve opens at top centre, and closes 25° past bottom centre.

(2) Exhaust valve opens at 40° before bottom centre, and closes at top centre.

These valve timing angles are usually marked on the rim of the engine's fly-wheel so that should the valve cam gear-wheels be dismantled at any time the engine can be re-timed without difficulty. The lower diagram of Fig. 16 shows the valve timing diagram for a typical three-port type of two-stroke engine. Here, the exhaust port is open for a period equal to 120°, and the transfer port to 95°, of crank angle. The port from the carburettor to the crankcase (namely, C., Fig. 7, diagram 3) is open for 70° during the period the piston is at the upper end of its stroke.

In passing it should be emphasized that it is the shape of the contour of the cam which determines the valve opening period, whilst the position of the cam, or its driving gear, relatively to the engine crank-shaft fixes the moments of opening and closing. Hence by altering the mesh of the valve gear-wheels, the time of opening and closing are altered, *not* the period during which the valve remains open. This is equivalent to shifting the shaded parts of (1) and (2) in Fig. 16, bodily around the centre of the crank circle.

The function of the *Piston* (Fig. 10) is to receive and transmit the force of the exploding and expanding gases to the crank-shaft, via the *Connecting Rod;* it must also act as a gas-tight plunger for the suction, compression and exhaust operations. The piston receives the thrust of the connecting rod; this causes it to press heavily upon the cylinder wall during the firing stroke. The piston is given a good bearing area for this reason and means are provided for lubricating the cylinder walls. Split rings, made of cast-iron, but which are flexible, known as the *Piston Rings*, are provided to ensure gas-tightness. The piston also houses the *Gudgeon* or *Wrist Pin*, upon which the upper end of the connecting rod rocks.

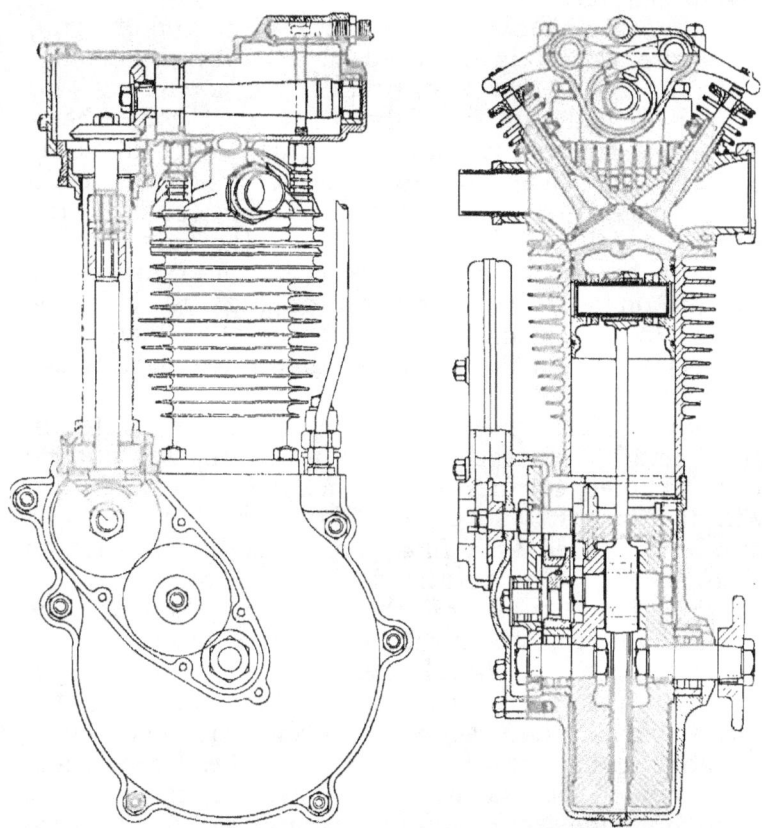

Fig. 17.—A typical Motor Cycle Engine.

The connecting rod, as we have mentioned, is the connecting link between the piston and the crankshaft, whereby the reciprocating motion of the former is converted into the rotary motion of the latter; the lower bearing of the connecting rod is on the *Crank Pin*, or *Journal*, of the crank-shaft.

The cam-shaft, which rotates at one-half engine speed, is also known as the half-speed shaft.

Referring to Fig. 10 again, on the extreme left is shown a gear drive for a high voltage electricity generating device, or *Magneto*, which produces the high voltage electric spark for the ignition of the charge; this is referred to again at the end of the present chapter.

Finally, the whole of the mechanism and gearing below the piston is enclosed in an oil-tight casing, known as the *Crank-case*, or *Crank Chamber*. This latter is usually in the form of a cast-iron, sheet-metal or aluminium alloy casting and not only contains the bearings for the crank-shaft, cam-shaft, and other moving parts, but confines the lubricating oil, or "oil spray," so that all moving parts can be adequately lubricated. The crank-case is also provided with projections or lugs to bolt the engine to the frame of the automobile and with brackets upon which the engine-driven accessories are mounted.

Before referring to the engine components in more detail, reference will be made to two typical single cylinder engines, shown in section, commencing with the motor-cycle engine shown in Fig. 17, this may be regarded as a final step from the illustration of Fig. 10. The left-hand diagram refers to the exterior front view of this engine, showing the ribbed cylinder, crank-case, and the vertical shaft provided with a bevel gear-wheel at its upper extremity, engaging with a similar wheel on the horizontal camshaft shown. In this case the valve-operating cams are on the latter overhead shaft, and are machined integral with it. Referring, next, to the right-hand side sectional illustration, it will be observed that the cams on the cam-shaft just mentioned actuate the valves by lifting up the inner ends of the two rocking levers, each of which is pivoted at about its centre. The inlet valve is on the left, and the exhaust on the right. Both valves are inclined; this arrangement gives a very efficient shape of combustion chamber, and provides easy paths for the inflowing and outflowing gases. The piston in this case is made of a copper-aluminium alloy, and has two rings above, and one ring (known as a *Scraper Ring*) below the hollow (*Gudgeon* or *Wrist*) pin shown The crank arrangement consists of a pair of cast iron fly-wheels, connected by a hard steel pin which

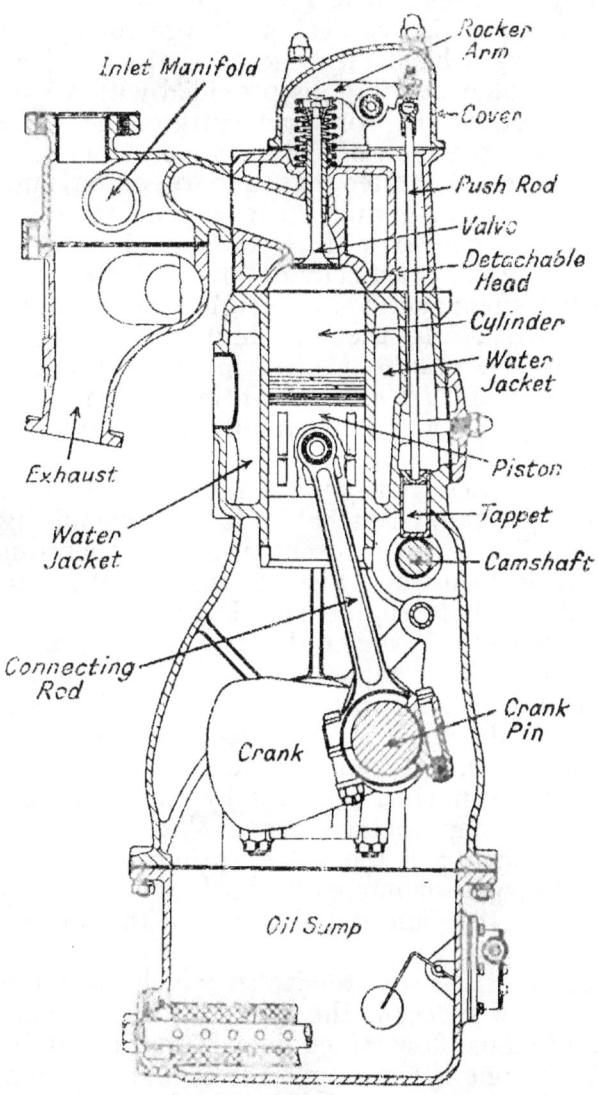

Fig. 18.—End Sectional View of Overhead Valve Water-Cooled Car Engine.

forms the crank-pin; this combination of crank-shaft and fly-wheels enables a good deal of space to be saved. A short coned shaft is attached to the centre of the left-hand fly-wheel, and rotates on roller bearings, at its other end it is provided with a driving gear-wheel, which engages with a train of gears, to drive the vertical shaft previously referred to. The right-hand fly-wheel shaft, which also rotates on roller bearings, is provided with a sprocket-wheel, keyed to it, for transmitting the drive, by chain, to the gear-box of the machine.

The engine illustrated has a bore of 69 mm. and stroke of 77 mm., giving a capacity of 347 c.c.

The valves shown in Fig. 17 should be especially noted. These are practically hollow, in order to reduce the heating effect and to save weight; this type of valve, which is frequently used on high speed engines, is known as the *Tulip Valve*, in distinction to the *Mushroom-Headed Valve* shown in Fig. 10.

Fig. 18 shows a typical water-cooled car engine in end sectional view.

The valves are arranged in the cylinder head, the latter being made detachable for decarbonising and valve grinding. The mechanism for operating the valves, clearly shown in Fig. 18, consists of a camshaft driven at one-half engine speed, the cams on which move the tappets above and the latter operate the rocker arms through long tappet rods having cups at their upper ends, engaging with ball-ended members screwed into the right-hand ends of the rocker levers. The latter can rock on central bearings so that as their right-hand ends are pushed upwards by the push rods the left-hand ends depress the valves. When the cam ceases to lift the push rod the valve springs—which are of the double-type—maintain the push rod and tappet in contact with the cam and also return the valve to its seating.

The cylinder has a water-jacket which communicates with the water spaces in the cylinder head so that there is a continuous flow of cooling water through both when the engine is working. The carburettor and inlet manifold are on the upper left-hand side and the exhaust manifold seen just below. The whole of the valve mechanism in the cylinder head is enclosed by a detachable cover so that there is ready access to the valve clearance adjustment gear. The lower part of the crankcase terminates in a sump which contains the oil for lubricating the engine. A submerged gear-type of oil pump (not shown) draws oil from this sump and delivers it to the various parts requiring lubricating. All surplus oil from these parts drains back into the sump again.

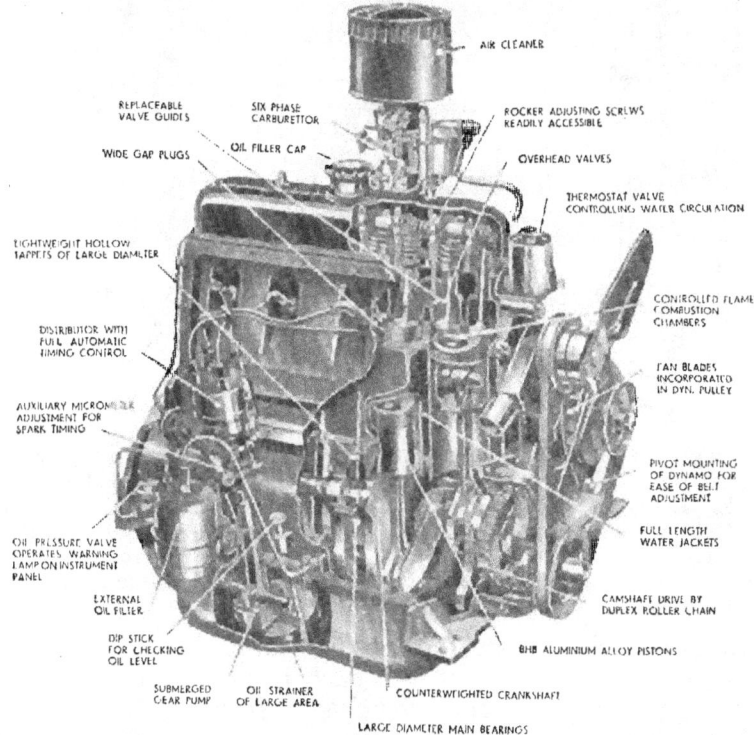

Fig. 19.—Showing Principal Features of the Vauxhall 10 H.P. Engine.

The Engine Components.—Having considered the general arrangement of one or two representative four-cycle engines, from the single-cylinder engine view-point, and thus obtained a broad idea of the purposes of the different parts, it is now proposed to consider each of the important components in more detail, commencing with the cylinder.

C

Figs. 20 and 21 illustrate the external fittings of a typical six-cylinder engine, viz., the Armstrong-Siddeley Special. The various components are indicated by the lettering, a key to which is given in the captions below.

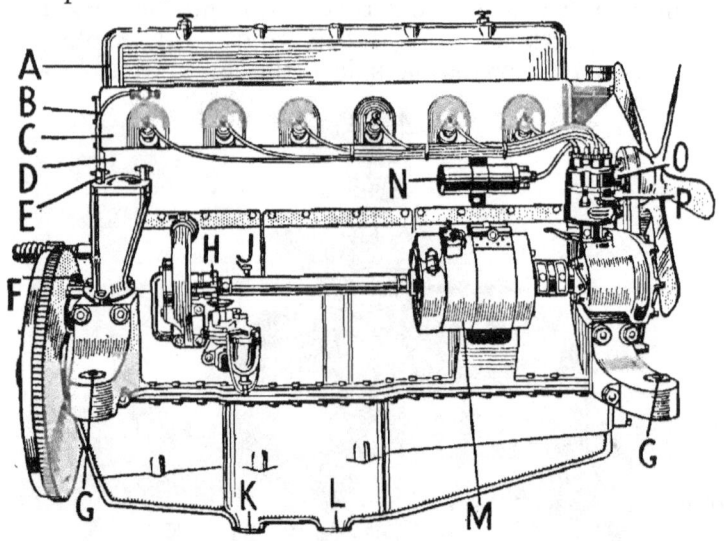

Fig. 20.—The Siddeley Six Engine, showing A—Valve cover. B—Oil pipe to rocker shaft. C—Cylinder head. D—Cylinder block. E—Oil filler. F—Connection for revolution counter drive. G—Rubber buffer mounting. H—Water pump gland adjustment. J—Oil level indicator. K and L—Oil sump plugs. M—Brush cover clip. N—H.T. ignition coil. O—Fan Belt Adjustment and P—Distributor.

In connection with the engine mounting lugs G, these are four in number and each has a rubber buffer inside, in order to insulate the chassis frame against any engine vibrations that may occur.

The Engine Cylinder: Air-Cooled Types.—The cylinder, we have seen, encloses the combustion chamber, and provides the working surface for the piston; it also contains the valves and sparking plug. In the preceding chapter mention was made of the high explosion pressures which occur in the case of modern engines, namely, from 300 to 600 lb. per sq. in. In the case of a cylinder of 4 inches bore, with a compression ratio of 5 to 1, the total force due to the explosion (for a pressure of 390 lb. sq. in.), and

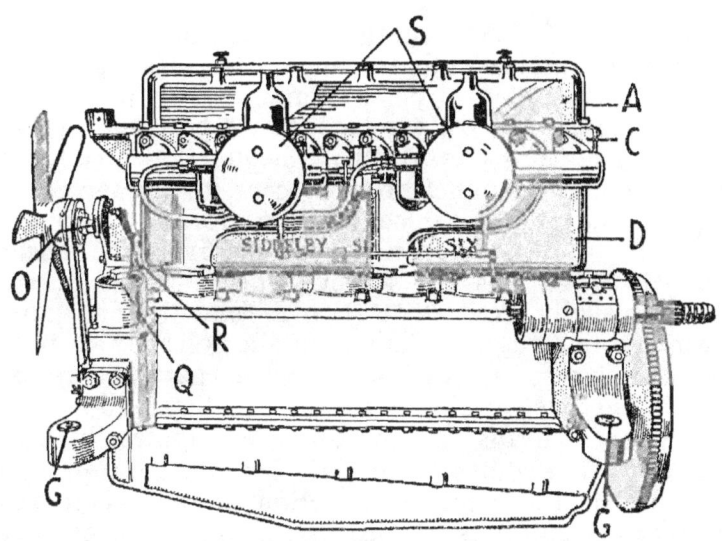

Fig. 21—The Siddeley Six engine, showing A—Valve cover. C—Cylinder head. D—Cylinder block. G—Rubber buffer mounting. O—Fan belt adjustment. Q—Oil filler cap. R—Oil relief valve and S—Downdraught carburettors.

which tends to blow the cylinder head off is no less than about 2¼ tons; the holding down bolts and nuts must be strong enough to withstand this pressure. A good example of an air-cooled engine cylinder is given in Fig. 17 and another in Fig. 22. The former has a detachable head, the whole valve unit and cylinder head being removable after the holding down nuts have been taken off; the cylinder barrel is held to the crank-case by four holding down studs and nuts. A turned extension of the barrel fits into a recess in the top part of the aluminium alloy crank-case; this ensures the barrel being in line with the crank-shaft, and enables it to be "located" again correctly whenever it has been dissembled. This locating spigot portion will also be observed in the example shown in Fig. 22. In Fig. 17 the cooling fins in the cylinder head are vertical, whereas in the case of Fig. 22 they are horizontal.

Cylinder Materials. — The majority of air-cooled cylinders are made of a hard grade of grey cast-iron the fins being chill cast, with the result that they are extremely hard.   An average composition of cast-iron suitable for motor cylinder and piston castings is as follows : Iron from 93 per cent. to 95·5 per cent., Carbon (combined) 0·5 per cent. to 0·8 per cent. (total) 2·7 per cent. to 3·3 per cent., Silicon 1·2 per cent. to 1·8 per cent., Manganese 0·6 per cent. to 1·0 per cent., Sulphur (max.) 0·12 per cent., Phosphorus (max.) 0·85 per cent.   This cast-iron gives an average tensile strength of 16 tons per sq. in.   The thickness of the walls is usually from $\frac{5}{32}$ to $\frac{7}{32}$ inch, and the fins taper down to a fine edge.   Aluminium alloys have also been used for air-cooled cylinders, more particularly on aircraft engines, although the use of these alloys for the cylinder heads of motor-cycle engines is now favoured.   The alloys themselves are not hard enough to withstand the hammering action of the valves and the wearing action of the piston, so that it is necessary to insert cast-iron or bronze valve seatings for the

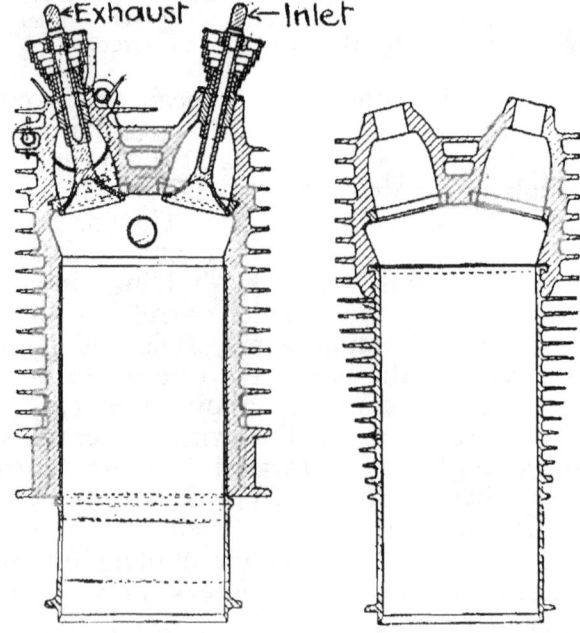

Fig. 22.—Examples of Air-cooled Cylinders.   Left— Aluminium Alloy with Steel Barrel and shrunk in alloy steel Valve Guides and Seatings.   Right—Steel Cylinder with Cast Aluminium Alloy Head, and Bronze Seatings.

valves (these are usually cast in place), and cast-iron, alloy steel or Nitralloy liners in the case of the cylinders. (Fig. 22.)

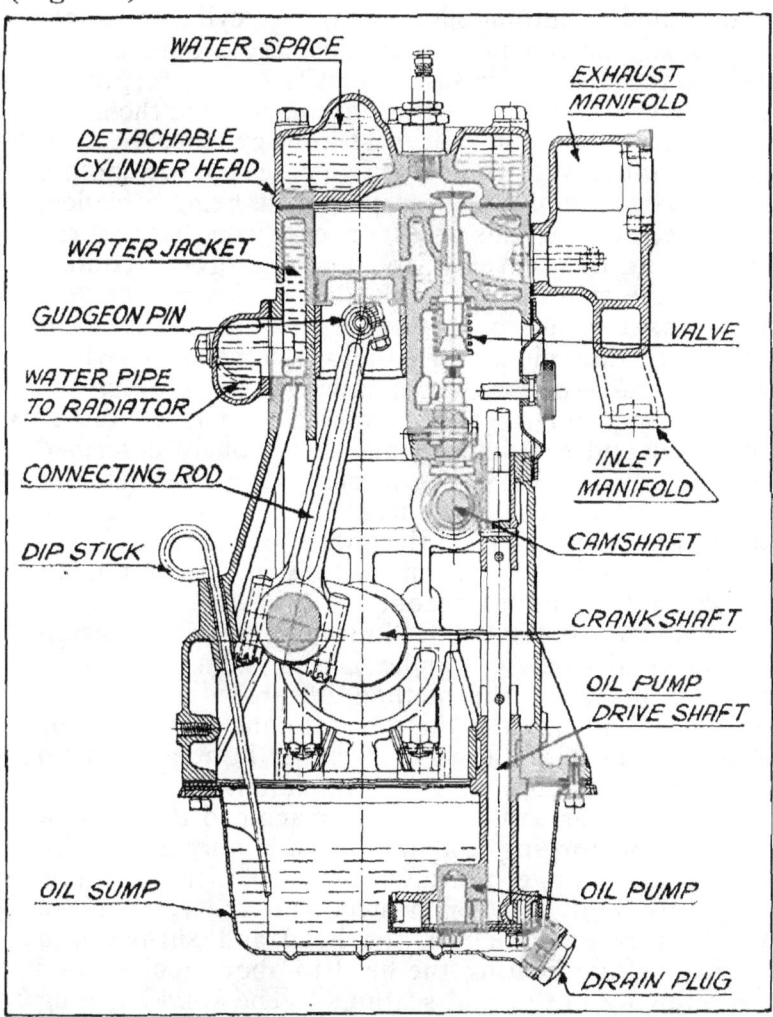

Fig. 23.—Side Sectional View of Water-Cooled Car Engine.

A typical aluminium alloy for cylinder and piston castings is as follows: Aluminium 91 per cent., Copper 7 per cent., Tin 2 per cent. Other well-known alloys are the British Engineering Standards Alloys 2L8, 3L11 and the "Y" and R.R. Hiduminium

Alloys. Elektron, a magnesium alloy, weighing only 60% that of aluminium, has been used for pistons, crankcases, etc.

Air-cooled automobile engine cylinders are occasionally made of steel. In this case a hard carbon steel (40 to 60 tons per sq. in. tensile stress) forging is used, the fins and barrel being machined from the solid. A separate aluminium or cast-iron cylinder head is generally employed. In one or two cases copper cooling fins have been employed, the separate fins being threaded, or pressed on to the steel or cast-iron barrel; the difficulty in this case is to obtain a good contact between the two metals; a film layer of dirt acts as a partial heat insulator. Copper is the better heat conductor, aluminium and its alloys come next, whilst steel and cast-iron are much inferior, but possess the best wearing and strength qualities. The combined aluminium and cast-iron cylinder previously described has the advantages of both metals. The cylinder barrel in this case is shrunk into place in the aluminium alloy. The thickness of the walls of steel cylinders is only about $\frac{1}{16}$ to $\frac{1}{8}$ in., as compared with $\frac{3}{16}$ to $\frac{1}{4}$ in. for cast-iron.

A Modern Air-Cooled Cylinder.—A modern design of aircraft air-cooled cylinder is shown in Fig. 24 in order to illustrate the method of providing a large area of cooling fins and the general construction. The cylinder head is of a high strength light aluminium alloy, such as duralumin, Hiduminium R.R. 50 or R.R. 53 alloy or Y-alloy. It is machined throughout from a solid forging and the fins are only about $\frac{1}{8}$in. apart. The valve seatings are of nickel-chrome-manganese high expansion steel and are screwed to suit the threaded holes in the head and shrunk into position by first heating the head to about 400° C. and then screwing in the cold seatings. The sparking plug adapter is fitted in a similar manner. It may be of interest to note that no less than 80 different machining operations are necessary from the forged to the finished condition of the head, during which the weight of the forging is reduced to less than one-third.

The lower part of the head is screwed internally to suit the thread on the cylinder barrel. The latter is

made of a special nickel-chrome-molybdenum steel
which is hardened, after machining, by the nitrogen
hardening process known as nitriding; this gives a
glass hard wearing surface.

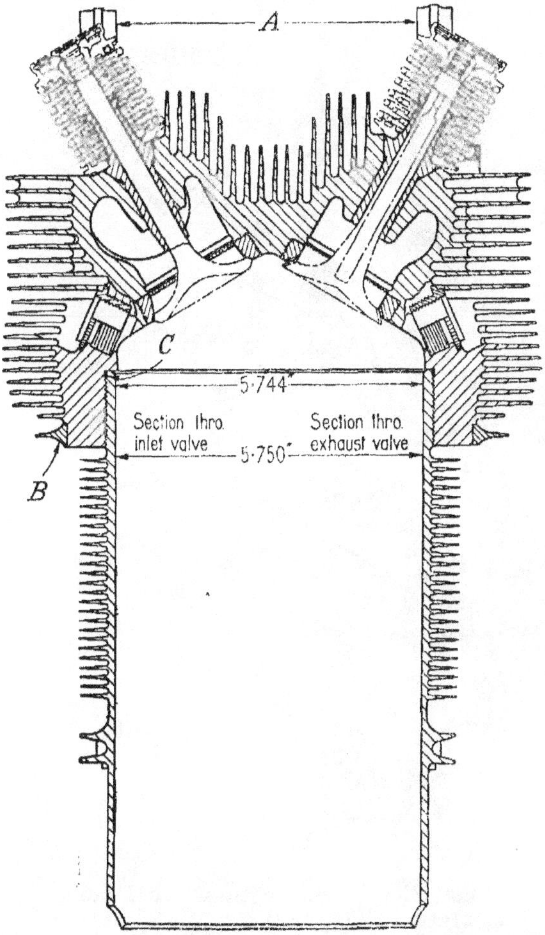

Referring to Fig. 24 a ring of steel B is shrunk on to the lower part of the cylinder head and a copper gasket ring is expanded into an undercut portion at the end of the thread at C. The rocker mechanism for actuating the valves is shown at A. The inlet valve in on the left and the hollow sodium-cooled exhaust valve on the right.

With this type of cylinder an output of over 45 B.H.P. per litre is obtained under normal oper-

Fig. 24.—The Bristol Air-Cooled Engine Cylinder.

ating conditions, with a compression ratio of about 7 : 1
and corresponding minimum fuel consumption of
0·40 lbs. per B.H.P. per hour. The mean effective
pressure (B.M.E.P.'s) at full power for supercharged
engines using this design of cylinder are 170 to 200 lbs.
per sq. in.

Water-Cooled Cylinders.—Water-cooled automobile engine cylinders are, generally speaking, more intricate in design, since the cooling water ports and passages

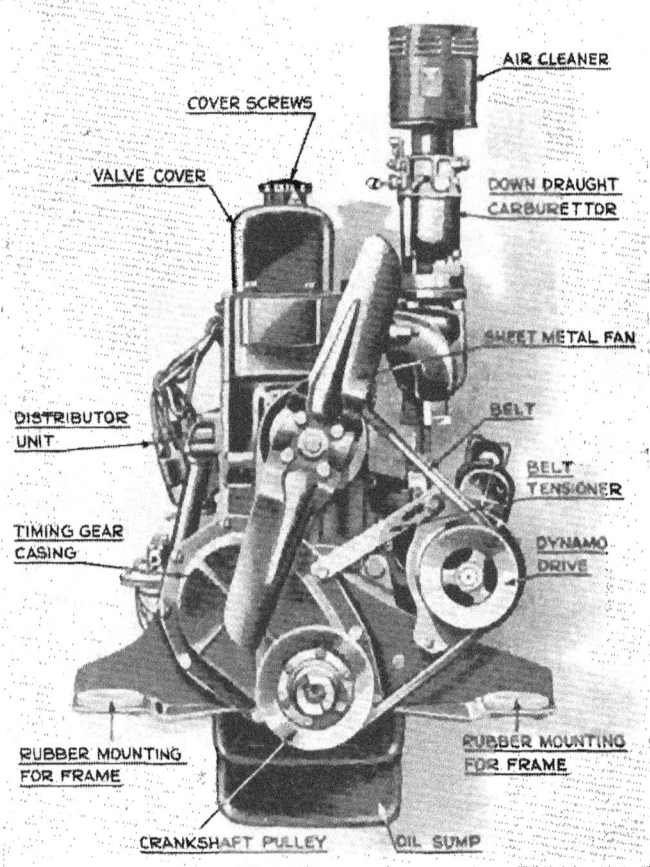

Fig. 25.—A modern Light Car Engine with Down Draught Carburettor (Vauxhall).

have to be arranged in the castings. The examples given in Figs. 18 and 23 will enable some idea to be obtained of the general type of casting employed.

Hitherto plain cast-irons have been used for cylinders, since they gave sharp clean castings in the case of the complicated multiple cylinder blocks. The irons suitable for such purposes had not, however, the best long-wearing properties. In order to obtain suitable cylinder castings, combined with much longer service, it is now becoming the practice to use an alloy cast-iron, containing nickel and chromium in small percentages. Cylinders made from such irons can be used for much greater mileages before reboring becomes necessary, than for ordinary irons.

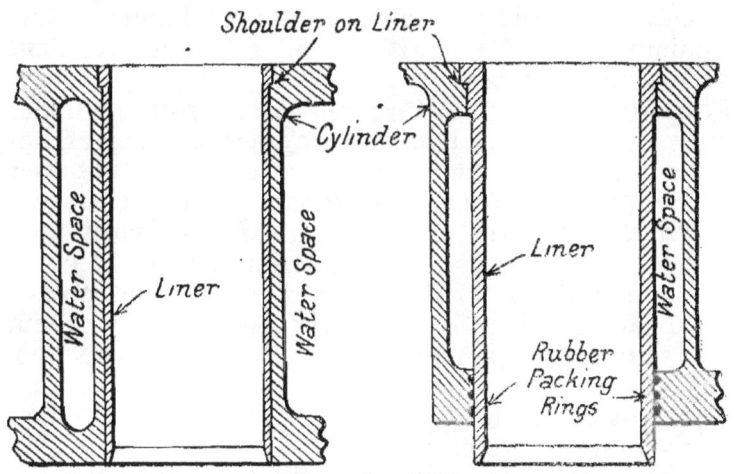

Fig. 26.—Showing (left) Dry Cylinder Liner and (right), Wet Liner.

Cylinder Liners.—It is now also customary to employ plain cast-iron or aluminium cylinders, and to fit the cylinder barrels with liners machined from an alloy cast-iron such as Chromidium, Centrard or Durocyl. These irons are of the oil-hardening type, and give considerably longer useful service. Moreover, it is unnecessary to scrap the pistons, since worn liners can be replaced with new ones to suit the existing pistons if not worn oval or scored. In this way the life of a cylinder block can be extended almost indefinitely. Most of the expensive car and all commercial compression ignition engines are fitted with hardened and ground alloy cast-iron liners.

When the liner forms the complete cylinder barrel with the water in contact outside, it is termed a *Wet Liner*, but if it is inserted into an existing cylinder barrel it is termed a *Dry Liner*, for it is no longer in

C*

contact with the water in the cylinder jackets (Fig. 26).
Another type of cylinder liner is that made from a
special alloy steel known as Nitralloy. This has the
property of becoming hardened when heated to redness
and exposed to ammonia gas. The surface then
becomes extremely hard and wear-resistant. Another
cylinder liner material is a nitrogen-hardening cast-
iron known as Nitricastiron; it contains aluminium and
chromium in small quantities, and gives an excellent
wear-resistant surface.

Chromium-plated Cylinders.—Cast iron cylinder
barrels of commercial vehicle engines are sometimes
given a hard chromium coating in order to obtain better
wearing qualities. A process, known as the " Listard "
used in this country enables the layer of chromium
to adhere perfectly to the cylinder wall metal and the
chromium surface can be given a mirror-like polish.
The surface obtained is extremely hard, namely, about
three times the diamond hardness of hardened tool
steel.

Tests made with different types of cylinder liner
materials on commercial engines showed that with cast
iron a wear of 0·010 occurred in the cylinder at the end
of 30,000 road miles; with hardened alloy cast iron
liners the same amount of wear occurred after 60,000
miles whilst with Listard treated cast iron only 0.006 in.
occurred after 120,000 miles.

Aluminium Cylinders.—Aluminium cylinders are
more expensive, but give better cooling than in the
case of cast-iron cylinders and jackets. Incidently, it
has been found possible to employ higher compression
ratios with aluminium cylinders and pistons than with
cast-iron ones, for the heat is conducted away quicker
in the former case and the same cylinder temperature
attained as with a lower compression ratio.

Fig. 23 is a side sectional view of a modern light
car engine having side-by-side valves, aluminium
piston, nickel-chrome steel connecting rods and a
detachable cylinder head.

An important feature of this engine is the shape
of the combustion chamber, which is practically flat
on the left side so that when the piston approaches the
top of its stroke the compressed mixture is forced out

of this part and swirls around in the space on the right. This gives the necessary turbulence to the mixture at the moment of ignition; in this way higher

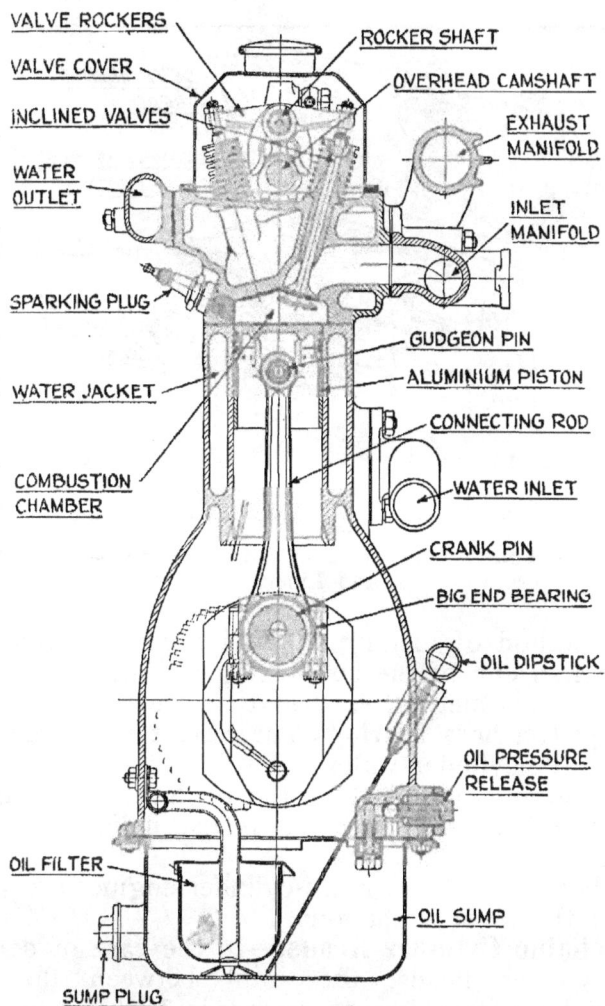

VALVE ROCKERS

VALVE COVER

INCLINED VALVES

WATER OUTLET

SPARKING PLUG

WATER JACKET

COMBUSTION CHAMBER

OIL FILTER

SUMP PLUG

ROCKER SHAFT

OVERHEAD CAMSHAFT

EXHAUST MANIFOLD

INLET MANIFOLD

GUDGEON PIN

ALUMINIUM PISTON

CONNECTING ROD

WATER INLET

CRANK PIN

BIG END BEARING

OIL DIPSTICK

OIL PRESSURE RELEASE

OIL SUMP

Fig. 27.—Typical Overhead Camshaft Engine, showing principal components.

compressions can be used without detonation. It will be noted that the sparking plug is placed centrally in the open combustion space.

Another important point is the vertical shaft drive, on the right, down to the gear-wheel oil-pump situated in the oil-sump.

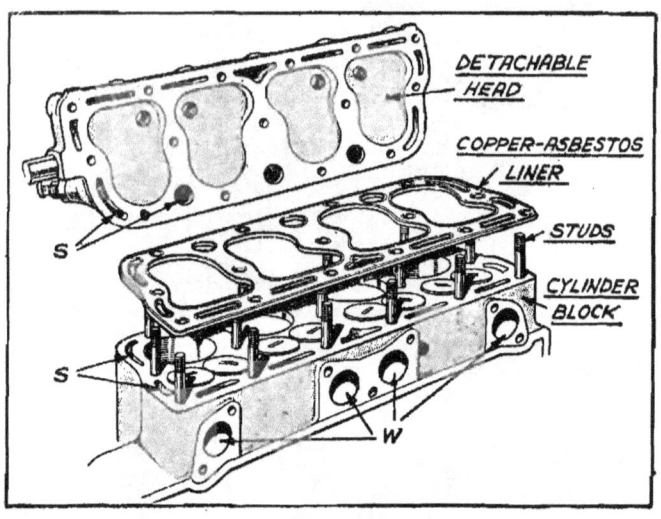

Fig. 28.—Four-cylinder Engine showing Detachable Head and Liner.

The method of locking the wrist, or gudgeon pin, to the small end of the connecting rod should also be observed; this method ensures that the pin shall rock in the piston boss bearings but cannot move endwise to score the cylinder walls.

The method of indicating the oil level by means of a dip stick (shown on the left) is also indicated clearly in Fig. 23.

Further examples of multi-cylinder engines are given later in the present chapter.

Detachable Cylinder Heads.—In the case of detachable cylinder heads, the joint between the two machined surfaces has to be made both water- and gas-pressure-tight, by means of a special gasket, or liner, made of two very thin sheets of copper, with asbestos packing between; this liner, when screwed up, beds down and conforms to the machined surface shape, thus making a good joint. Such a joint can frequently be used two or three times if it is not dented or damaged in removal; the application of a little gold-size, or boiled linseed oil, to the surfaces enables a better joint to be made.

A typical detachable cylinder head is shown in the upper part of Fig. 23 in cross-sectional view, the holding down nuts and sparking plug socket being clearly illustrated, whilst Fig. 28 shows a four-cylinder engine with its detachable head removed. The copper-asbestos liner is also shown ready to drop over the studs on the top of the cylinder block.

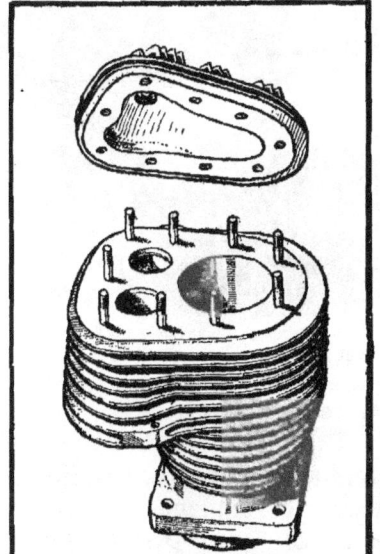

Fig. 29.—A Typical Motor Cycle Engine Detachable Cylinder.

The cylinder heads of water-cooled engines are made hollow for the water to circulate round. For this purpose a number of slots and holes, marked S in Fig. 28 are arranged to correspond with one another in both the cylinder head and block.

The cylinder head, it should be noted, carries the sparking plugs, top water-pipe connection to radiator and, as in the case illustrated, the cooling fan pulley boss.

The four holes marked W. in Fig. 24 represent the exhaust manifold connections.

In the case of motor-cycle engines it is the practice to fit cast-iron cylinders with aluminium alloy detachable heads. This arrangement not only ensures better conduction of the heat but also enables a higher compression to be employed. Fig. 29 shows a typical example of a detachable cylinder head for a motor cycle.

The advantages of a detachable cylinder head are as follows:—(1) It enables the cylinders to be machined uniformly, and also the heads; and (2) it obviates the necessity of removing the whole (and usually heavy)

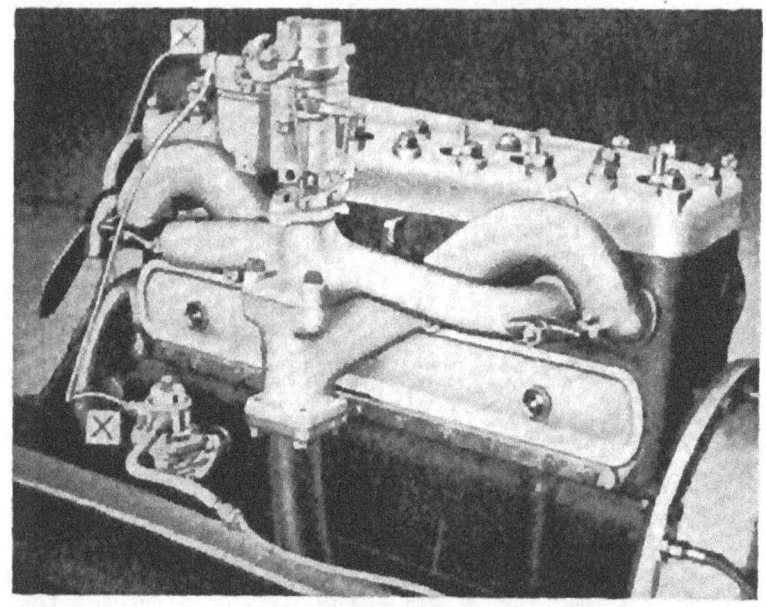

Fig. 30.—The Austin Aluminium Alloy Cylinder Head.
The Carburettor and Fuel Feed Pump are shown at **X**
above and **X** below, respectively.

cylinder block for decarbonizing and valve regrinding.
Those who have endeavoured to replace a four-
cylinder monobloc casting on to the four pistons and
the crank-case, single-handed, will appreciate the
advantages of the detachable head.

In connection with the subject of cylinder head
gaskets, a recent tendency is to use thin sheet metal
gaskets instead of copper- asbestos ones, since the latter,
being flexible, cause variations in the compression
ratios when they are tightened by different amounts.

Linerless Aluminium Cylinders.—The Cross linerless aluminium alloy cylinder is made from a special
hard and strong alloy and machined all over.   The
piston used in this cylinder is also of aluminium alloy
and is prevented from actually touching the cylinder
walls by three rings of a specially hardened type (Wellworthy); these rings fit the bottoms of their piston

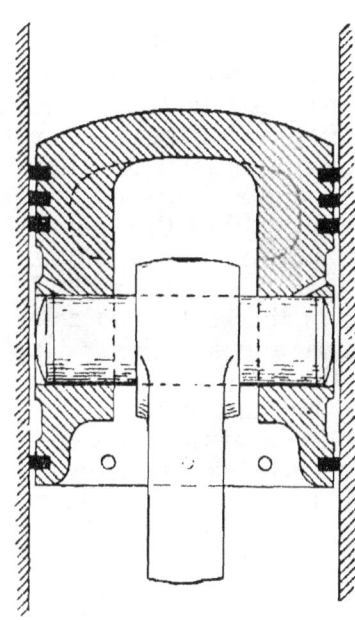

Fig. 31.—The Cross Liner-less Aluminium Cylinder.

grooves (Fig. 31). The advantages claimed for this system are that there is appreciably less cylinder wear; the cylinder and piston have the same co-efficient of expansion and therefore require very little working clearance, thus giving good gas-tightness, low oil consumption and freedom from cold piston slap; better heat conductivity so that a rather high compression can be used, and reduced weight.

Tests have been made on motor cycle engines fitted with this system, over hundreds of thousands of miles and the results obtained appear to justify the claims made for it.

Motor Car Engine Cylinder Construction.—In the case of most mass-produced car engines the cylinder block is made of cast iron. The complete casting comprises the cylinder barrels, water jackets, crankcase and main bearing housings, as shown in Fig. 30, which are rigidly designed to take the working loads transmitted from the pistons, connecting rods and crankshafts. The more recent models of car engines use an alloy cast iron containing chromium or chromium and nickel; such alloys have much better wearing qualities than ordinary cast irons used hitherto.

The castings for cylinder blocks are usually normalized and then stored for a period of some weeks in order to allow the metal to " age " before machining. This ensures that there will be no subsequent volume changes. Cast iron detachable cylinder heads have been widely used on such cylinder blocks, but more recently aluminium alloy heads (Fig. 30) have been employed; these enable higher compression ratios to be used, due to their better heat conduction. In one or two cases copper cylinder heads have been used with satisfactory results; copper has a better heat conductivity than aluminium. The lowest part of the crankcase usually terminates in a flat machined face provided with a relatively large number of studs for the purpose of attaching the pressed steel or cast aluminium alloy oil sump. The crankcase of the cylinder block casting also includes the bearer arms for bolting the engine to the chassis frame; the front portion of the clutch housing is also made integral with the cylinder block casting. The front end section of the latter also contains the timing gears or chains and sprockets. In one or two instances, however, these are at the rear or flywheel end of the crankcase, but this arrangement does not give the same accessibility as at the front end.

In some instances the cylinder block is made of aluminium alloy with inserted alloy cast iron liners for the cylinder barrels, following the method employed in aircraft engine construction where minimum weight is essential. Magnesium alloy, such as Elektron, is also used for automobile crankcases and this arrangement gives a lighter construction than for the aluminium alloy.

Cylinder Shapes.—The shape of the combustion chamber has an important bearing upon the output and thermal efficiency of the engine. This shape depends upon the disposition of the valves. If the valves are in the cylinder head, it is possible to approximate to the ideal spherical shape shown in Fig. 32 (A). Examples of efficient cylinder designs are given in Figs. 17 and 23. The inclined overhead valves and dished piston shown in Fig. 32 (B) are the nearest practical approach to (A). A more common arrangement, which is also efficient, is that shown in (C); this design can be constructed more cheaply, and is often used on high speed and racing engines.

Most overhead valve engines give more turbulence to the fresh charge than in the other designs shown in Fig. 32, and also give less heating to the incoming charge. An arrangement, which represents a compromise between the overhead and the side-by-side valve shape, is that illustrated in (D) Fig. 32 and in which the inlet valve is over the exhaust. This arrangement enables a single cam-shaft to be used; the inflowing mixture also cools the exhaust valve.

The side-by-side valve engine can be constructed economically, and is accessible; it is very widely employed in light and medium cars. The turbulence promoted in this type is not as satisfactory as in the overhead types, but the special combustion chamber

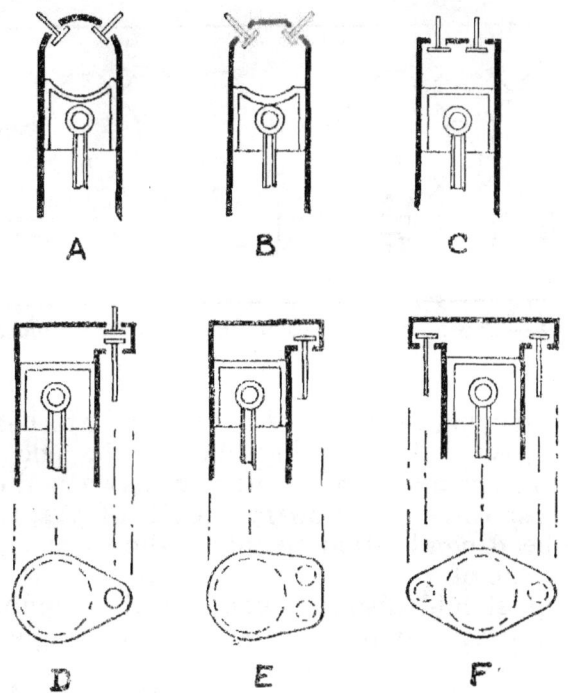

Fig. 32.—Showing various Combustion Chambers and Valve Dispositions.

design illustrated in Fig. 35 improves the degree of turbulence considerably, and renders this type almost as efficient as the overhead one. The T-headed engine shown at F (Fig. 32) is now obsolete; it is bulky, inefficient, and expensive to produce.

In regard to the overhead valve type of engine with inclined valves in the head as shown in Fig. 33, the results of research work carried out by Ricardo with different positions of the sparking plug and, in one instance with the cylinder head fitted with two sparking plugs as shown at B (Fig. 33), show that the highest useful compression ratio (H.U.C.R.) for the typical grade of petrol used was 6·0 : 1 for the arrangement shown at A, 5·9 : 1 for that at B; 5·3 : 1 for that at C, and 5·6 : 1 for that at D. The corresponding value for the side valve engine head with sparking plug over the inlet valve at E as was found to lie between 4·3 : 1 and 4·6 : 1.

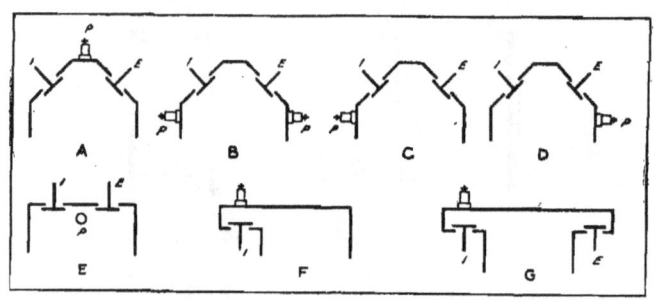

Fig. 33.—Types of Combustion Chamber, with different Sparking Plug Positions.

The plain cylindrical type of head, similar to Diagram E, Fig. 33, with the sparking plug on the side of the combustion chamber wall, centrally between the inlet and exhaust valves gave an H.U.C.R. of 5·2 : 1. For the Tee-head combustion chamber, shown at G, gave the low value of 3·6 : 1.

Mention should also be made of the symmetrical pocketless combustion chambers of the single-sleeve valve, the Aspin and Cross rotary engines which are described in Chapter III; each of these combustion chambers enable relatively high values of the H.U.C.R. to be utilized without detonation effects.

Non-Detonation Type Combustion Chambers.—A considerable amount of experimental work has been carried out in recent years with the object of improving the combustion chamber so that higher compression ratios can be employed with fuels which previously required lower compressions in order to prevent detonation. Now that the causes of detonation are more fully understood the shape of the combustion chamber can be designed accordingly and improved performance thus obtained.

The general principle of the method employed is illustrated in Fig. 34,* which shows an efficient design of car engine cylinder head. It will be seen that the chamber is divided into three zones. The first zone, on the left, is the ignition area which is likewise the heat loss area. In this area the metal is exposed to burning for the longest time and it is the region of the highest temperature. This area should therefore be protected against heat loss and the exhaust valve placed within this zone.

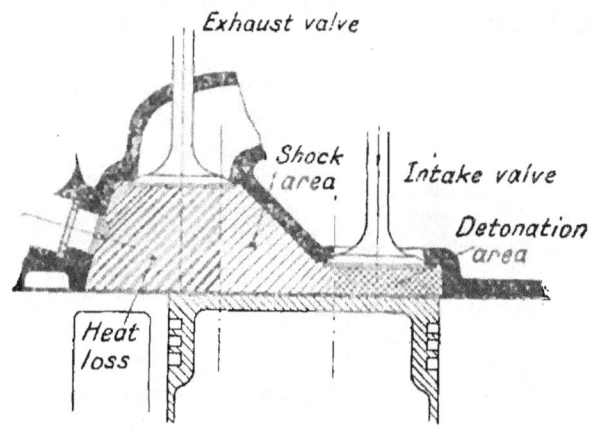

Fig. 34.—Combustion Zones.

The second zone is the shock area, because during the time this area is burning the crankpin and piston are passing through top-dead-centre and thus the structure is given the maximum effect of the pressure rise. It is desirable to reduce the pressure-rise rate in this area and hence a portion of the volume must be displaced from this point. If this is placed in the third area, which is the detonating zone, more volume-to-surface results than is satisfactory for detonation control. The relatively cooler inlet valve when placed in this area helps to absorb the heat of this super-compressed last portion of the charge to burn.

This is added to the last third, and, providing the depth of the section through this area is not over $\frac{5}{16}$ in.,

* " Carburation." A. Taub. Proc. Inst. of Autom. Engrs., March, 1938.

detonation may be adequately controlled for compression ratios up to 6·5 : 1. In this manner it is possible to get a lower heat absorption at the first area allowing a lowering of the flame front in the shock area and yet providing additional heat in the detonating area.

It may here be mentioned that a mechanical method has been devised by the Research Department of Messrs. Vauxhall Motors Ltd. to obtain the best shapes of combustion chamber for utilising the highest compression ratios for a given grade of fuel.

In the Whatmough design of combustion chamber for side-valve engines shown in Fig. 34A, the sparking plug is arranged over the exhaust valve so that the hottest part of the mixture is the first to be ignited, the flame then spreading to the relatively cooler parts of the mixture and combustion chamber. The latter is also designed so as to allow the mixture to flow into it on streamline principles. In order to secure as even a distribution of temperature as possible the cooling water is arranged to give the most efficient removal of the surplus heat from the combustion chamber possible.

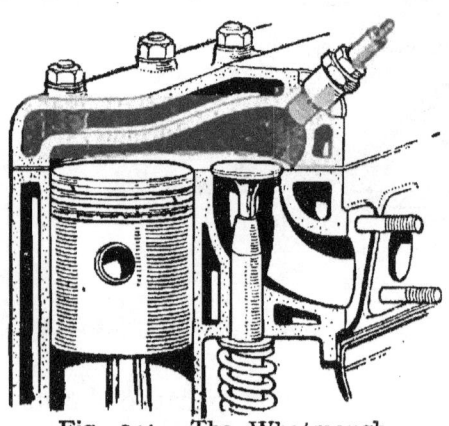

Fig. 34A.—The Whatmough Combustion Chamber.

Cylinder Dimensions.—The cylinder walls should have a minimum thickness of $\frac{3}{16}$ in. when cast in iron or aluminium alloy and 1-16 in. in steel (machined forging).

A useful design formula is as follows :—

$$t = \frac{p\,d}{2f}$$

where t = minimum wall thickness in inches, p = maximum explosion pressure in lb. per sq. in. absolute, d = cylinder diameter in inches, and f = safe working stress for the metal used, i.e., 2,000 lb. per sq. in. for aluminium and cast-iron, and 3,500 for steel.

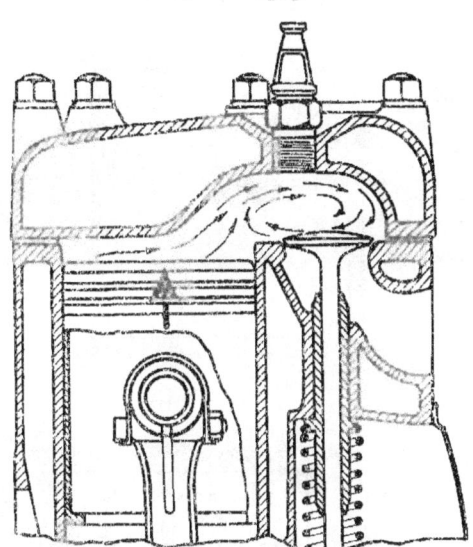

Fig. 35.—The Ricardo Turbulence Combustion Chamber Design for Side Valve Engines.

Another empirical formula is

$$t = \frac{d}{32} + \tfrac{1}{8} \text{ (inch)}$$

It is usual to allow a water space of $\frac{5}{16}$ to $\frac{7}{16}$ in. for natural circulation cooling, and for $\frac{3}{16}$ to $\frac{5}{16}$ in. for pump circulation.

Engine cylinders are generally tested hydraulically before assembling on the engine. The water jackets are tested to 40 to 60 lb. sq. in., and the cylinders themselves to 500 lb. sq. in. (minimum).

The Piston.—The piston has to maintain a gas-tight chamber for the cylinder charge, at all times, to transmit the explosion and expansion pressure loads to the connecting-rod, and to form one member of the connecting rod *Small-End Bearing*.

Pistons are now made either in (1) Cast-iron, (2) Alloy or Mild Steel, or (3) Cast Aluminium alloy.

At one time the cast-iron piston was universally employed, but subsequently was replaced by the

aluminium alloy piston which was appreciably lighter—
thus reducing the value of the reciprocating loads—
and could be more accurately made by the die-casting
process.

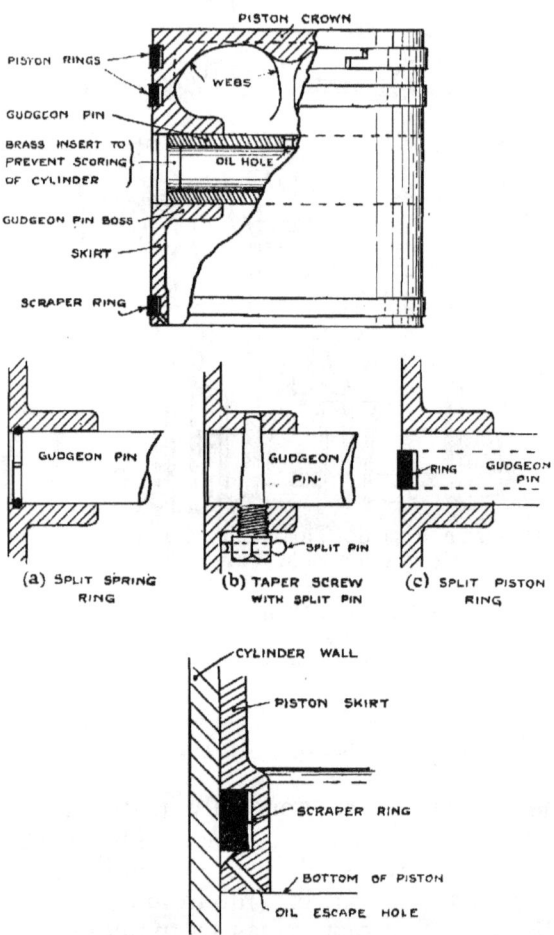

Fig. 36.—Cast-iron Piston, and some earlier methods of
securing the Gudgeon Pin, and (below) Oil Scraper Ring.

There has, however, been a tendency, more recently,
to revert to the cast alloy-steel or alloy-iron piston for
commercial engines, designed for operation over long
periods, since these types show reduced cylinder wear.
A notable example is the Ford steel piston which is

cast very accurately, and is made relatively much
lighter than the earlier cast-iron pistons.

In addition to its better wearing qualities, which
result in much longer piston and cylinder life, before
re-grinding becomes necessary, owing to the fact that
it works in a similar metal, the clearances with the
cast-iron pistons can be made quite small. On the
other hand aluminium alloy, which is very much lighter
is a far better heat conductor; it therefore runs much
cooler, and more free from carbon deposit. Owing
to its higher co-efficient of expansion than cast-iron,
it has usually to be made a "looser" fit when the
engine in cold; this extra clearance used to result in a
knocking noise, known as *Piston Slap*, which dis-
appeared, however, as soon as the engine warmed up.
To overcome this objection, it is usual to split the *Skirt*
or lower end of the piston, and to slightly spring the
segments. Other designs of non-slap aluminium alloy
pistons are described later in this section.

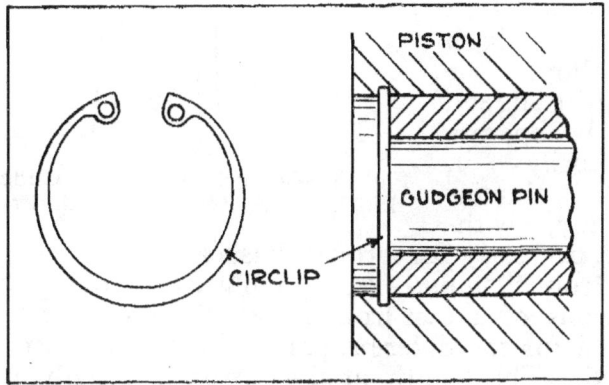

Fig. 37.—Circlip method of securing Gudgeon Pin.

The Gudgeon Pin.—The gudgeon pin is usually
made hollow to save weight, and to give as much
bearing surface as possible. The material used is a
high-grade carbon or low-nickel content steel, case-
hardened and ground to size.

Various methods have been used, in the past, to
secure the gudgeon pin in the piston bosses; some
of these are illustrated in Fig. 36. Of these, only those
at (*a*) and in the top illustration have survived.

The usual method employed in modern aluminium alloy pistons is to make the gudgeon pin a push fit in the small-end bearing and piston bosses. It is secured against end movement by means of spring clips, such as those shown at (*a*), or with a special design of clip known as a Circlip (Fig. 37); this has lugs at its ends for facilitating its removal with a pair of round-nosed pliers. The Circlip springs into a groove turned in the piston boss in the same manner as the device shown at (*a*).

In certain designs the gudgeon pin is actually clamped tightly in the small end of the connecting-rod and allowed to rock in the piston bosses. (Fig. 38).

When the piston is made of high ium alloy, the floating gudgeon pin will operate satisfactorily in strength aluminthis alloy, without the necessity for bronze bushes.

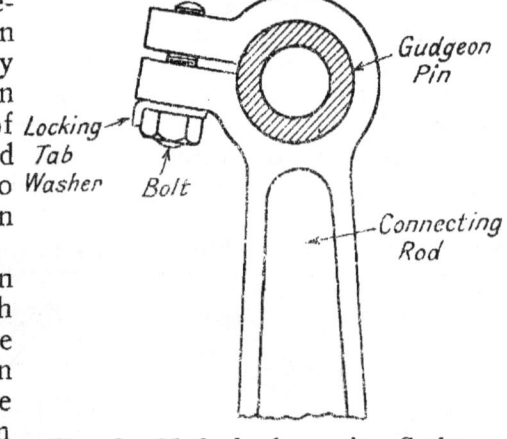

Fig 38.—Method of securing Gudgeon Pin in Connecting Rod Small End.

Cast-Iron Pistons.—Fig. 36 illustrates the general proportions of a cast-iron piston; the common terms employed for the different parts of the piston are also indicated. The length of the piston is usually made equal to the diameter, except in the case of slow-running engines (for commercial purposes), when the length is often from 25 to 45 per cent. longer. The average stroke of modern engines is from 1·3 to 1·4 times the cylinder bore.

The pressure of the piston on the cylinder walls, due to the thrust of the connecting-rod on the explosion stroke, should not exceed about 25 lb. sq. in. The gudgeon pin diameter is usually from a quarter to one-fifth of the piston diameter; it is an advantage to make it hollow, for weight saving reasons.

The thickness of the crown of the piston is usually from $\frac{3}{16}$ in. in the case of pistons of 2 to $2\frac{1}{2}$ in. diameter up to $\frac{1}{4}$ in. for 4 in. diameter. A useful formula for the thickness t is as follows : —

$t = \frac{1}{32} (D + 3)$ inches, where D = diameter in inches.

The number of piston rings employed varies from 2 to 4. A common arrangement is to use 2 rings above the gudgeon pin, and 1 ring below, but it is not unusual to fit 3 rings only, above the pin. The lower ring, in the former case, is termed a *Scraper Ring*, and its purpose is to "scrape" off any surplus oil from the cylinder walls, and thus to prevent over-oiling of these, which otherwise would result in much carbon deposit. Very often a bevel is arranged on the lower piston slot, and holes are drilled right through the metal, with the object of conducting the surplus oil away (Fig. 36, lower diagram).

Cast-iron pistons are made slightly smaller in diameter than the cylinder bore, in order to work properly, without seizing. It is usual to allow a working clearance between the piston and cylinder of $\frac{1}{1000}$ in. for every 1 in. diameter of piston; thus a 3 in. diameter would have a clearance of $\frac{3}{1000}$ in. (·003 in.). Further, since the top of the piston is much hotter than the bottom, the clearance is made rather more at this end, by about $\frac{2}{1000}$ to $\frac{4}{1000}$ in. as a rule.

Steel Pistons.—Previous to the general adoption of aluminium pistons, the light steel piston was used in racing car engines, but it was generally a more expensive piston to manufacture.

More recently the cast steel piston has come to the fore as a rival to the aluminium type on account of high strength and better wearing qualities.

The Lincoln Zephyr engine, made by the Ford Company, uses the copper steel piston design shown in Fig. 39. This piston has a skirt thickness of only ·035 inch, and a crown thickness of ·090 inch. It is, therefore, an excellent heat disperser as compared with the heavier cast-iron pistons. It is so light that the piston shown ($2\frac{3}{4}$ ins. diameter) weighs only 11 ozs., which is about 1 oz. more than the same diameter of aluminium piston. *The clearance* between the piston and cylinder wall can, of course, be made much smaller than for aluminium alloys, so that the oil consumption

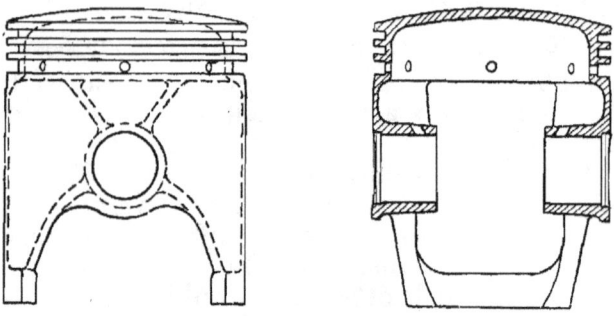

Fig. 39.—A modern Copper-Steel Piston.

is reduced and no piston-slap occurs at starting. The steel used contains 1·35 to 1·70 Carbon; 2·50 to 3·00 Copper; 0·6 to 1·00 Manganese; 0·9 to 1·3 Silicon; and 0·15 to 0·20 per cent. Chromium, the rest being Iron.

The pistons are cast in " gates " of eight pistons, the latter being afterwards separated by sawing.

Aluminium Alloy Pistons.—In the case of aluminium alloy pistons, the clearances are greater,* as we have mentioned before.

The average clearance for plain aluminium alloy pistons working in cast-iron or steel cylinders is $\frac{3}{1000}$ to $\frac{6}{1000}$ in. per inch diameter of cylinder. An appreciably greater clearance is given at the top of the piston, however. Thus in the case of a 3 in. diameter cylinder the skirt clearance is about $\frac{3}{1000}$ in. ; that of the first land† $\frac{15}{1000}$ in.; of the second land $\frac{10}{1000}$; of the third land $\frac{10}{1000}$ in.; and of the top or fourth land $\frac{60}{1000}$ in. The compensated types of aluminium alloy pistons described later in this section have relatively smaller clearances compared with those of the plain cylindrical ones.

* Due to the fact that aluminium has an expansion co-efficient about three times that of cast-iron.

† The land is the band of metal between or adjoining the piston ring slots.

Typical examples of the clearances for different parts of two different kinds of aluminium alloy pistons, namely, plain and split skirt ones, are illustrated in Fig. 40. The former illustration refers to a piston ground elliptical in shape to have a total clearance of 0.010 to 0.012 in. on the portion beneath the gudgeon pinholes—in the shaded area, and 0·002 to 0·0025 in. total clearance on the thrust faces.

The piston shown in Fig. 40 (B) is of a similar pattern to that given in Fig. 42. It permits close fitting, with

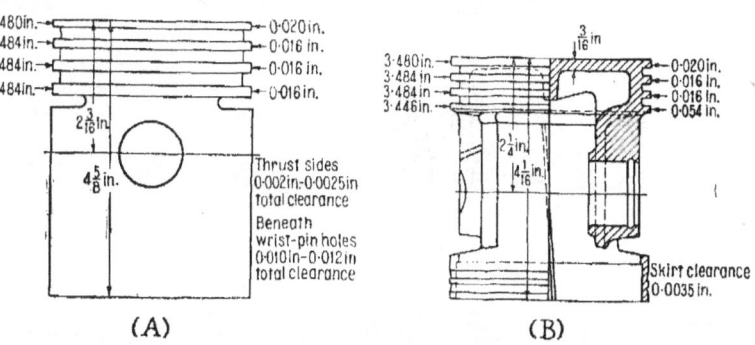

(A)   (B)

Fig. 40.—Illustrating Aluminium Piston Clearances.

skirt clearance of 0·0005 to 0·001 in. per inch piston diameter. The diagonal slit is arranged on the thrust face.

The materials used for aluminium alloy pistons include Aluminium-Copper (B.E.S.A. L.8 Alloy), Y—Alloy and R.R. Alloys Nos. 53 and 59.

Fig. 41 illustrates a good example of light aluminium piston design, in which the weight has been reduced to an absolute minimum by the removal of all redundant or surplus metal. It is known as the *Slipper Type Piston* from the fact of the " skirt " portion having the bearing surfaces only on the two piston thrust side; that on the explosion thrust side is of the larger area. In the case of a 4 in. diameter slipper piston, complete with its rings and gudgeon pins, the weight is only about 1·4 lb., whereas that of the ordinary light cast-iron piston is about 3·3 lb. The slipper piston has been shown to reduce piston friction by

25 to 35 per cent., giving an increase of power at
normal speeds of 5 per cent. and up to 15 per cent. at
the highest speeds. The gudgeon pin in the slipper
piston illustrated (Fig. 41) actually bears on the

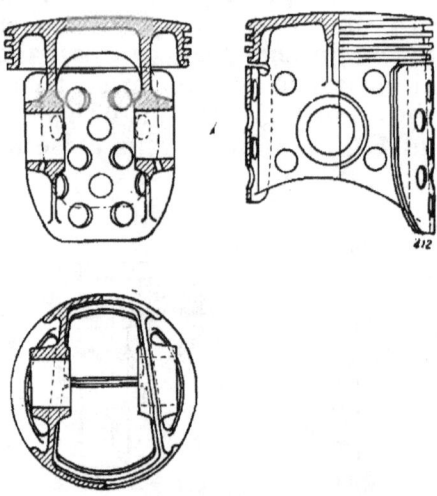

Fig. 41.—The Ricardo Slipper Piston.

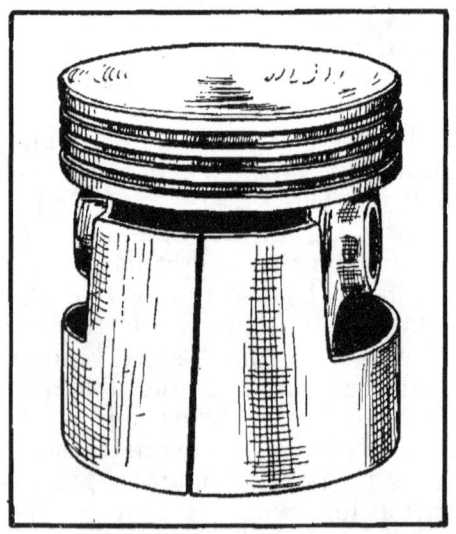

Fig. 42.—A Typical Aluminium Alloy Piston.

aluminium alloy; there is no bronze bush. This arrange ment works perfectly well, and there is very little wear.

Fig. 42 illustrates another popular type of aluminium piston used on car engines. In this case there are three ring grooves above the gudgeon pin bosses, and the piston head is separated from the skirt by horizontal slots; this slot, apart from reducing the heat conduction, allows the oil to drain away from the cylinder walls to the crank case through the interior of the piston. It will be observed that the skirt is cut away between the bottom land and for about one-half the length of the full skirt shown in the centre; this is done for lightness. The piston skirt is split in an oblique direction on one face, as shown; this is arranged on the thrust face of the piston. With this type of piston skirt the spring effect obtained by splitting it enables the clearance to be reduced so that there is no piston slap when cold.

A typical aircraft engine piston of aluminium alloy as used on certain Rolls Royce engines is illustrated in Fig. 43. Its special features include general lightness of construction; short length in relation to its diameter; three narrow compression rings above the gudgeon pin; one slotted oil scraper ring near base of skirt; large diameter hollow gudgeon pin with spring clip retainers at its ends and dished piston crown.

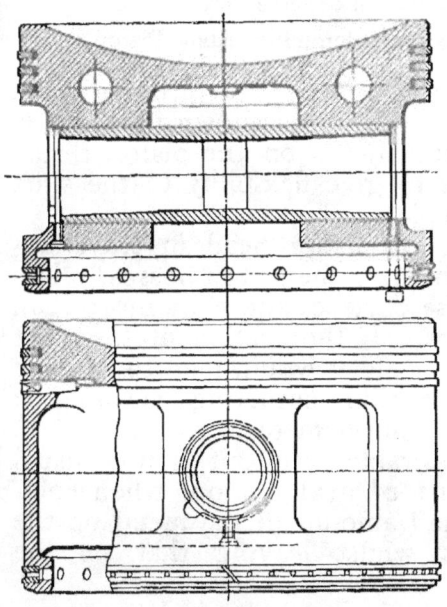

Fig. 43.—Rolls Royce Aircraft Type Piston.

Oval Section Pistons.—In order to insulate, so far as possible, the skirt from the hotter crown portion of the piston, and thus prevent distortion of the former

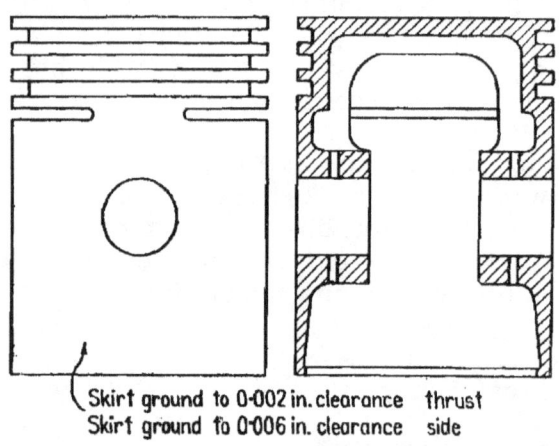

Skirt ground to 0·002 in. clearance   thrust
Skirt ground to 0·006 in. clearance   side

Fig. 44.—An Oval-Section Aluminium Alloy Piston.

part, it has become the practice to provide slots between the head and the skirt, as shown in Figs. 42 and 44. These slots are always on the piston thrust faces. They also serve to give flexibility to the skirt when the latter is also split.

The heat from the hot piston head in travelling down the un-cut portions near the piston bosses, raises the temperature of these parts of the skirt more than that of the thrust faces. If the section of the skirt were truly circular when cold it would tend to become oval, the greater expansion occurring across the piston boss, or gudgeon pin diameter.

To obviate this undesirable result, it is now usual to make the piston skirt of oval section, when cold, by grinding away a small amount of the metal on the minimum thrust faces, whilst leaving that on the thrust faces.

Fig. 44 illustrates an oval piston of 3 inch diameter, and shows the clearances usually allowed. These pistons are known as "constant clearance" ones, since they become circular in section when hot.

Another example of an oval section type piston is the Aerolite one shown in Fig. 45. The skirt of the

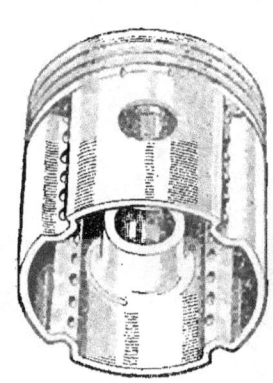

Fig. 45.—The Aerolite Piston.

piston is cam ground and is divided into four faces by means of vertical flutes. The gudgeon pin faces are designed to allow for the correct amount of expansion and the thrust faces have a fine cylinder clearance. In this design of piston the heat from the piston crown is conducted by internal webs to the gudgeon pin faces. It is restricted by slots in the lower ring grooves and by the reduced cross-sectional area of the flutes from materially affecting the thrust faces; this ensures the correct cylinder clearances at all temperatures.

Another well-known design of aluminium piston is the Invar Strut type, in which four strips of a practically non-expansible nickel-iron alloy known as Invar are cast into the piston bosses and the skirt of the piston as shown in Figs. 46 and 47. With this type it is

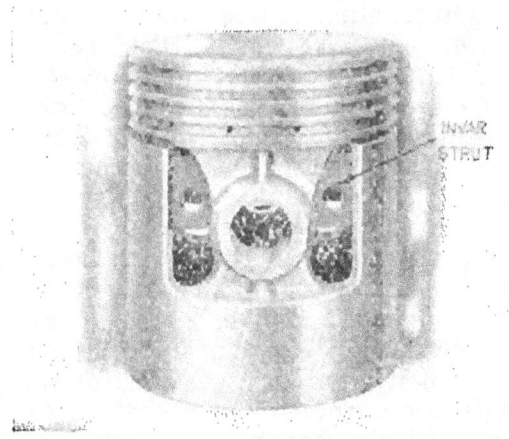

Fig. 46.—The Invar Strut Piston.

not necessary to allow so much clearance either to the lower part of the head or to the skirt. These pistons are made of a strong aluminium alloy called Birmalite, which has a tensile strength of 16 to 19 tons per sq. in. and is therefore almost as strong as wrought iron.

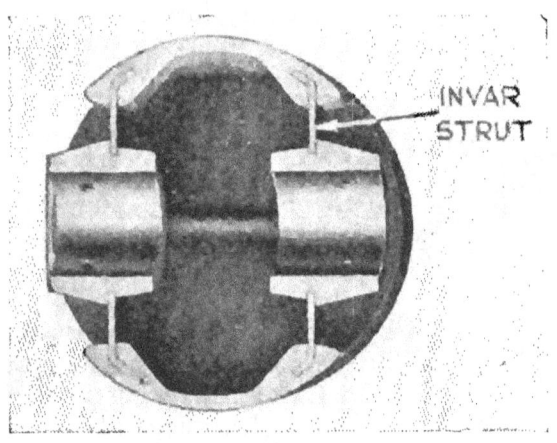

Fig. 47.—Plan View of Cut-Away Piston
to show the Invar Struts.

Another and more recent design of compensating piston, known as the Nelson Bohnalite Autothermic one, shown in Fig. 48, employs a pair of low carbon steel plates P, with their ends anchored to the skirt but not to the gudgeon pin bosses; the skirt, bosses and head are a one-piece casting which relieves the steel inserts of thrust. Since the aluminium alloy when heated expands more than the steel the two metals form a kind of bi-metallic element which tends to bend outwards as shown by the arrows with the result that the expansion of the skirt is decreased in the direction perpendicular to the axis of the gudgeon pin and is increased in the direction parallel to the pin. In this way the expansion of the piston is effectively controlled and by suitably apportioning the dimensions of the steel members practically any degree of expansion can be obtained. Usually the net expansion of the piston when heated is about the same as that of cast iron. It is, however, slightly less in the direction perpendicular to the gudgeon pin at both top and bottom of the skirt and slightly greater at the bottom of the skirt in line with the gudgeon pin. The piston is thus oval when cold and cylindrical when hot, whilst it requires a much smaller cold clearance than a plain piston of the same alloy. The latter is of high-silicon aluminium and has a lower expansion coefficient than most other commercial aluminium alloys.

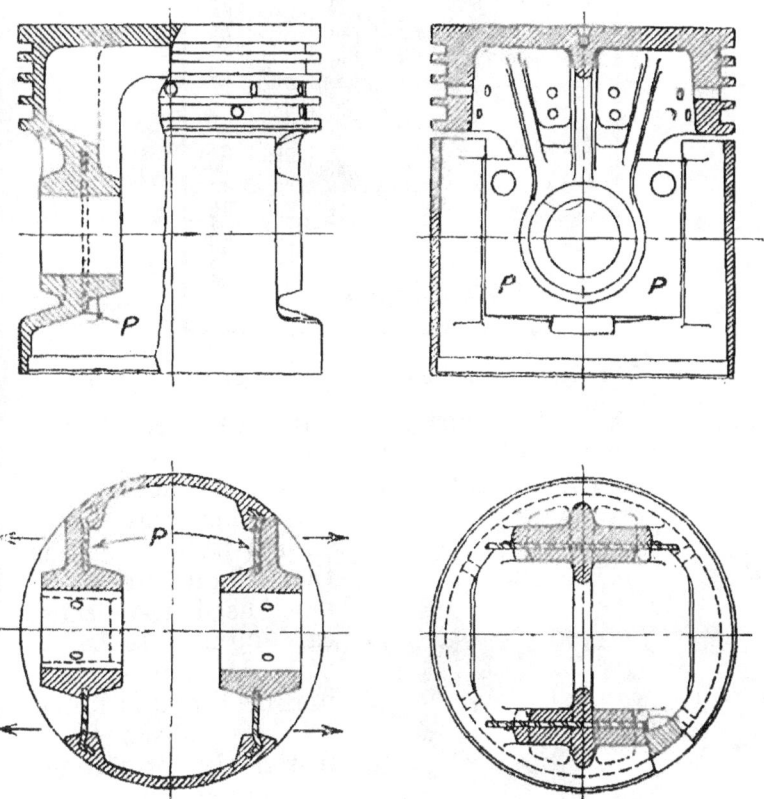

Fig. 48.—The Nelson Bohnalite Autothermic Piston.

The piston shown in Fig. 48 is of the solid skirt type; the same method of compensation for expansion can be applied to split skirt pistons.

Another class of piston employs an aluminium alloy head and piston bosses with a steel skirt, thus combining the excellent heat-conducting properties and lightness of the former material with the good wearing qualities of steel for the skirt.

Fig. 49 illustrates the Zephyr composite piston in part cross-section showing how the two metals are

D

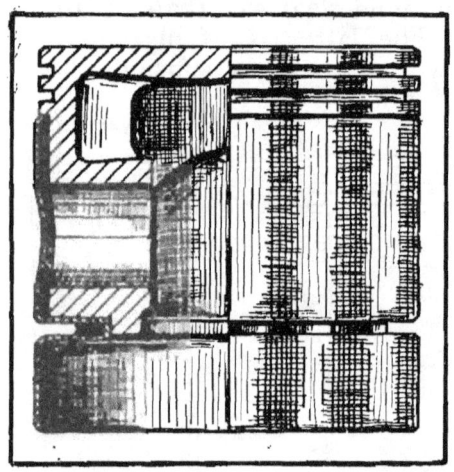

Fig. 49.—The Zephyr Combined Aluminium and Steel Piston.

employed. The steel skirt gives a much longer life and requires only a small clearance. These pistons are very light and are used both for commercial and racing cars.

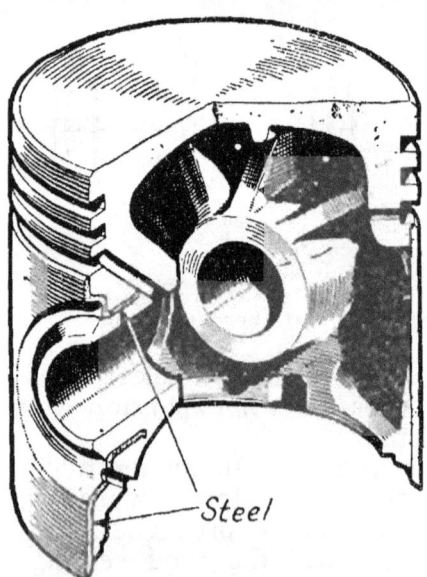

Steel

Fig. 50.—The Flowerdew Composite Piston.

Another example of composite piston is that of the " Flowerdew" used in Wolseley car engines. It has a steel skirt and aluminium head and is noted for its absence of piston slap and long-wearing qualities.

Coatings for Aluminium Pistons.—The ordinary aluminium alloy pistons are more liable to wear and " scuffing " action than cast iron ones, so that in more recent models the cylindrical surfaces have been given a protective coating of aluminium oxide or tin.

When tin is used as a coating it not only prevents "scuffing", or partial seizure during the running-in period of the engine, but forms a highly polished bearing surface of low frictional resistance.

The oxide-coating method results in an extremely hard fine texture surface which resists wear and prevents seizure under normal running-in conditions. The pistons are immersed for about 18 minutes in a sulphuric acid solution at 70° Fah. While in solution the piston, which is carried on a conveyor, is the anode for the electro-coating process and it emerges from the solution with an aluminium oxide coating of about 0·00025 inch; this coating is so hard that further machining cannot be done to the piston. The piston ring slots and the gudgeon pin holes are treated as well as the piston itself.

It is considered advantageous to treat the coated pistons with a solution of *colloidal graphite*, such as Aquedag, in water allowing the solution to dry off, when the microscopic grains of graphite remain embedded in the pores of the metal and reduce the frictional coefficient appreciably.

In connection with the *anodizing process* an alternative method is to immerse the pistons in an electrolyte containing 20 per cent of sulphuric acid and 5 per cent. oxalic acid. A current of 13 amperes per sq. ft. at a voltage of 14 is passed through the bath for about 30 minutes to give the proper thickness of coating.

Piston Rings.—These are now always made of cast-iron, although spring steel ones have been used. The principal requirement of a piston ring is that when compressed into the cylinder it shall bear evenly all round, and shall not have too large a gap. If a concentric type of ring is sawn through at one place and inserted into the cylinder it will not expand to a cylindrical shape, but will bear on the portions near the cut end A and on the diametrically opposite part of the ring D as shown in Fig. 51. There will be no contact with the cylinder on either side at C and B. The type of ring illustrated is therefore unsuitable for the purpose of petrol engines, and must be modified either by reducing the radial thickness near the ends A or by suitably shaping the ring when free so that when it closes within the cylinder it is perfectly circular.

A method which is sometimes employed for attaining the latter result is to turn the rings to slightly larger diameter than the cylinder bore and then to cut out a small portion leaving a parallel sided gap. The ring is then closed by pressure and in this condition is clamped between circular discs of rather smaller outside diameter and the exterior surface is then ground down to exactly the same diameter as the cylinder bore. This gives a circular outside form when the ring is placed in the cylinder, but it is necessary, in addition, to increase the gap between the ends of the ring (when cold) so as to give a clearance of ·002 in. per inch of cylinder diameter when placed in the cylinder; thus a ring of 3 in. nominal (cylinder) diameter would have a gap of ·006 in. when in place. The object of this clearance is to allow for expansion when the engine is at its normal working temperature. For high speed air-cooled engines the gap should be from ·003 to ·004 in. per inch of diameter. The vertical, or slot clearance should be ·0015 in. for rings up to 6 in. diameter for cast iron pistons and ·002 to ·003 in. for aluminium pistons. In order to make this circular ring exert an equal pressure all around, it is usually hammered inside, the hammer blows being greatest opposite the gap, and decreasing in intensity to nothing at the gap. Another type of ring is the *Eccentric* one : this, if properly made, and hammered, gives a uniform wall pressure, so that leakage of the gases past the piston is minimized.

The disadvantage of the eccentric ring is that it is difficult to ensure that there is not an excessive pressure outwards at the two joint ends—for the examination of rings taken from engines that have been in service for appreciable periods always shows that the greatest wear occurs at these places. It has also the drawback of providing more space behind the ring for carbon deposit to accumulate. For these reasons the eccentric ring has practically been superseded by the parallel or concentric type.

Wall Pressure of Piston Rings.—In order to prevent the escape of gases past the rings it is necessary to

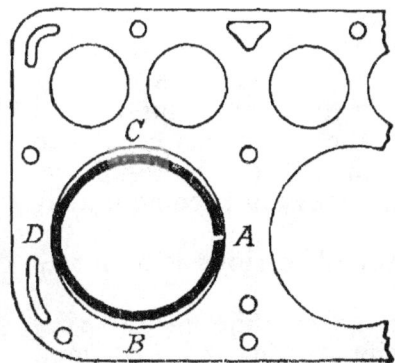

Fig 51.—Effect of inserting a Plain Concentric Ring in Cylinder Barrel.

arrange for the latter to exert a certain radial pressure o u t w a r d s against the cylinder walls. This pressure must only be sufficient for the purpose during the normal working conditions for excessive pressures reduce engine power and also give rise to greater cylinder wear.

For car engines up to about 5 ins. cylinder bore the correct value of the radial pressure should be about 8 to 10 lbs. per sq. in. For larger rings such as those used in Diesel and gas engines, experience indicates that an average pressure of 8 lbs. per sq. in. is the most satisfactory.

Piston rings are often made with a ratio of 1 : 25 between the radial thickness and the cylinder diameter and these give radial pressures of 14 to 17 lbs. per sq. in. for a gap of 12 per cent. of the diameter with cast iron having a modulus of elasticity of 14,000,000 lbs.

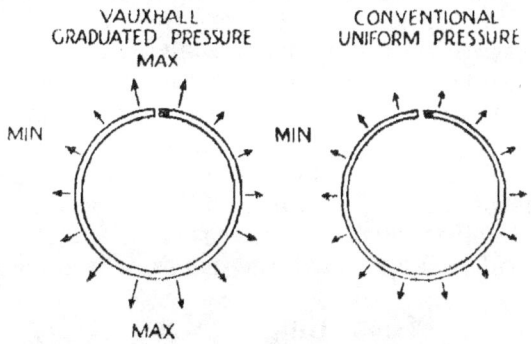

Fig. 52.—Piston Ring used in Vauxhall Engines.

per sq. in. To obtain higher pressures the above ratio must be increased and the size of the gap made greater at the same time. To reduce the radial pressure the reverse procedure must be adopted.

The free gap, i.e., the distance between the ends of the ring before placing in the cylinder should be about 3½ times the radial thickness. When the ring is stretched to go over the piston in order to insert it in the ring grooves of the latter the gap opens usually to about 8 times the thickness, but with the free gap previously mentioned the material does not become unduly stressed.

The usual thickness of automobile rings is about $\frac{1}{30}$ of the diameter.

A typical example of an automobile piston ring illustrating the usual dimensions is that of a 75 mm. (2·953 in.) one which has a maximum width of ·0938 in. and minimum width ·0928 in.; radial thickness of ·098 to ·106 in. and nominal free gap of ·34 in. Such a ring will give a radial pressure of 9 to 13 lbs. per sq. in. according to the width and thickness and would be suitable for use in aluminium alloy pistons.

A special type of piston ring developed by the Vauxhall engineers enables a liberal amount of oil to be delivered to the cylinder walls, in order to reduce cylinder bore wear, whilst preventing escape of this oil into the combustion chamber; thus adequate lubrication with minimum oil consumption has been the aim in view.

Instead of using the conventional pattern piston ring, shown on the right, in Fig. 52, the Vauxhall ring exerts a pressure on the cylinder walls graduating from maximum values at the two ends of the ring and at the place diametrically opposite, to minimum values between these two positions, as shown by the lengths of the radial arrows. This method, it is claimed, prevents piston ring vibration and gives better sealing of the combustion chamber charge and gases; at the same time oil leakage past the rings is reduced to a minimum.

Materials for Piston Rings.—Most of the piston rings employed in commercial automobile engines have been of cast iron of the practically " mottled " pearlitic variety. The castings from which these rings are machined are either sand or centrifugally cast; the latter method of casting the metal in rotating moulds give a more uniform structure free from gas bubbles or holes and other defects. The following are the standard specifications for the compositions of cast irons recommended for the two casting methods:—

Composition.	Sand Castings.	Centrifugal Castings.
Combined carbon ...	0·55 to 0·80*	0·45 to 0·80
Total carbon ...	3·5 (max.)	3·5 (max.)
Silicon	1·8 (max.)	1·8 to 2·5
Sulphur	0·12 (max.)	0·12 (max.)
Phosphorus ...	1·0 (max.)	1·0 (max.)
Manganese ...	0·4 to 1·2	0·4 to 0·2

* Percentages.

Apart from these ordinary cast irons, alloy cast irons are used in high performance engines. A typical alloy iron used for high performance rings is one containing from 3 to 3·5 per cent. carbon and ·25 to ·45 per cent. chromium. Examples of such alloy irons used commercially are Brico[1] and Thermocrom.[2] It is usual to heat treat such irons in order to obtain the desired strength and hardness qualities.

Molium Piston Rings.—A new piston ring material, known as Molium, developed by Messrs. Wellworthy Ltd., is of the self-lubricating kind and softer than the cast iron used for cylinder walls, giving less wear than ordinary piston rings. It is used for multi-edged piston rings which give good sealing effects against oil loss and blow-by of the gases. The rings are made up in individual sections of $\frac{1}{32}$ in. thickness, each element being flat in cross-section but spiral in form. When a number of these elements are assembled in the piston ring groove the side pressure effect obtained prevents blow-by. With this type of ring, breakages, filing and fitting are stated to be eliminated.

Coated Piston Rings.—In order to reduce the wear or scuffing action on piston rings which is usually most marked during the new engine running-in period, rings are now coated by special processes, the more common of which are the tin, iron-oxide and iron-manganese-phosphate methods.

Tin coating consists of an electrolytic deposition of tin on the surface of the piston ring. Iron-manganese-

[1] British Piston Ring Co., Ltd. [2] Wellworthy Piston Rings Ltd.

phosphate coating consists of an iron phosphate with a high percentage of manganese phosphate. This coating is softer than grey cast-iron and sufficiently porous to absorb an appreciable amount of oil. The treatment necessary to supply iron-oxide coating consists in oxidising the machined surfaces of the iron to a predetermined depth in an enclosed chamber heated to a temperature of approximately 1,000° Fah. in the presence of a suitable gaseous oxidising agent.

Test results are reported showing that treated piston rings wear only half as much as untreated rings.

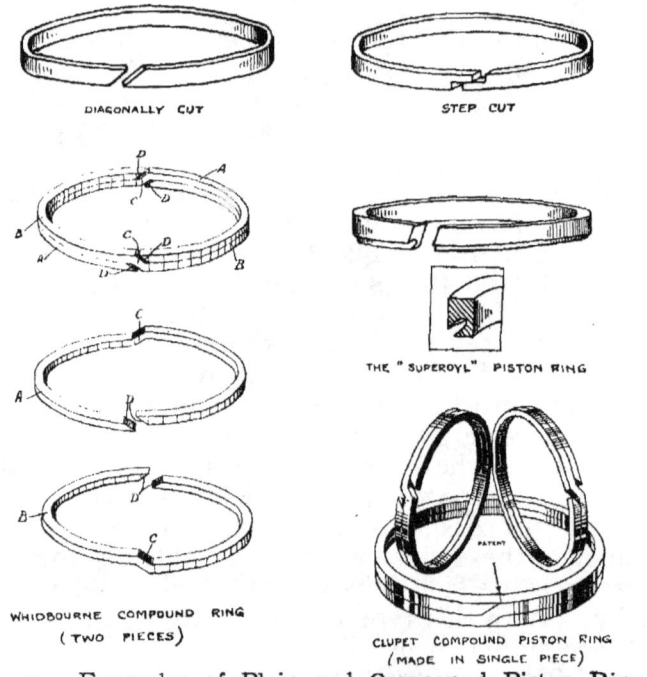

Fig. 53.—Examples of Plain and Compound Piston Rings.

There is a number of compound piston rings on the market, in which the gases are given long and tortuous leakage paths. These rings, of which the Clupet, Wellworthy, and McQuay-Norris, are typical examples, are much used for replacements on worn engines.

Oil Control or Scraper Rings.—The object of these rings is to reduce excessive oil consumption by scraping off the surplus oil on the cylinder walls *on the* *down stroke of the piston;* they have little or nothing to do with the maintenance of compression.

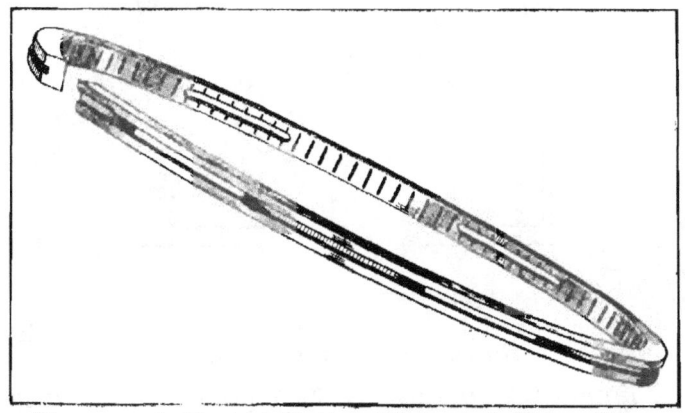

Fig. 54.—A Typical Slotted Oil-Control Piston Ring.

Fig. 55.—*A, B* and *C*—illustrates three typical oil scraper rings, the sections shown being taken through the piston wall. In all cases the bearing area of the scraper ring on the cylinder wall has been reduced, as

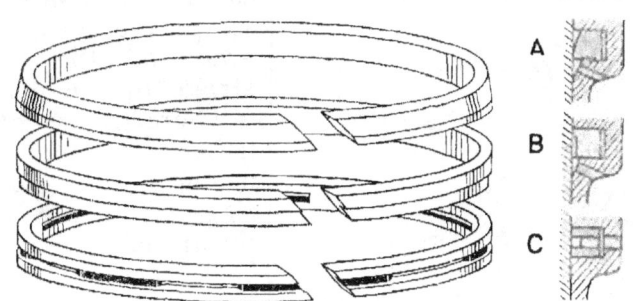

Fig. 55.—Oil Scraper Piston Rings.

compared with that of an ordinary compression ring. The ring shown at *A* has its upper side bevelled away; ring *B* has its lower portion turned away so as to remain clear of the cylinder wall, whilst ring *C* has a central groove turned away and slots made at intervals through the bottom of this groove.

Owing to their reduced bearing areas these rings give a greater pressure on the cylinder walls and, therefore, exert a satisfactory scraping action.

D*

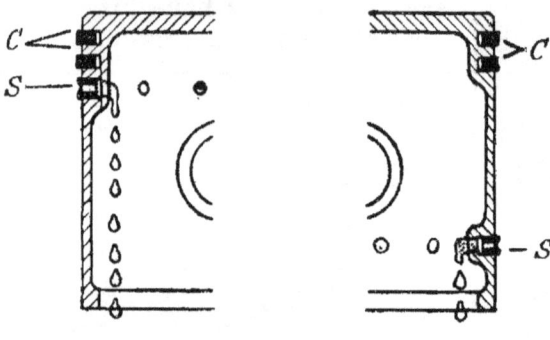

(A)   (B)

Fig. 56.—Methods of Fitting Scraper Rings. C—Compression Rings. S—Scraper Rings.

It is important to observe two points when fitting scraper rings, viz.: (1) Oil holes must be provided through the piston wall, as shown in the right hand sectional views (Fig.55), to conduct away the surplus oil scraped off the cylinder walls. (2) The rings must be fitted with their scraping edges downwards, as shown, in order to scrape the surplus oil away on the down stroke of the piston.

There are two alternative methods of fitting oil control rings. These are illustrated in Fig. 56. The method shown at A leaves the piston skirt quite clear, but the oil scraped off the cylinder walls is hotter than for the arrangement shown at B; the latter method gives a slightly heavier piston but there is a reduced tendency for the oil to gum and carbonise in the scraper ring groove.

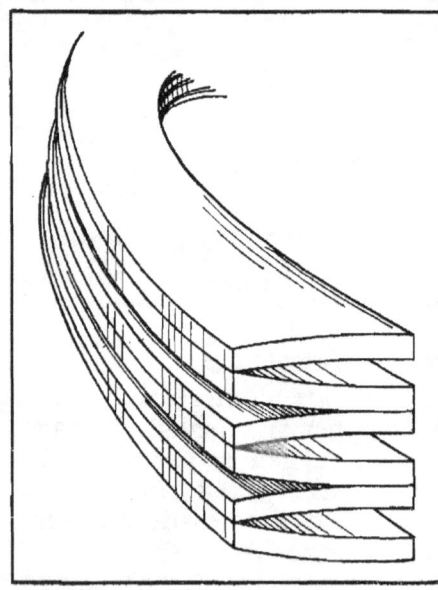

Fig. 57.—The Cord Flexible Piston Ring.

" Scraypoil " rings of the type illustrated in Fig. 55 are made by the British Piston Ring Co., Ltd., Coventry.

Fig 57 illustrates the Cord compound piston ring made up of two single sheet-metal annular members, welded at the common junctions. This flexible ring acts. as a compression and scraper ring. It has sufficient lateral spring to seal the groove and to compensate for wear in the latter; moreover, it will compensate for any irregularity in the cylinder bore. Usually two sets of these rings are fitted in the two lower piston grooves and a plain compression ring in the top groove.

Piston Rings for Worn Cylinders.—If a standard pattern ring is fitted to a cylinder that has worn oval or is tapered as a result of wear, it will not fit the cylinder correctly, and leakage of gas will occur. In order to overcome this disadvantage piston rings of a flexible nature, having an internal steel expanding device are frequently employed. Owing to the combination of flexibility and internal expansion thus obtained the ring will adapt itself to the inequalities of

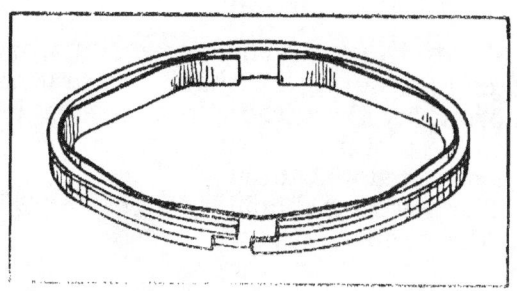

Fig. 58.—The Wellworthy "Simplex" Ring.

bore and produce a better gas-tight fit than is possible with a plain ring. A typical example of such a ring is the Wellworthy " Simplex " one illustrated in Fig. 58. It has a number of thin slots to give side pressure in the piston ring grooves and is provided with an internal spring steel expander having eight points of contact with the interior surface of the ring. It is claimed that this ring will compensate for cylinder inequalities up to ·015 in. and that owing to the design of the slots the space behind the ring becomes filled with oil, thus providing adequate lubrication for the ring; all surplus oil is scraped by the ring through holes drilled below its groove so that it is returned to the crankcase.

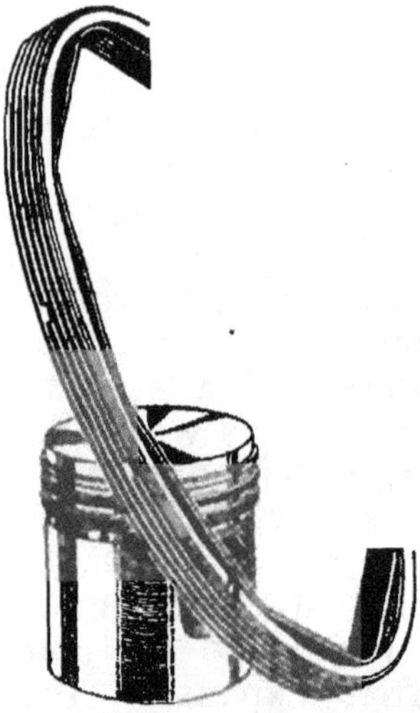

The expander ring, illustrated in Fig. 59, is the Wellworthy Molium multi-edge type with individual sections of $\frac{1}{32}$ in. thick and of spiral form.

The Ramco expander ring shown in Fig. 60 can be designed to exert any radial pressure from zero to several times that of a plain split ring. It is usually necessary for expander rings to exert slightly more pressure than plain compression ones. It is claimed that this expander ring will prevent piston vibration, reduce piston slap and give more effective oil control.

Fig. 59.—Wellworthy Molium Ring.

Fitting Piston Rings.—When new rings are to be fitted to an engine, during overhaul, it is first necessary to clean thoroughly the carbon from out of the piston grooves, and to turn down any ridges on the sides of the piston slots caused by wear. The new ring should be tried in the groove, before fitting, by holding it in the reverse way (that is, with the outside of the ring in the groove). The ring should have a side clearance of about ·002 to ·003 in. for aluminium pistons. The ring must be ground down to give this clearance. It should next be tried in the worn part of the cylinder, and the slot filed carefully (and parallel) until the ring fits the cylinder with a gap clearance of

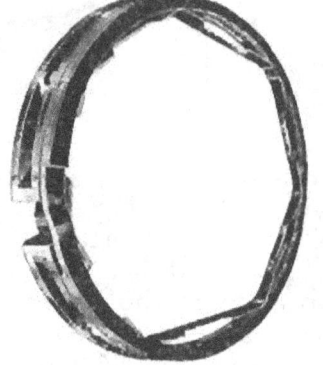

Fig. 60.—The Ramco Expander-type Ring.

·002 in. per inch diameter of cylinder. To file the slot, a small strut of wood can be used to hold the gap open. The ring can now be sprung on over the piston.

When three rings per piston are used, the slots should be spaced equidistantly before insertion of the piston in the cylinder. A thin metal clip, with a thumb-nut and screw can be used with advantage to compress the rings in position on the piston prior to inserting the piston in the cylinder; this clip will slide off as the piston is pushed in.

Piston and Cylinder Wear.—After an automobile engine has run about 25,000 to 35,000 miles on the road, (for ordinary cast-iron cylinders) it will be found that the piston clearance has increased to about $\frac{10}{1000}$ to $\frac{15}{1000}$ in. (the average new values ranging from about $\frac{3}{1000}$ to $\frac{4}{1000}$ in. for a $3\frac{1}{2}$ in. piston). This somewhat excessive clearance causes the oil to leak past, and results in more rapid carbonization; it also causes a loss of gases past the piston. The general result is a falling off in the power and efficiency, and an increase in the fuel and oil consumptions. It is usually reckoned that the average clearance increases by $\frac{1}{1000}$ in. for every 3,000 to 4,000 miles of road running; most of the wear is due to road-dust drawn in with the carburettor air. The use of air-cleaners on carburettors obviates much of this wear, whilst the harder alloy cast-iron cylinders also have a longer life.

In such cases the remedy consists in having the cylinders re-bored or re-ground by a specialist firm, and new pistons (and rings) fitted ;the engine will then take on a new lease of life, if the other parts are adjusted at the time of overhaul.

New pistons are frequently "lapped," or worked up and down, in their respective cylinders, using an extremely fine abrasive and oil for the purpose. It is inadvisable to run a car with an engine which has had its cylinders re-ground and new pistons fitted, at a speed exceeding about 30 m.p.h. for the first 300 to 500 miles, otherwise the wearing surfaces will not bed in properly.

The Valves.—Altnough many designs of rotary, piston, cuff, sleeve and other types of valve have been tried in the past, present-day practice is confined to the sleeve and poppet valve examples; we shall refer to the former again. The poppet, or mushroom-head valve, illustrated in the cases of the engines shown in Figs. 10, 17 and 22, is made of high-tensile alloy steel, since it is subjected to severe impacts, tension and high temperatures. Whilst the inlet valve is kept fairly cool by the passage of the wet mixture the exhaust valve is exposed to the intense heat of the outflowing exhaust gases, so that whereas in water-cooled engines the average temperature of the head of the inlet valve seldom exceeds 250° C. to 275° C., that of the exhaust valve is generally about 700° C. to 760° C. The material for the inlet valve need not therefore be so strong at its working temperatures as the exhaust valve. The latter must also be made of a suitable steel to withstand the hot corrosive effects of the exhaust gases under its head and around the upper part of the stem. When fuels containing tetraethyl lead are used the corrosive action is accentuated.

Valve Materials.—Various alloy steels are used for the inlet valve, including plain nickel ones, nickel-chromium and chrome molybdenum.

The steels employed for exhaust valves include nickel chrome high tensile steels, stainless steels, silicon chrome steel, cobalt chrome steel, high speed steel, high nickel chrome (known as *austenitic* steel) and tungsten steels. Of these steels those possessing the necessary tensile strengths and hardnesses at the operating temperatures are the cobalt chrome, silicon chrome and the high nickel chrome *austenitic* steels. A typical cobalt chrome steel has a tensile strength of 45 tons per sq. in. cold, and 25 tons per sq. in. at 700° C. Similar properties are given by certain silicon chrome steels. The high nickel chrome steel, having 0·40 to 0·45 per cent. carbon; 1·5 to 1·8 per cent. silicon; 0·6 to 1·5 per cent. manganese; 12 to 13 per cent. chromium and 18 to 22 per cent. nickel with tungsten up to 3 per cent. gives a cold tensile strength of about 43 tons per sq. in.; 34 tons per sq. in. at 700° C. and 24 tons per sq. in. at 800° C. The cold hardnesses of the three valve steels mentioned lie between 250 and 300 Brinell. All of these steels possess good anti-scaling properties.

A group of valve steels, often used for aircraft engines, is the "Silchrome" one, having from 1 to 4 per cent. silicon and 8 to 24 per cent. chromium, with 1 to 4 per cent. nickel and sometimes up to 3 per cent. molybdenum. The Silchrome, cobalt chrome and tungsten steels are not so good in their corrosion resistance properties to the action of leaded fuels as the high nickel chromium austenitic steels. The latter, however, have a higher thermal expansion coefficient, so that a greater allowance must be made in the valve clearances when such steels are employed.

With austenitic steels it is necessary to provide the valve stem with a hardened tip to resist impact effects; it is now usual to provide a welded button of Stellite on the valve stem end for this purpose.

Sodium-cooled Valves.—Although employed to a limited extent only on automobile engines this type of valve is standard on aircraft engines. It is made with a hollow stem and head and part of the interior space is filled either with the metal sodium or a mixture of salts, such as potassium nitrate and lithium nitrate. Sodium melts at 97.5° C. and the latter mixture at 130° C. Sodium boils at 880° C. In each case the

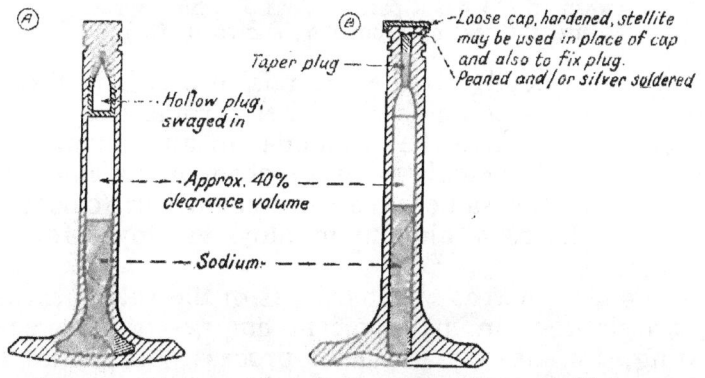

Fig. 61.—Sodium-filled Hollow Stem Exhaust Valves.

molten salt and its vapour in the hollow exhaust valve stem conduct the heat from the valve head to the cooler part of the stem, thus cooling the head. Fig. 61 illustrates typical sodium-cooled exhaust valves as used on aircraft engines.

Valve Seatings. — Although the ordinary valve seatings of cast-iron cylinders give satisfactory service, a greatly increased life, before re-seating is necessary, is given by special seatings, which are either screwed or shrunk into position. The materials used for valve seating insert rings include aluminium bronze (used in aluminium alloy cylinders and heads) and alloy steels. The best of these steels appears to be the one having the requisite strength and hardness combined with a coefficient of expansion equal to that of

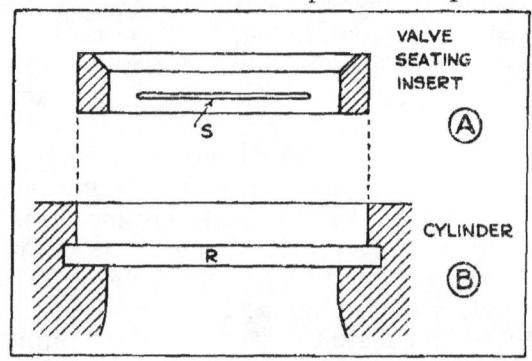

Fig. 62.—Showing one method of inserting Hard Valve Seating into Cylinder. The insert ring shown in (A) has a pair of slots S, which spring into the seating groove R, shown in (B.).

the cylinder metal. If the latter is of cast iron then a stainless or austenitic (high nickel chrome alloy) steel can be used. When the cylinder is of aluminium alloy it is now usual to employ a nickel-chromium manganese steel having the same expansion coefficient (0·000022) as that of the usual aluminium alloys employed for the cylinder.

In order to increase the hardness of the valve bearing or seating face in many recent engines it is given a coating, by the oxy-acetylene process, of Stellite or Brightway (80 per cent. nickel and 20 per cent. chromium). Stellite is an extremely hard synthetic cutting alloy containing cobalt, tungsten and carbon, used also to tip the cutting portions of machining tools in order to provide much higher cutting speeds than for carbon or high speed tool steels. The valve seatings are often shrunk into position by first immersing them in liquid oxygen or in dry ice, or by first heating the cylinder or head.

Valve Guides.—Usually, the valve guides are of close-grained cast iron and are made a press or shrink fit in the cylinder or its head, for side or overhead valves, respectively.

When the cylinder is made of aluminium alloy it is usual to heat it to a temperature of about 350° C. in order to expand the valve guide hole and then to insert the cold guide. When the cylinder cools down the guide is held very firmly in position; for this purpose the diameter of the guide is made slightly greater than that of the guide hole when both are at the normal air temperature, so as to give what is known to engineers as an *interference fit*.

Another method that is used for aluminium and also cast iron cylinders is to make the guides to an interference fit in their holes and to first cool them down in

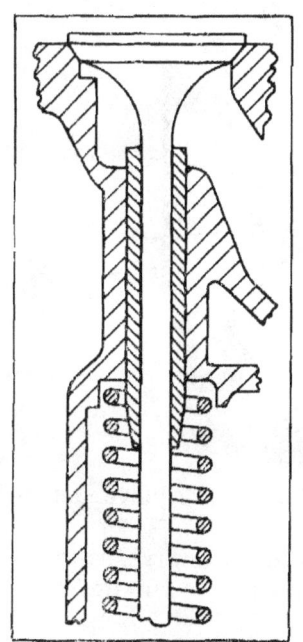

Fig. 63.—The Ferrocite Valve Guide.

a mixture of dry ice (solid carbon-dioxide) and alcohol, which has a temperature of about minus 60° Fah. (if liquid oxygen is used the temperature is minus 182° Fah.), before inserting them in their holes in the cylinder. When in contact with the cylinder metal the guides expand and become a very secure fit.

Exhaust valve guides are occasionally made of phosphor bronze as this metal has been found to give less wear and better heat conduction; this type of guide is used on Rolls Royce " Merlin " aircraft engines. After being shrunk into place the holes are burnished by forcing hardened steel balls through them.

Another material for valve guides is that known as

Ferrocite. It is made of powdered pure iron by a die pressing process and has an oil absorption capacity up to 30 per cent. of its volume, according to density. The guide as die pressed is given a mirror finish and requires no further machining. It is claimed to extend the life of the valve stem considerably owing to its low coefficient of friction and excellent oil-retaining properties. Fig. 63 shows a typical Ferrocite valve guide in position.

Valve Components and Dimensions.—The components of a typical side-by-side valve example are shown in Fig. 64. Commencing at the top, the diameter of the head should be as large as possible consistent with the combustion chamber design. It is good practice to make the valve diameter d=0.414 (piston diameter), or alternatively to so choose d that the gas velocity through the port does not exceed 120 feet per second; this consideration leads to the formula :

$$d^2 = \frac{2D^2 l N}{86,400} \text{ ins.}$$

where D is the piston diameter, l the stroke (both in inches) and N the r.p.m.

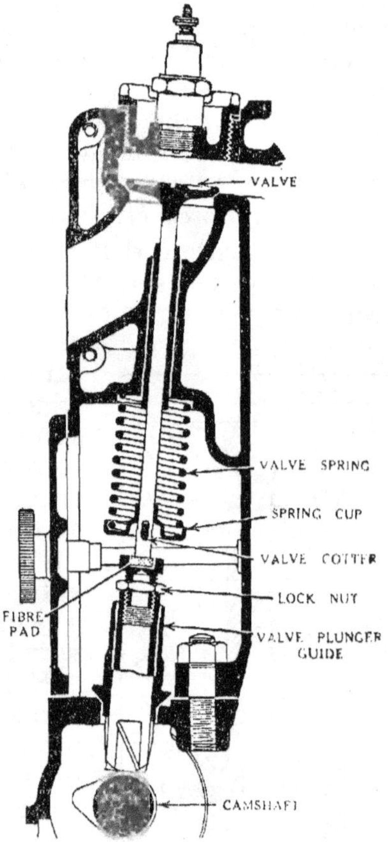

Fig. 64.—A typical Valve and its Components.

The lift of the valve is usually about a quarter to one-fifth of the valve diameter dimension at the smaller part of the conical seating.

The angle of the seating is usually 45°; 60° angle and also flat seatings have also been used.

The stem of the valve should be about one-fifth of the diameter of the valve; it should be a good sliding fit in its cast-iron guide, but without allowing any leakage past the stem.

The valve spring, which usually has a compression of 25 to 45 lbs. per in., should be sufficiently strong to keep the usual size of valve and tappet on the cam after the latter has ceased to lift them. The tappet, or plunger, is interposed between the cam and the valve stem. At its lower end it carries a roller (or ball) which obviates the rubbing action of the cam. The top of the tappet is provided with a screw adjustment (with its lock-nut) for the purpose of adjusting the distance between it and the valve; it is essential to maintain the correct clearance amount.

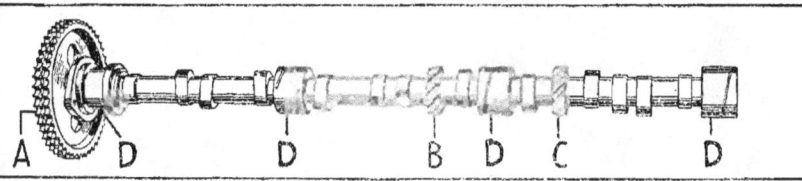

Fig. 65.—A typical Camshaft. *A*—Timing gear wheel. *B*—Helical gear for driving oil pump. *C*—Ignition gear drive. *D*—Camshaft bearing surfaces.

A Typical Camshaft.—Fig. 65 shows a typical six-cylinder engine camshaft. The sprocket *A* is driven at one-half engine speed by means of a duplex roller chain. There are four bearings, *D*, in the engine casting for the camshaft, phosphor bronze or white-metal-lined bushes being employed.

An examination of the illustration will show that there are twelve cams in all, arranged in pairs (inlet and exhaust). *B* is a skew gear for driving the oil-pump shaft, whilst *C* is another skew gear for driving the ignition distributor and contact-breaker operating shaft.

Valve Clearances.—The clearance allowed depends upon the design of the engine, that is, upon the relative heating of the valve and cylinder; most manufacturers specify the appropriate clearance values for their engines. In the absence of such information, it is usual to give side valve engines a clearance of $\frac{3}{1000}$ in. for the inlet and $\frac{5}{1000}$ in. for the exhaust, the adjustment being made whilst the engine is hot. Overhead camshaft operated valves have clearances of $\frac{3}{1000}$ to $\frac{4}{1000}$ in.

Large Clearance Valves.—In order to obviate the necessity of frequent attention to tappet clearances and to avoid any possibility of burnt-out valves due to insufficient clearance or stretch, it is becoming the practice, in the case of side-valve engines to employ special designs of valve cams, which will enable the engines to run quietly with valve tappet clearances of the order $\frac{15}{1000}$ to $\frac{25}{1000}$ inch. Examples of engines having large clearance valve tappets are the Daimler (poppet-valve), Morris and Wolseley.

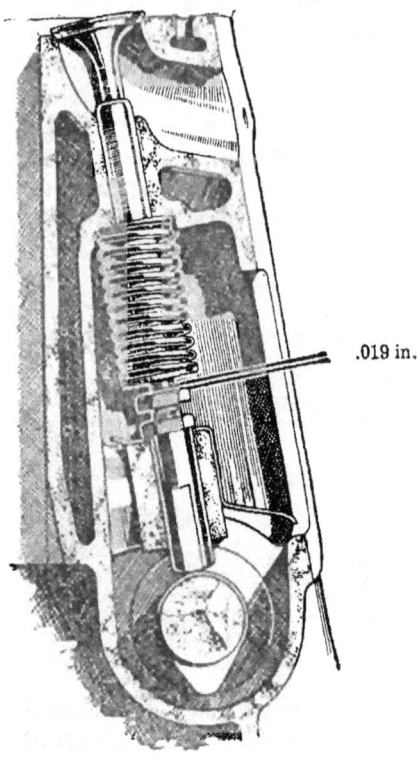

.019 in.

Fig. 66.—Large Clearance Valves used on some Morris Engines.

Automatic Tappet Adjusters.—In order to do away with the need for making regular adjustments of valve tappet clearances, some manufacturers have fitted automatic adjustment devices, thus ensuring maximum engine efficiency at all times.

The method usually adopted is to utilise the oil pressure of the engine lubricating system to hold an inner plunger member of the tappet against the valve stem. Fig. 67 illustrates a popular American automatic adjuster. The valve tappet is made hollow, and has a plunger member which abuts against the end of the valve stem. Oil under pressure is delivered to the reservoir R, whence it passes through a hole H into the hollow tappet, up the central tube and past a non-return ball valve into the adjusting chamber below the

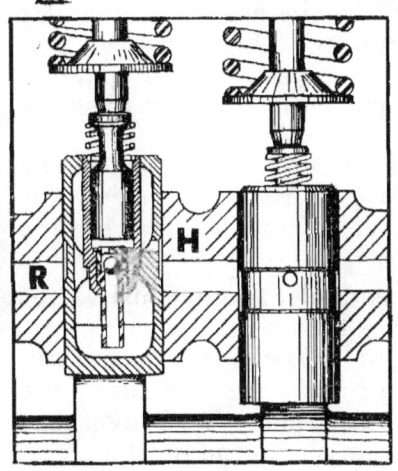

Fig. 67.—Automatic Valve Tappet Adjusting Device.

end of the plunger. Oil also passes from the reservoir through a hole in the upper end to lubricate the tappet guide, and to assist in getting rid of any air in the system, for it is just as important to eliminate air in the oil system, in this instance, as with hydraulic brakes. The hydraulic adjuster shown maintains zero clearance under all conditions; it is known as the Wilcox Rich "Zero Lash."

Another type of automatic tappet clearance adjuster operating on a purely mechanical principle is illustrated in Fig. 68.

The spindle of the tappet is drilled throughout its length to accommodate a coil spring and is threaded externally at the top. On to this threaded section screws a phosphor bronze nut drilled to accommodate a cross pin which is held in position by a spring circlip. The steel mandrel in the centre of the coil spring is slotted at each end, the lower slot engaging the crossbar at the bottom of the spring and the top slot engaging the cross pin. It will thus be seen that the phosphor bronze nut is spring loaded, the action of the spring tending to unscrew the nut in an upward direction. In the top of the nut is located the hardened steel tappet head resting on a number of

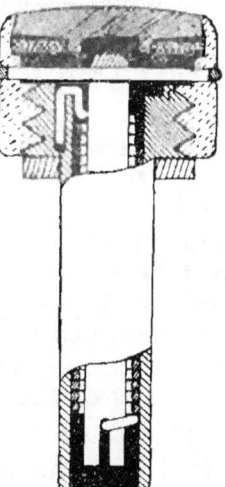

Fig. 68.—Mechanical type of Automatic Valve Tappet Adjuster.

spring steel discs which deflect to a pre-determined amount in accordance with the recommended tappet clearance, which is governed by the degree of chamfer on the underside of the tappet head. This forms the complete assembly for " located " tappets, but for spinning tappets a flat spring clip is placed in position over the tappet head.

In practice the action of the coil spring keeps the tappet in contact with the tip of the valve stem, the deflection of the internal steel discs allowing for the normal clearance.

Overhead Valves.—Overhead inlet and exhaust valves are often given more clearance than side ones, viz., $\frac{4}{1000}$ and $\frac{8}{1000}$ in. respectively to allow for the greater difference in the expansion of the cylinder and push rods. With overhead cam-shafts the former clearances are employed. In adjusting the clearances, it is advisable to employ a " feeler-gauge." The feeler of the appropriate thickness corresponding to the clearance recommended by the makers of the engine is selected from the hinged batch supplied. The tappet adjustment is unscrewed, the feeler inserted between the valve stem and tappet,

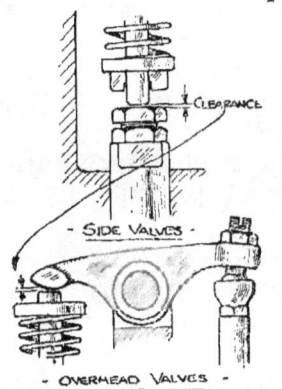

Fig. 69.—Showing Clearances of Side and Overhead Valves.

and the adjustment screwed up until the feeler can only just be pulled out. The adjustment locking screw is tightened, taking care to hold the other screw head securely, otherwise the clearance will be altered. It is advisable to check the valve clearances at regular intervals of about 1,000 miles.

The Ford Valve.—The modern Ford engines are fitted with the non-adjustable type of side valve shown in Fig. 70. In this case the end of the valve stem is expanded, as at E (Fig. 70), and a split form of spring retainer or valve collar is employed. Since it is not possible to use ordinary valve guides with this type of valve, the valve guide is made in two halves, which,

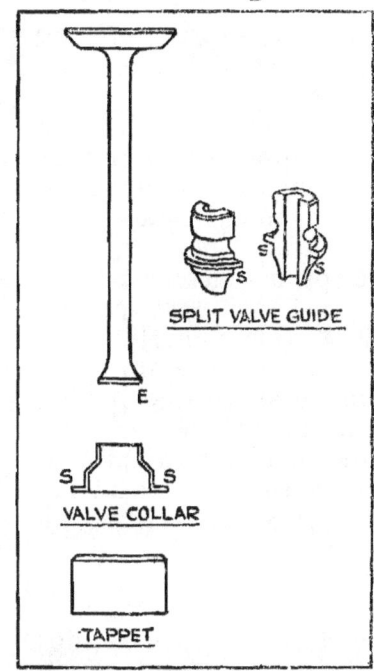

SPLIT VALVE GUIDE

E

VALVE COLLAR

TAPPET

Fig. 70.—The Ford Valve.

together, fit into a corresponding hole in the cylinder casting, and are retained in position by means of the valve spring which presses on the faces S, as well as on the corresponding faces S of the valve collar

To remove a Ford valve it is necessary first to compress the spring with a suitable tool (supplied by the makers), then to lift and remove the collar and, after lifting the valve a little, to take off the spring. The split valve guide can then be driven out, downwards. The valve should be pushed up as high as possible whilst driving the guide out with a piece of hardwood or copper drift.

The correct valve clearance is $\frac{13}{1000}$ inch; this can only be obtained by grinding the valve stem end, if the clearance is too small.

Valve Springs.—In many instances of earlier designs of automobile engine the valve springs fractured and caused engine breakdown. This was due to the unsatisfactory grade of steel and the heat-treatment employed; to some extent, also, it was due to the design of the spring.

In later engines the trouble has been cured by the use of better steels for the springs and by the use, in certain cases, of compound springs, i.e., one spring within another; occasionally, triple valve springs are employed. The object of compound springs is to avoid the surging or wave travel effects when the engine is running at higher speeds for with a single spring it sometimes happens that the natural frequency of the spring's vibration coincides with the normal operating frequency of the valve and resonance effects then occur which give rise to surging effects; by using compound springs of different natural frequencies these effects are obviated.

Owing also to the liability of springs to fail through initial surface defects it is now the practice to employ ground steel wire for making springs or to shot-blast the surfaces after coiling so as to give a work-hardening effect.

The steels employed for valve springs include a hard-drawn carbon steel with 0·7 to 0·8 per cent. carbon and 1·0 per cent. manganese (maximum). Pure Swedish iron is often used in the manufacture of carbon steel springs. Another spring material is a steel having 0·4 to 0·5 per cent. carbon and 1·0 to 1·5 per cent. chromium with 0·15 per cent. of vanadium; when suitably heat-treated it has a tensile strength of 90 to 115 tons per sq. in. The carbon spring steels when oil hardened and tempered give 80 to 90 tons per sq. in. tensile strength.

Valve Spring Retainers.—There are several methods in use for retaining the lower end of the valve spring, i.e., for transmitting the downward spring pressure to the valve stem; typical examples are given in Fig. 71.

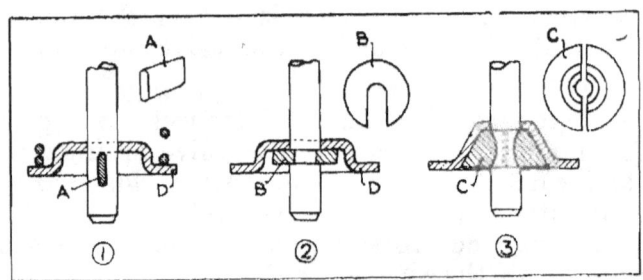

Fig. 71.—Valve Spring Retainers.

Diagram 1 shows the valve cotter method. In this case the valve stem is slotted near its lower end, and a steel cotter is used to secure the valve collar *D*.

Diagram 2 illustrates the slotted collar method, in which the horse-shoe collar *B* slides in a recess turned for the purpose in the valve stem.

Diagram 3 shows the split collar scheme, whereby the somewhat deeper section valve spring collar *D* is retained by a pair of similar split cotters *C* which fit into a recess turned in the valve stem. A similar method of holding the lower end of the valve spring on certain Austin engine models is illustrated in Fig. 72.

Types of Valves.—Confining our attention for the present to poppet-valves, these follow certain definite designs. For normal and slow-speed engines the valves take the form illustrated in Figs. 64 and 66, whilst for high-speed engines it is necessary to reduce

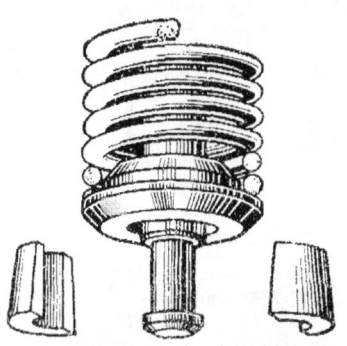

Fig. 72.—Split Conical Cotter Method of Spring Retainer.

the weight, and to design the valve so that it keeps as cool as possible. In this case, the valve stem above the valve guide (which is a separate cast-iron or bronze sleeve forced into the casting) is made of tapering form, and is hollow for part of the way from the head, as shown in Figs. 17 and 22. This type is known as the *Tulip Valve*.

Valves have been made with cast-iron heads, secured firmly to hard steel stems by screwing and riveting, or by welding; no pitting or corrosion occurs with this type.

Overhead type valves are, of course, shorter than the side types. They can thus be made much stiffer, and the more exposed valve-springs can be cooled much better.

Overhead Valve Operation.—Overhead valves are operated either (*a*) by push-rods and rocker arms or (*b*) by an overhead cam-shaft and rockers. An example of the former method is shown in Fig. 73 (B). The fulcrum for the rocker arm is screwed to the cylinder head in some cases, whilst the valve end of the arm is usually rounded, or provided with a ball or roller, and bears direct on the valve stem end. An example of the

roller-ended rocker arm is shown in Fig. 73 (B).
The push-rod is now invariably provided with ball-and-
cup ends, to minimize the friction constraint in the
working parts. Sometimes the cup is in the rod end,

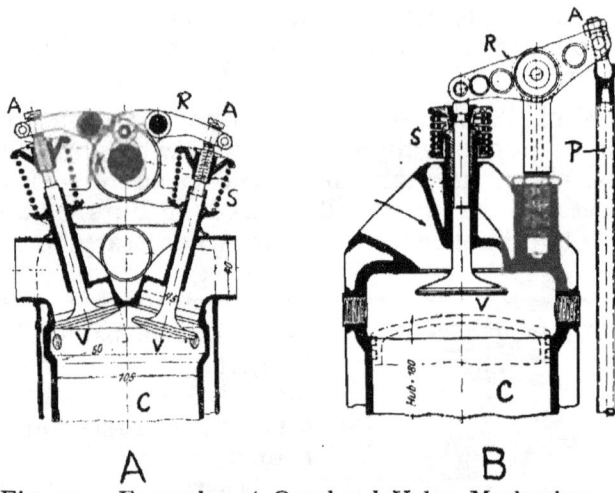

Fig. 73.—Examples of Overhead Valve Mechanism.
A—Valve clearance adjustment.   C—Cylinder.   R—Rocker Arm.
P—Push rod.   S—Valve spring.   K—Operating cam.   V—Valve.

and sometimes in the rocker or tappet. Means are
provided for adjusting the valve stem clearance, either
by means of a screwed portion on the cam tappet, on
the rod itself, or at one end of the rocker arm
[Fig. 73 (A)]; the latter is the more accessible type.
Tappet rods are often made hollow, for lightness.
Another example of overhead valve gear is given in
Fig. 18.

The overhead cam-shaft method of operating the
valves is employed upon the more expensive engines;
it requires a greater number of working parts, but is
usually quieter, better lubricated and more efficient.
The more popular method is to employ a single overhead shaft having all the cams cut on it in their
appropriate positions; Figs. 17, 27 and 73 (A) show
typical examples of single overhead cam-shafts.
In the case of larger engines, two cam-shafts are
often used, one for the inlet valve, and one for the
exhaust, as shown in Fig. 76, this method enables

the valves to be inclined to a greater degree, and leaves a convenient "trough" in the cylinder head for the sparking plugs.

The example of a push rod and rocker arm valve mechanism shown in Fig. 73 B is intended merely to illustrate the principle of this method of valve operation and in connection with the modern interpretation of this principle the general practice is to enclose the push rods within the cylinder block unit or its cover plate, as shown in Fig. 18. The push rod is sometimes arranged to work through a hole bored in the metal of the cylinder block so that it will attain the same working temperature as the cylinder unit and thus obviate differences in expansion effects which are experienced with the exposed type of push rod; in the latter instance the valve stem clearance increases as the cylinder block heats up.

The overhead camshaft engine requires a longer drive between the crankshaft and camshaft; for this reason a roller chain drive is generally preferred to timing gear wheels. The valve stem clearances do not alter between the cold and hot engine conditions and so can—and generally are—made smaller, namely, about 0·003 to 0·004 in. for both valves, than for side valves or push-rod operated overhead valves.

In connection with the decarbonizing of overhead camshaft engines this is a lengthier procedure as it is necessary to disturb the camshaft drive gear and usually the engine has to be retimed when the cylinder head is replaced.

Side Valve Operation.—The principle of the valve-operating mechanism for a side-valve engine is well illustrated by the example shown in Fig. 23 and also in the case of the Triumph motor cycle engine valve gear shown in Fig. 74.

The smaller pinion wheel seen in the centre is the crankshaft drive gear. This meshes with two larger wheels each having twice the number of teeth, shown dismantled below. Each of these gear wheels has its own cam made integrally with it. The cam wheels are bushed and rotate on fixed pins secured to the crank chamber. The cams operate direct on to the large mushroom-ended tappets seen above the cam wheel pins, so that no intermediate rocker gear is employed to operate the valves.

The gears are drilled, as shown, for lightness. The spiral gear behind the engine drive pinion is for operating the oil pump below.

Fig. 74.—Motor Cycle Side Valve Operating Mechanism.

High Camshaft Engines.—It is an advantage, in the case of push-rod type overhead valve engines to place the camshaft as high as possible, in order to lighten the components of the operating mechanism and to reduce thermal expansion effects to a minimum. An example of a high camshaft engine is shown in Fig. 216. A feature of the usual high camshaft arrangement is that the valve gear comes away with the cylinder head when the latter is removed from the engine.

Many commercial vehicle engines, both of the petrol and Diesel types, now employ the high camshaft arrangement.

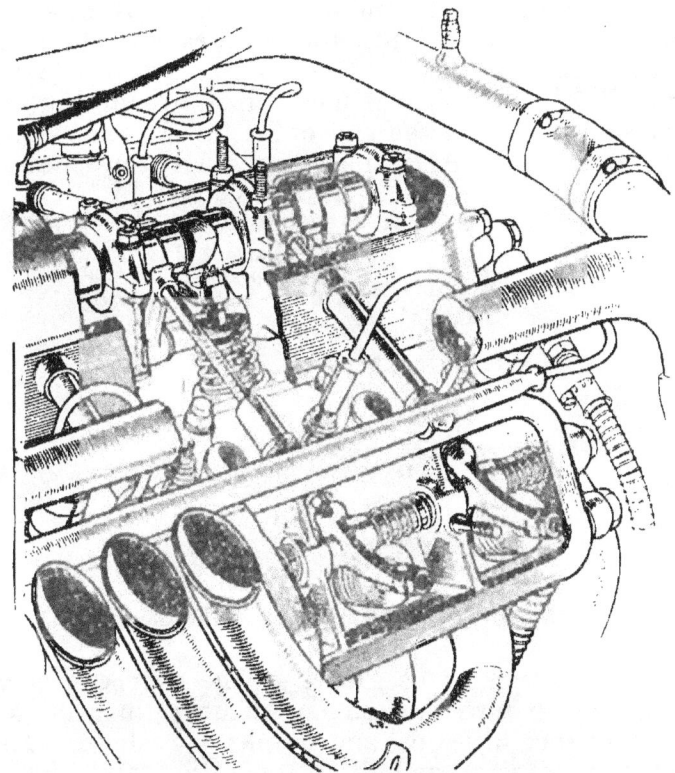

Fig. 75.—Auto Union Engine Valve Operating Mechanism.

Single Camshaft Operating Four Sets of Valves.—

In order to obviate the use of two separate camshafts in the case of a Vee-type engine of narrow angle a single camshaft can be employed and the four sets of inclined valves actuated by means of rocker arms and levers. An example of such a mechanism is that of the Auto Union sixteen cylinder racing engine shown in Fig. 75.* The inlet valves which are on the inside of the cylinders are operated direct from the rocker arms shown, whilst the exhaust valves which are on the outside parts of the cylinder are actuated by means of rocker arms and push-rods working in suitable guides.

Two Camshaft Drive.—A unique method of driving the two camshafts in the case of one particular commercial engine consists in providing a gear-wheel

* The Motor.

above the crank-shaft, and driven off it at one-half speed. This gear-wheel, together with each of the camshaft ends, has a small crank; each crank has the same " throw ". The Y-shaped member shown in Fig. 76 has roller-bearings at the end of its three arms, which engage with the three cranks mentioned; the rotation

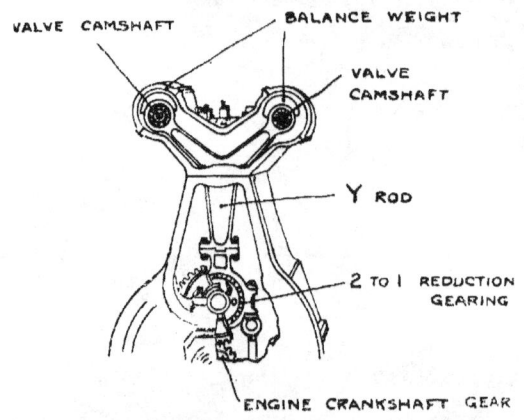

Fig. 76.—A Novel Method of Driving the Camshafts.

of the 2 to 1 crank causes each of the overhead camshaft cranks to turn in a similar manner, just as two pairs of connecting-rods and cranks would do. The advantages of this method of operation are that it avoids the backlash and wear of the usual trains of gearing, and gives a direct and inexpensive method of operating the camshafts. Actually there is an eccentric and strap (similar to steam engine practice) at the 2 to 1 shaft, and not a crank.

A somewhat similar method, following this "link-plate" idea, had previously been employed by Ricardo in the case of the Tylor-Ricardo engine.

Cam Shapes.—The amount of lift, and the time during which the valve is lifted, is determined solely by the shape of the valve cam. The common practice is to lift the valve fairly rapidly, and then to maintain it at this constant lift for a certain period, afterwards closing it fairly quickly. Fig. 77 (R.H.) illustrates the shape of such a cam, and also the type of valve-lift diagram given. The (L.H.) curve refers to the now almost obsolete gradual lift cam. The shape of the cam has a marked influence on the power output of the engine; the best shape is that which gives the quickest (practicable) opening and closing, consistent with the maximum allowable period of opening. Examples of typical valve-timings are given in Fig. 16, it may be added that the method of giving " overlap " to the exhaust valve at its closing period is frequently used in high speed engines. The sides of the cam must

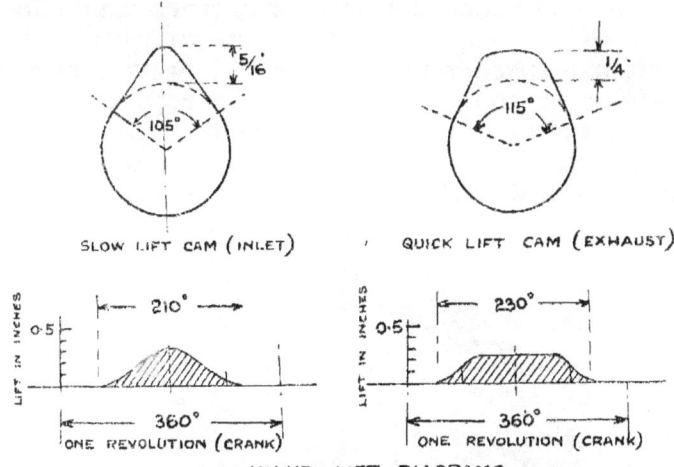

SLOW LIFT CAM (INLET)   QUICK LIFT CAM (EXHAUST)

VALVE LIFT DIAGRAMS

Fig. 77.—Cam Shapes and Lift Curves.

not be too steep, however, or there will be too much noise at opening and closing, due to impact, too much side force on the tappet or rocker arm, and the tappet will tend to "jump" or leave the cam during the opening period, thus necessitating the use of stronger valve springs to keep it in contact. The use of large area valves, with small lift, has to some extent obviated the necessity for steep-sided cams in high efficiency engines.

Number of Valves.—For engines up to 3 or 3½ ins. diameter of cylinder, it is the common rule to employ one inlet and one exhaust valve per cylinder. For larger sizes, high speed, and racing engines, two exhaust and one inlet valve (or two inlets) per cylinder are sometimes employed. It is considered better practice to use two valves of smaller diameter than one valve of larger diameter, although the operating mechanism is more complicated.

Driving the Cam-Shaft.—Since the cam-shaft must open and close each pair of valves of any cylinder once every two revolutions of the main crankshaft it must be driven at one-half engine speed by gearing down from the crank-shaft.

In earlier types of engine it was usual to employ a pair of gear-wheels—sometimes an intermediate wheel was used to avoid having the cam-shaft too near to the crank-shaft. The cam-shaft wheel had twice the number of teeth of the engine shaft pinion to give it the required half-speed ratio. Unless, however, special herringbone or spiral gear-wheels are used this form of camshaft drive is apt to become noisy in service. The later practice with many side-type and push-rod operated overhead valves was to employ a silent chain

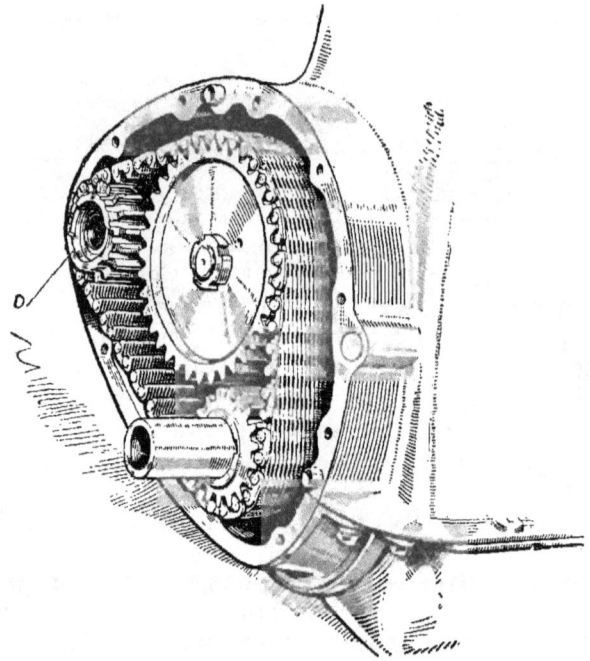

Fig. 78.—Illustrating the Silent Chain Driven Camshaft Arrangement.

drive for the cam-shaft as shown in Fig. 78. The lower shaft is the engine crank-shaft extension. The silent chain-wheel sprocket on this drives the larger chain-wheel attached to the cam-shaft above at one-half engine speed. The sprocket marked D on the left drives the lighting dynamo at engine speed.

The inverted-Vee type of silent chain has more recently been replaced by the roller chain, since the latter is lighter, less bulky and rather more efficient, mechanically.

The more recent examples of car and commercial vehicle engines employ the double or triple roller chains, according to the size of engine. These chains may be regarded as two, or three, roller chains placed side by side, but with common links for the inside members.

A typical arrangement for a medium-size car engine is shown in Fig.79. This has a double roller chain

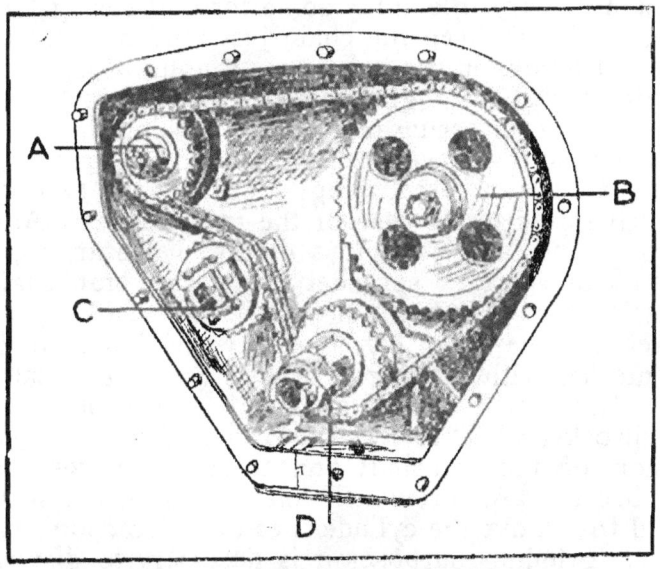

Fig 79.—Double Roller Chain Drive to Camshaft.

for driving the cam-shaft B and dynamo A. It has also an automatic chain tensioning sprocket C. The latter consists of an idle sprocket, the fixed axle of which is subjected to a gentle spring pressure all the time, in order to take up any slackness that may develop. A ratchet device is incorporated in the slide—which has a straight-line motion—so that as the

E

sprocket moves inwards it is always retained in its inward position by the pawl of the ratchet. Another type of tensioning sprocket uses an eccentric device and a clock-spring tensioner, with ratchet device.

These chains operate silently and only require replacement about once every 40,000 miles or so.   A typical double-roller chain suitable for a 15 h.p. rating engine has a pitch of ⅜ in., roller diameter of ¼ in., and total width of 1 in.; the breaking load is 4,000 lbs.

Overhead Camshaft Drive.—Owing to the relatively big distance between the crankshaft and the camshaft in overhead camshaft engines it is not convenient, nor desirable, to employ a train of gear wheels to drive the camshaft so that the roller type of chain drive is employed. This gives a simple and direct transmission but it is necessary to employ some form of spring tensioning device to keep the chain just taut in order to prevent variations in the valve timing and knocking of the chain against the engine casing.   In some instances a simple cantilever spring tensioner is employed, whilst in others a spring-loaded tensioner sprocket, running idle on its shaft, is used.   Fig. 80 shows the type of chain drive used in the case of the twin-camshaft Alfa Romeo engine, which employs the 45° valve arrangement.   The chain sprocket on the crankshaft has half the number of teeth of the upper sprocket and the latter has a gear wheel bolted to it which meshes with equal gear wheels on the two camshafts.   In the case of a single overhead camshaft the upper sprocket wheel is keyed to or bolted to a flanged extension of the camshaft and therefore drives the latter direct.   As previously mentioned, when it is required to remove the cylinder head for decarbonising and valve-grinding purposes it is necessary to disturb the timing of the engine or to provide for the bodily removal of the upper sprocket in an endwise direction so that it will clear the camshaft; it can then be rested in a bracket provided for the purpose so that in this instance the chain need not be detached from the sprocket.

Usually a ⅜ in. double roller chain is employed for engines up to about 15 H.P. (rated power) and a ⅜ in. triple one for higher powers.

The Crank-Shaft.—The crank-shaft and connecting-rod (or rods) convert the reciprocating motion of the piston (or pistons) into one of rotation. It is made very stiff, since it is subjected to severe and varying twisting stresses due to the explosion pressures, and also to the " inertia " effects of the reciprocating parts;

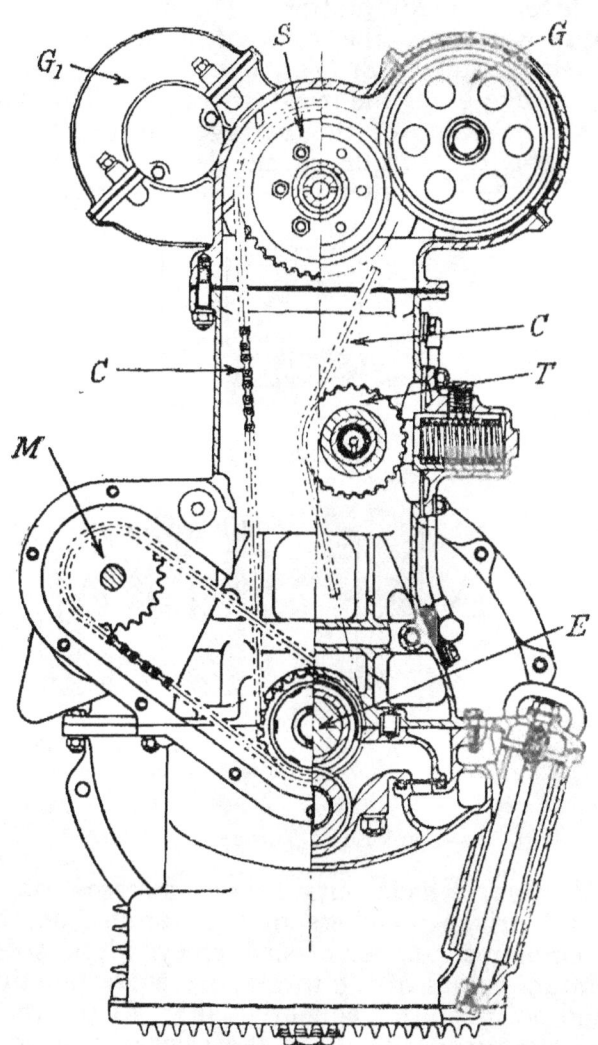

Fig. 80.—Alfa Romeo Camshaft Drives.
C—Chain.   E—Engine sprocket.   G and G—Camshaft sprockets.
S—Common sprocket.   T—Tensioner.   M—Magneto drive.

the latter effects are really the forces due to the acceleration and deceleration of the piston and connecting-rod in their strokes.

The twisting or turning action on the crank-shaft, which is generally spoken of as the *Torque*, is constantly changing; this fact necessitates a stronger shaft than for a steady motion. The manner in which the torque varies in the case of a single cylinder engine is shown in Fig. 81 (*a*). It will be observed that the firing stroke gives the greatest torque. The dotted line shows the value of the average torque. In this case the greatest torque is no less than 8 times the mean value. In order to reduce this great varia-

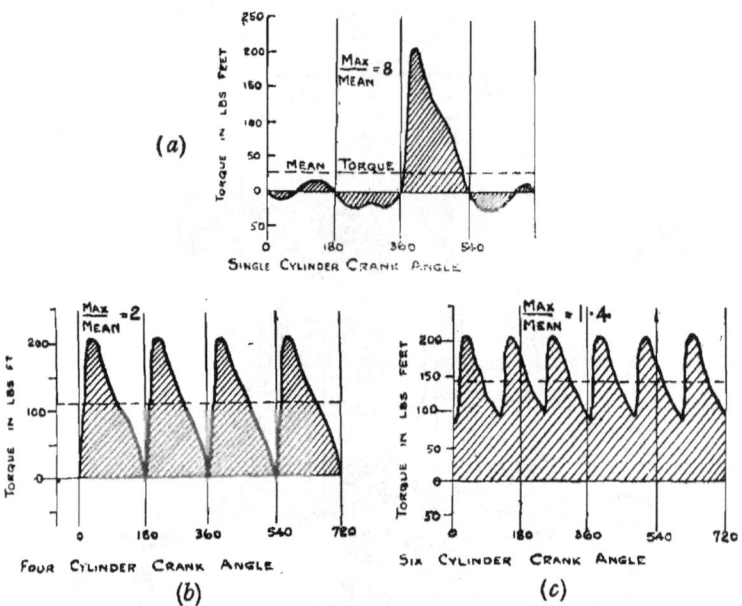

Fig. 81.—Crank-Shaft Torque Diagrams.

tion of torque, which puts severe stresses on the driving and driven members of the automobile, it is usual to employ a number of smaller cylinders, so that the maximum values of the torque are correspondingly lower, and occur more frequently. Fig. 81 (*b*) and (*c*) illustrates the torque values in the case of a four and a six-cylinder engine; it will be noticed from this that the greater the number of cylinders the smoother becomes the torque curve.   Interpreted practically, this means a smoother running engine, with a lighter chank-shaft, and incidentally lower minimum top-gear and higher maximum speeds.   The single cylinder engine is, relatively speaking, very jerky in its running, on account of the wide torque variations, and it is necessary to provide heavy fly-wheels to reduce the speed and torque variations. Crank-shafts for motor-cycle engines are usually of the combined fly-wheel crank-pin type illustrated in Fig. 17.

Fig. 82.—A Two-Cylinder Opposed Engine Crank-shaft, with Connecting-rods and Slipper Pistons.

For four-cylinder engines, the crank-shaft is usually made in one piece by a forging process, known as " drop-forging " (Fig. 83); in this process a powerful steam or pneumatic hammer is provided with a pair of dies, having internal impressions corresponding to the required shape of the crank-shaft.   The almost white

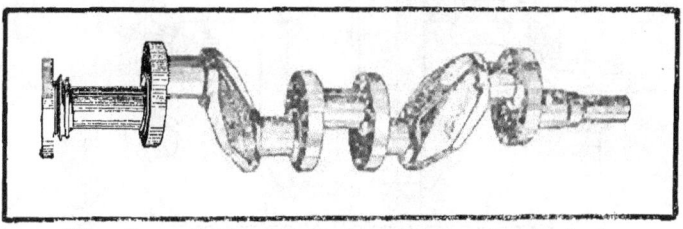

Fig. 83.—A Four-Cylinder Engine Crank-shaft made from a Drop Forging.

hot steel is pressed into these dies, and assumes the
desired shape.   The bearings and crank-pins only are
machined.

In some cases the crank-shaft was planed, milled and
turned from a solid slab of steel; this was a more
expensive process, however.   Another method, which

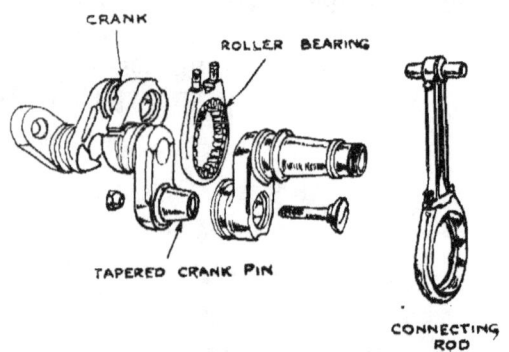

Fig. 84.—Example of a Built-up Crank-shaft.

has been developed, is to build up the crank-shaft from
a number of components, as shown in Fig. 84.   This
method, although more complex, enables roller bearings to be employed for the big ends of the connecting-
rods. Crank-shafts for single and twin cylinder engines
are provided with two main bearings, one at either
end.   In the case of the popular four-cylinder engine,

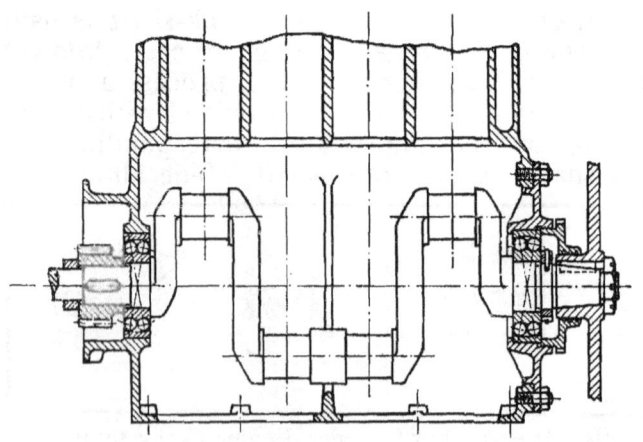

Fig. 85.—A Two-bearing Four-Throw Crank-shaft.

of the light car and cheaper touring car types, the use of the compact monobloc system of cylinder design enables a short but stiff crank-shaft with only two bearings to be used, as shown in Fig. 85. It is better practice, however, to employ a centre bearing whenever possible, since this prevents deflection or whipping under load.

In the case of six-cylinder engines, the best practice allows one main bearing between each pair of cranks, although the cheaper models employ in all three bearings only.

A modern eight-cylinder crank-shaft is illustrated in Fig. 86. As with most well-designed crank-shafts this has a damping device to smooth out the torsional vibrations, and balance weights for each crank-pin. The crank-shaft in question has five main bearings. The fly-wheel is securely attached to one end, and on its periphery are gearwheel teeth for the starter electric motor pinion to engage.

Crank-Shaft Materials.—Inexpensive and slow-speed engine crank-shafts are frequently made of a medium carbon steel known as " 40 ton steel," unhardened. The majority of modern crank-shafts are made of alloy steels, including 3 per cent. nickel-steel, chrome-vanadium and nickel-chrome. After machining to approximate dimensions the crank-shaft is hardened, and the bearing surfaces finally ground to size. In the hardened condition these alloy steels have tensile breaking stresses of from 50 to 90 tons per sq. in.

More recently high grade crankshafts have been made of alloy steels containing molybdenum; typical steels of this class include the nickel-chromium-molybdenum and somewhat similar steels with the addition of a small proportion of vanadium When suitably heat-treated such steels give tensile strengths of 65 to 75 tons per sq. in.

Another class of high tensile steel that has been used for automobile engine crankshafts is the nitriding one containing chromium, molybdenum, vanadium and aluminium. After machining, the crankshafts are heated to about 500° C. and exposed for a period of 40 to 90 hours to a stream of ammonia gas, the nitrogen

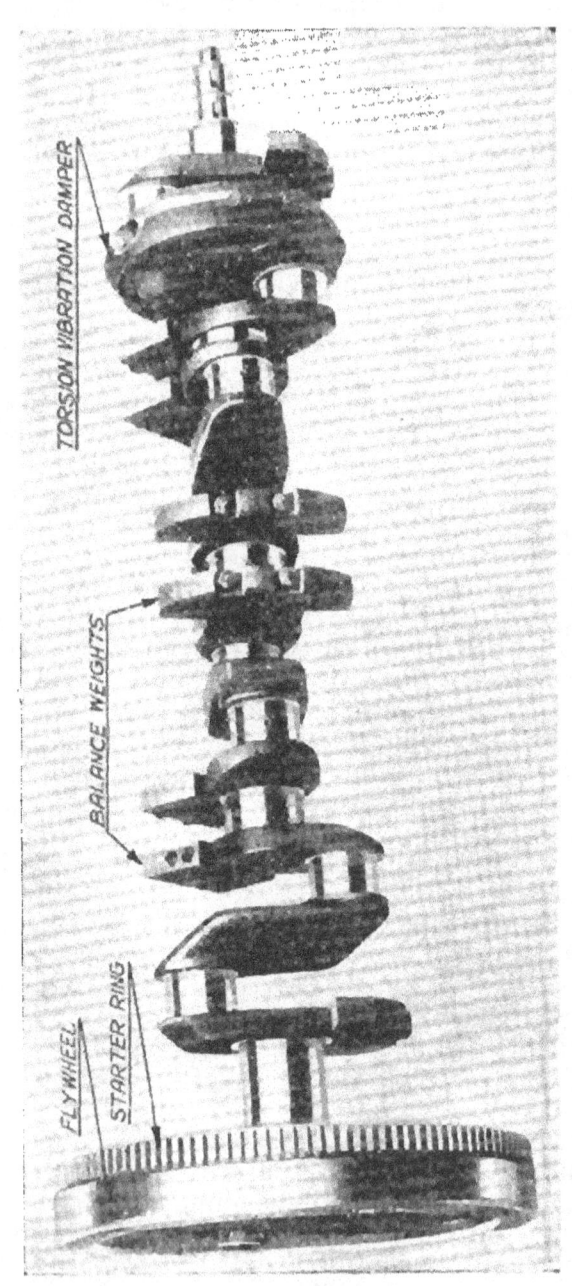

Fig. 86.—A Straight Eight Engine Crank-shaft with Torsion Vibration Damper, Balance Weights and Flywheel attached.

of which takes part in the surface hardening process and finally produces a glass hard surface capable of extreme resistance to wear effects. Such steels do not distort when hardened as the temperature is well below that of ordinary case-hardening processes. They possess tensile strengths of 60 to 90 tons per sq. in. according to the composition and heat treatment. In instances where the maximum resistance to wearing is required, the crank pin and main journals are chromium plated. The result is an extremely hard surface which will give at least 80,000 miles of useful service before re-grinding and replating become necessary.

Cast Crankshafts.—Hitherto all motor car engine crankshafts have either been built-up or made from forgings. More recently the Ford Company has adopted the cast crankshaft, using a special composition of alloy iron, containing Carbon (1·35 to 1·60), Manganese (0·5 to 0·6), Silicon (0·85 to 1·10), Chromium (0.4 to 0.5), Copper (1.5 to 2.0) per cent., with very small percentages of sulphur and phosphorus.

The steel has a breaking stress of 60 to 65 tons per sq. in.

The previously used forged shaft for the 8-cylinder Vee engine weighed 90 lbs. in the rough and 66 lbs. when machined. The cast crankshaft weighs 69 lbs., and only 9 lbs. of metal is removed by machining.

The forged shaft shows measurable wear after 10,000 miles, whereas the cast shaft shows less than two-ten-thousands of an inch.

Other alloy irons that have been or are being used for cast crankshafts include copper-chromium ones, inoculated cast irons, chrome-molybdenum and nickel-chromium ones.

In general, provided the necessary strength and fatigue resistance are obtained, cast crankshafts are cheaper to manufacture, require fewer machining operations, can be made lighter than forged alloy steel ones and possess hard bearing surfaces having a lower coefficient of friction.

A typical chrome-molybdenum iron crankshaft was fitted to a four-cylinder 75 H.P. commercial vehicle engine running at a maximum speed of 2,500 R.P.M

E*

which covered a distance of 40,000 miles in $2\frac{1}{2}$ years. At the end of this period the wear on the main crank-shaft journal averaged 0·001 in. to 0·002 in.; that on the crank pins was found to be negligible. The big-end bearings were made of R.R. 56 light aluminium alloy and had a running clearance of 0·003 in.

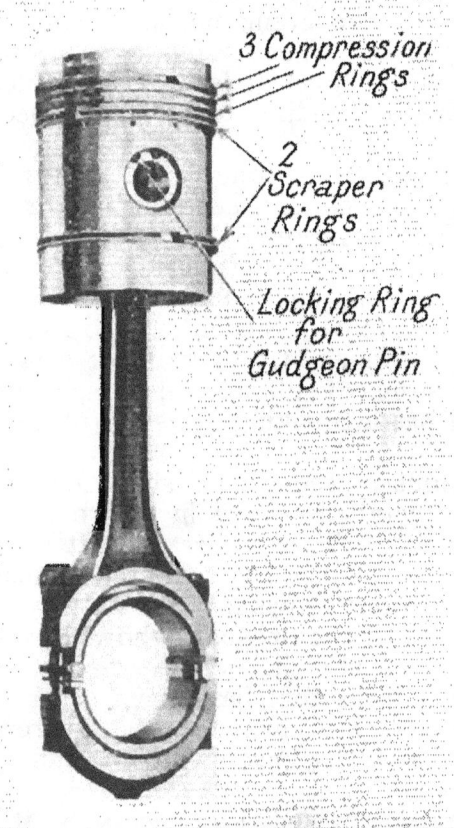

Fig. 87.—Piston and Connecting Rod Assembly for Commercial Engine.

Connecting-Rods.—These rods are subjected to heavy loadings of a variable nature. To resist such forces, the connecting-rods should be designed as struts subject to side bending. The tubular and H-sections are now invariably used, since these give the greatest strengths in compression and bending. Connecting-rods are usually made in one operation as drop-forgings; they are afterwards machined, and the bearings fitted. The small-end bearing, when the

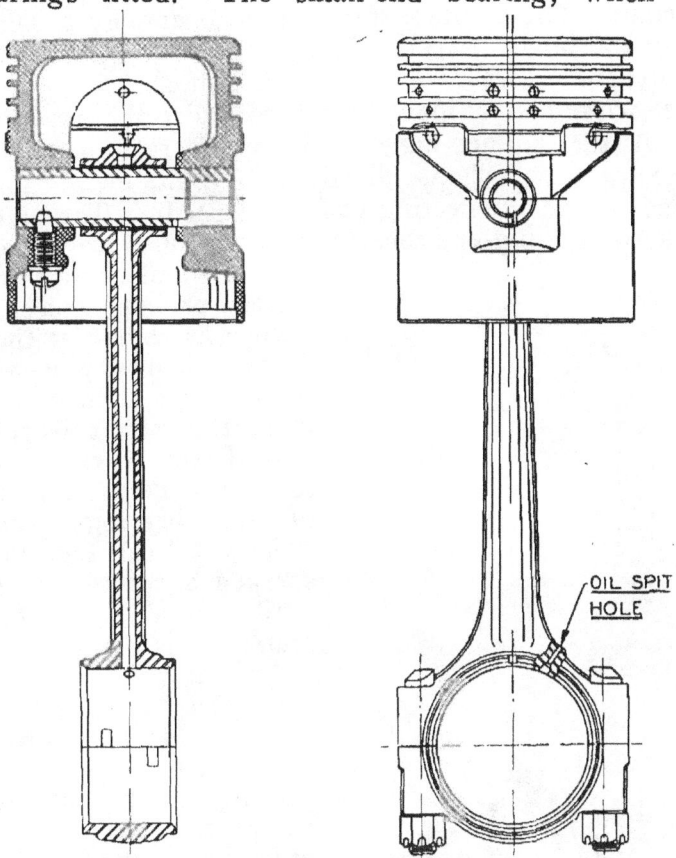

OIL SPIT
HOLE

Fig. 88.—A Modern American Car Engine Piston
and Connecting Rod Assembly.

gudgeon pin is fixed in the piston bosses, takes the form of a phosphor-bronze bush, which is pressed into the small-end of the rod. An oil hole above, or two holes below, and oil grooves are arranged for the purposes of lubrication. The gudgeon pin is sometimes secured to the small end and rocks in the piston bosses. Another method is to bush the small end, and to allow the gudgeon pin to " float " (i.e., to be free to rotate) in it and also the piston boss bearings; in this case, means are provided to prevent end-movement of the gudgeon pin; otherwise it would score the cylinder. The big-end bearing is very important. The common practice is to employ a split white-metal-lined bearing, bolted together with two or more bolts, as shown in Figs. 87 and 88.

The white-metal may be affixed directly to the steel of the connecting-rod or a separate pair of bronze or steel shells, lined with white-metal and secured to the connecting-rod parts used. The two halves of the big-end bearing are clamped together with a number of thin metal packing pieces known as shims, at the joint; when any wear occurs, one or more of these thin strips is removed from each joint to reduce the diameter of the bearing; the white-metal can then be scraped away by hand until a good fit on the crank-pin is obtained.

Although it has become the normal practice to split the connecting-rod big end bearing at right angles to the length or axis of the rod, the Vauxhall connecting-rod is now split at the angle shown in Fig. 89. This arrangement possesses certain advantages in regard to accessibility and is especially suited to the use of the white-metal lined steel shells used. The bearing cap has serrated faces which mate with

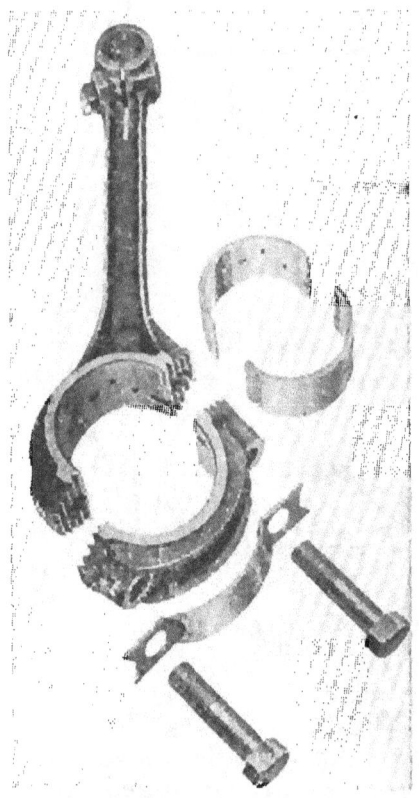

Fig. 89.—Vauxhall Connecting Rod.

corresponding ones on the connecting-rod so that any shearing action due to the piston loads is taken off the bolts holding the cap to the rod.

The use of the inclined type of big-end joint results in a more compact crankcase and enables the piston and rod to be withdrawn upwards through the cylinder barrel. Usually, with the right-angle connecting-rod joint the big-end part of the connecting-rod is too large to pass through the cylinder barrel.

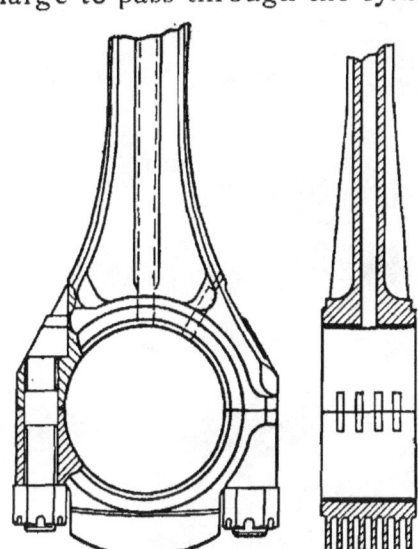

Fig. 90.—Connecting-Rod with Ribbed Cap.

Fig. 90 shows a modern big-end design of light yet very strong construction. Special features of this arrangement include the connecting-rod cap which is provided with fins for assisting the cooling of the bearing; light cap securing bolts with bevelled heads and reduced section, namely, to that of the bottom diameter of the nut thread and H-section rod with central oil hole in the enlarged web to lead oil to the small-end bearing.

The lubrication of the big-end bearing is important. Oil grooves radiating from the oil supply holes are cut in the white-metal, and the latter is recessed near the joints. In earlier car engines oil scoops, on the ends of the connecting-rod pick up oil from special troughs provided in the crank-case, and feed it to the oil hole (or holes) in the lower bearing, whence the oil travels along the grooves and floods the bearing. Occasionally oil holes are drilled in the upper half, and receive oil drained down the connecting-rods; it is not good practice, however, to feed the oil direct to the loaded side of the bearing.

Mention has already been made of the use of roller-bearings for the big-end (Fig. 84). These are perfectly satisfactory, if properly designed, and their use enables a compact and narrow crank-pin bearing to be obtained; these bearings give a rather lower frictional

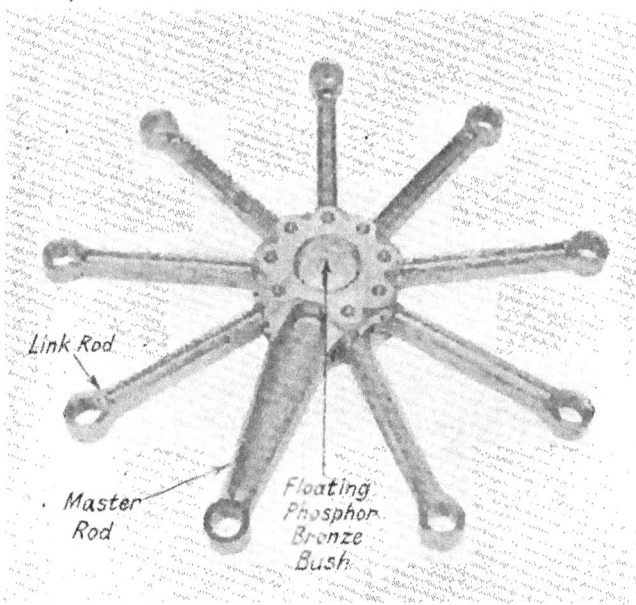

Fig. 91.—Connecting Rod arrangement of Bristol Radial Engine.

resistance, also. In the case of radial engines, in which only one crank-pin is used for the whole number of cylinders, the usual method is to employ one main, or "master," connecting-rod bearing on the crank-pin, and the remainder as secondary or link rods hinged to the master rod.

Connecting-Rod Proportions.—Connecting-rods in automobile engines are usually made equal in length to from $1\frac{3}{4}$ to $2\frac{1}{2}$ times the stroke of the piston. Where H-sections are used the approximate proportions are: *Depth* from $\frac{1}{5}$ to $\frac{1}{4}$ the cylinder bore, *Width* from $\frac{1}{7}$ to $\frac{1}{10}$ the cylinder bore, and *Web Thickness* from $\frac{1}{25}$ to $\frac{1}{30}$ the cylinder bore. The diameter of the gudgeon pin should be from 0·20 to 0·25 times the piston diameter and its length from 0·8 to 0·9 times the latter.

Connecting-Rod Materials.—Connecting-rods are usually made in 3 per cent. nickel steel, nickel-chromium and chrome-vanadium steels, heat treated. Aluminium alloy connecting-rods are now being used, and have given satisfaction. The alloys used are forged Duralumin, and " Y " alloys, which enable the connecting-rods to be made about 30 to 40 per cent. lighter than the drop forged steel ones of equal strength.

Connecting-Rods of Vee-type Engines.—In connection with the Vee-type of engine the crankshaft has one-half the number of crank pins or journals as the

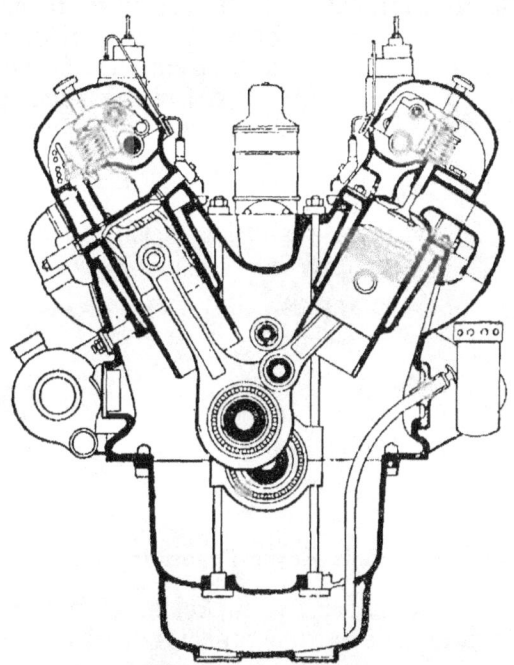

Fig. 92.—The Maybach Connecting Rod Arrangement.

number of cylinders; each pair of cylinders facing one another uses the same crank journal. The corresponding pairs of connecting-rod big-ends therefore work on the same journal. Two alternative arrangements for the rods are as follows, namely :—(1) Similar designs of rod working side by side on the journal, and (2) one forked rod and one central rod with its bearing between the forked portion. Another method, used in the case of the twelve-cylinder Maybach engine (Fig. 92) is to have a master connecting-rod running on roller bearings and a link rod working on a pin attached to one side of the master rod for the opposite cylinder. The link rod has a phosphor bronze bush type of bearing; this arrangement is somewhat similar to that of the connecting-rods of radial engines.

In the Rolls Royce Vee-type engines a forked rod is used for one cylinder unit and a plain rod for the facing cylinder one. The forked rod works on the crank journal and the plain rod on the big-end portion of the forked rod; in each case the bearing surfaces are of white-metal and of equal bearing area. Fig. 93 illustrates the connecting-rod arrangement of the engine in question. The material used for the rods is 3½ per cent. nickel steel, machined all over and heat-treated.

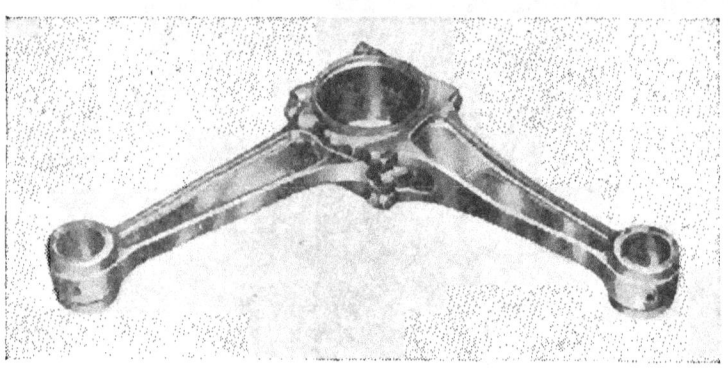

Fig. 93.—Rolls Royce Connecting Rods.

The forked rod carries a nickel steel bearing block lined inside and out with a special lead-bronze alloy. The bearing block which is divided across its bore bears directly on the crank-pin and is secured to the forked rod by four bolts, whilst the outer lining of the block forms the bearing for the plain rod.

Recent Bearing Improvements.—Although white-metal is a satisfactory material for the main and big-end bearings of petrol engines, provided the bearing pressures are not excessive and that oil filters or cleaners are fitted, when high compressions are used, as in high output petrol and compression-ignition engines, this material does not always give good service. In the latter type engines, some trouble was experienced owing to cracking and breaking down of the white-metal, so that it became necessary to seek stronger bearing metals having low frictional qualities. Many of the recent designs of engine are now fitted with *lead-bronze* main and big-end bearings. Lead-bronze is an alloy of copper, lead and tin having a distinct bronze colour; it is very much stronger in compression than white-metal, and has a comparable friction coefficient. As it is a more expensive material to employ, some makers only use it for the more highly loaded halves of the bearings, viz., the upper half of the big-end and lower half of the main bearings; white-metal is employed for the other halves. Lead-bronze bearings can be run for mileages of 80,000 to 100,000 before replacement or attention is necessary; white-metal under similar circumstances will not last longer than about 25,000 to 30,000 miles.

Another strong bearing alloy is one containing cadmium and nickel with small proportions of copper, magnesium, silver, etc. This is about twice as strong as the best white-metal.

The more recent tendency in regard to big-end bearings is to employ detachable steel shells lined with white-metal or lead bronze, these being of semi-cylindrical shape (see Fig. 89). The replacement of worn bearings thus becomes a comparatively simple matter. The result of using a very thin lining of bearing metal in the steel shells is not only to prolong the life of the bearing considerably but also to obviate any risk of the bearing metal breaking down when in service, if properly lubricated.

In certain 1941 engines the white-metal linings have been reduced to about 0·003 in. to 0·005 in. in thickness and it is claimed that the increase in bearing fatigue strength has resulted in the bearing life being prolonged to four or five times that previously obtained. The bearing metal is bonded to the steel shell by means of a special porous matrix.

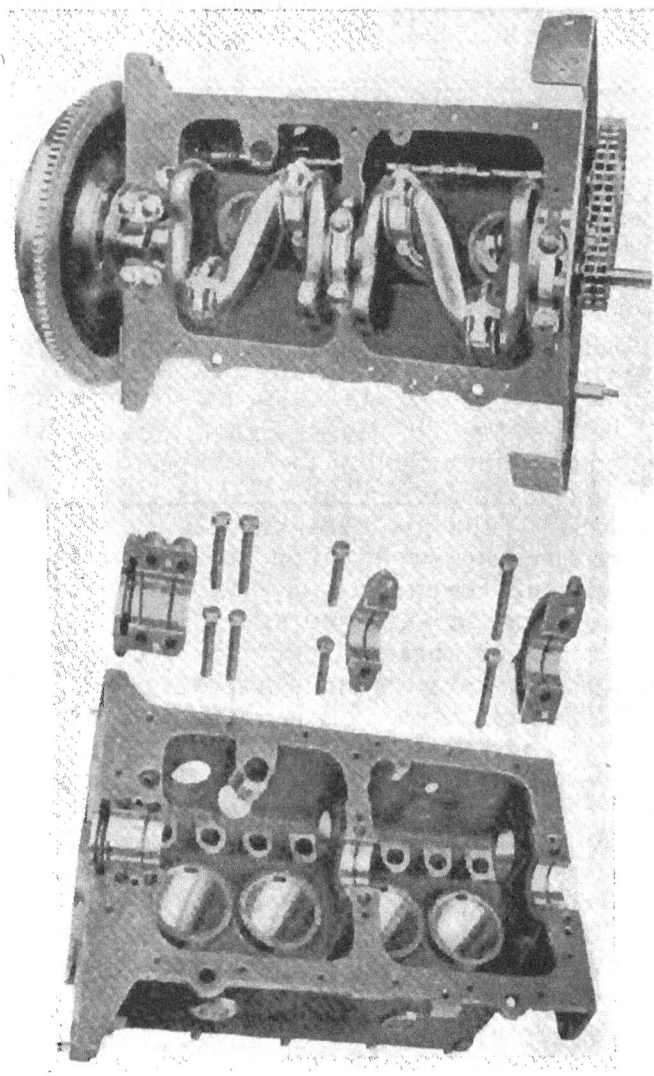

Fig. 94.—Vauxhall Crank-shaft and Crank-case, showing Main Bearings.

Where white-metal is employed it is generally of tne high tin content grade having from 80 to 90 per cent.

tin and 4 to 7 per cent. copper with from 4 to 10 per cent. of the hardening element, antimony.

A typical lead bronze alloy used for bearings contains 70 per cent. copper, 28 per cent. lead and 2 per cent. tin. It is nearly twice as hard and as strong in compression as the best white-metals and at 200° C. is nearly 12 times as hard as white-metal at 200° C.

Main Bearings.—The main crank-shaft bearings are usually carried in an extension of the cylinder block casting so that the bearing caps have to withstand the principal loading due to the combustion pressures of the charge. A typical example of a four-cylinder engine (Vauxhall) cylinder block and crank-case unit is shown in Fig. 94. In the upper illustration the crank-shaft is seen in place, the view being taken from underneath, looking towards the cylinder bores. In the lower illustration the crank-shaft has been removed to show the arrangement and components of the bearings. The bearing caps are provided with dowel pins which engage with corresponding holes in the opposite parts of the bearing in order to ensure correct location.

Other Components and Accessories.—The fly-wheel is provided primarily for the purpose of storing up part of the energy of the explosion period, and utilizing this stored up energy to carry the crank-shaft, rods and pistons over the three idle strokes; the fly-wheel thus tends to re-distribute the energy received, so as to give more uniform running. The essentials of a fly-wheel are a relatively heavy rim, situated as far away from the axis as possible; this is equivalent to stating that the " moment of inertia " of the fly-wheel should be as great as possible. Practical considerations fix the size and weight of the fly-wheel; thus for a motor-cycle engine the diameter should not exceed from 3 to 4 times the piston bore, and the weight about 1 to 2 lb. per cubic inch of cylinder capacity. For four-cylinder engines, owing to the more even torque and to the overlapping of the cycles, the fly-wheel can be made much lighter in proportion, namely from ½ to 1 lb. per cubic inch capacity. There is hardly any need for a fly-wheel in a six or eight-cylinder engine, owing to the evenness of torque. The fly-wheel serves, however, to form one member of the clutch, and also in most cases it is "toothed," to engage with the pinion of the electric starting motor.

Although electric motor starters are now the rule, a starting handle is usually supplied. Its use enables the engine to be cranked over on cold days, or after standing for a period, in order to fill the cylinders with mixture prior to using the starter; this relieves the batteries of much load; for valve timing and adjustments the starting handle is also very useful.

The forward end of the crank-shaft is provided with a pair (or more) of teeth or dogs, each having one straight and one sloping tooth. The starting handle has a pair (or more) of similar teeth, such that when the handle is put into engagement, the straight sides of

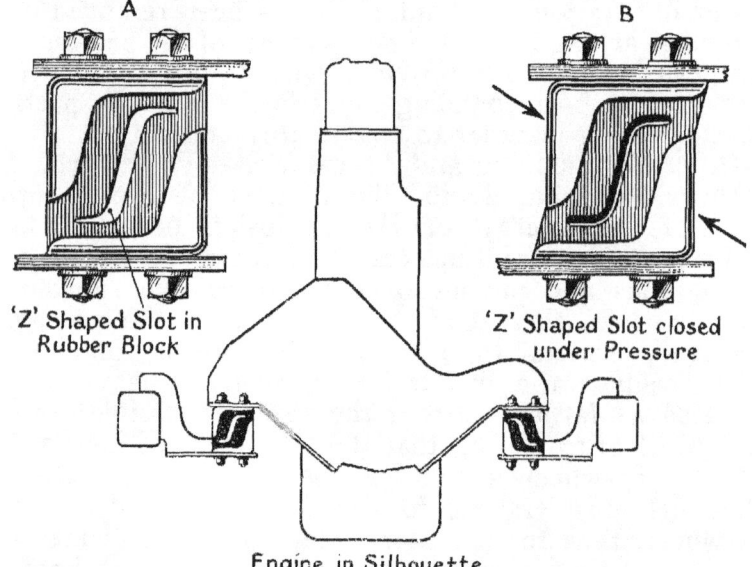

A   B

'Z' Shaped Slot in   'Z' Shaped Slot closed
Rubber Block   under Pressure

Engine in Silhouette
showing Front Mounting.

Fig. 95.—Vauxhall Flexible Engine Mounting.

the teeth engage, and the handle can then rotate the engine crank-shaft in its normal direction of rotation. When the engine fires, the crank-shaft teeth move the faster, and the sloping sides force the starting handle's teeth out of engagement. A compression spring sometimes holds the latter away from the crank-shaft's teeth. Should a backfire occur, the engine rotates in the reverse direction, and there is then a positive drive to the starting-handle. If the latter is held rigidly in the closed fist, it is very easy for the wrist to be broken. The handle should therefore be held on the under-side only, when pulling up, and without the fingers encircling it. *To obviate back-firing* a strong mixture and retarded ignition should be arranged for. Patent starting handles, having pawls and ratchet devices, have been marketed; these enable the risks of backfire to be avoided.

Engine Mountings.—Hitherto, it has been the practice to bolt the engine unit solidly to the chassis frame at three or four places. The " three-point " mounting consists of a front central fixing and two side ones; this method enables the engine to accommodate itself to small frame movements, without undue stresses on the engine casing. The modern method of mounting the engine dispenses with solid fixings and employs springs or rubber blocks between the engine bearers and the frame. In this way any vibrations due to misfiring or lack of correct engine balance, etc., are not transmitted to the frame and bodywork. From the car occupants' viewpoint the flexible mounting method is both quieter and free from transmitted vibrations.

An improved method of engine mounting employing rubber mounting blocks having " Z "-shaped slots is shown in Fig. 95. The effect of the slots is to allow for a very flexible slight amount of engine movement but to provide for increasing resistance to the larger move-ments; this occurs when the slots close as shown in diagram B (Fig. 95). The central illustration shows the two rear engine mountings of the Vauxhall car, in which the method described is employed.

The type of rubber used for engine mountings is of a special kind having a marked resistance to ageing (or perishing) whilst being proof against the action of petrol and oil. More recently, synthetic rubbers such as Buna and Neoprene have been employed as these materials offer increased advantages in connection with their resistance to the effects of sunlight, ozone, oil, petrol and grease and can be made to give excellent strength properties.

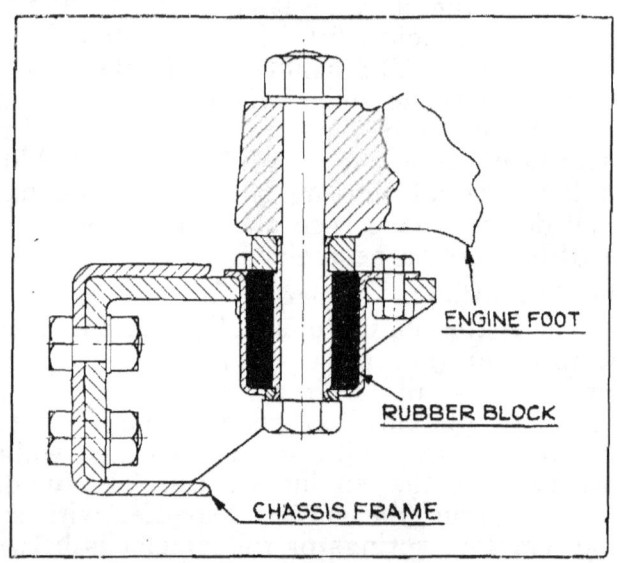

Fig. 96.—Flexible Engine Mounting.

Fig. 96 illustrates the Silenbloc type of engine mounting, the stretched rubber member (shown in black) affording correct insulation. When the flexible type of mounting is used it should absorb vertical vibrations, but should not permit of any appreciable side movements; in this respect the device shown in Fig. 96 is a good example. (See also Figs. 20 and 21.)

It is important to note that where an engine is mounted flexibly, the exhaust pipe should also be given a flexible support on the frame; otherwise it will restrain the engine's movements. Another point is the possible electrical insulation of the engine by the rubber blocks. As the engine forms the earth return of the ignition system, it must be connected electrically with the chassis frame.

The Silencer.—The exhaust gases leave the engine at a pressure of 30 to 50 lbs. per sq. in. at first, and at a temperature of between 600° C. and 800° C., with a high velocity, namely, about 150 feet per second in the case of a high speed engine. At each discharge, *i.e.*, on every exhaust stroke, a compression, or sound wave, is sent out, giving the characteristic exhaust note of the engine. It is necessary in public interests to reduce this noise considerably, and for this purpose the exhaust gases expand into a silencing chamber, where the strong sudden discharges are broken up into a more or less continuous one, emitting little

Fig. 97.—The Austin Flexible Front Engine Mounting.
E—Engine Unit. R—Rubber Block. F—Frame Member.

noise. The best design of silencer enables the exhaust gases to cool down and expand into the atmosphere continuously and to emerge as a more or less uninterrupted stream. Usually a number of baffles are fixed in the silencing chamber, but careful proportioning of these is necessary in order to avoid back pressure and overheating of the engine. Fig. 98 illustrates some typical examples of automobile engine silencers, and shows how the gases are broken up before emergence. The internal diameter of the exhaust pipe should not be less than one-quarter to one-third of the cylinder diameter, whilst the ordinary design of silencer should have a capacity of about 14 cubic inches for every B.H.P., at the maximum value. It should not weigh more than about 0·1 lb. per B.H.P. (maximum) developed by the engine.

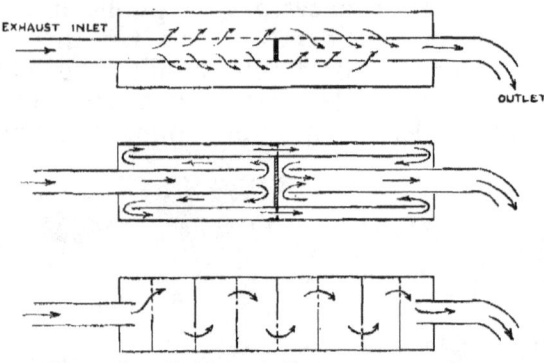

Fig. 98.—Types of Exhaust Silencer.

A well-designed silencer will not reduce the power of the engine appreciably; the results of tests show that very good silencing can be obtained with a loss of power of about 2 to 3 per cent.

Results of Silencer Researches.—The results of researches on petrol engines show that the exhaust noise which originates from the fluctuating pressures at the exit is due to a combination of notes of various frequencies, i.e., the sound wave is of complicated form made up of fundamental notes and overtones, etc. For practical purposes the exhaust noise may conveniently be regarded as consisting of two main parts, namely, a *low-to-medium frequency band* of about 50 to 600 cycles per second and, disregarding an intermediate frequency band from 600 to about 3,500 cycles per second, of little noise, a *high-frequency band* of about 3,000 to 10,000 cycles per second.

The lower frequency part appears to be due to a resonance effect between the varying capacity of the cylinder and exhaust system as the piston moves in the exhaust stroke and the area through the exhaust valve as it opens and closes. This conjecture has actually been confirmed by tests made upon a petrol engine driven by an electric motor at the same speeds as when running under its own power, when the low-pitch noises were shown to be almost as great as when the engine was working normally.

The high-pitch part of the noise is due primarily to
the release of high-pressure gas—at pressures of 50
to 70 lbs. per sq. in.—through the exhaust ports and
silencing system. Alterations of engine speed whilst
affecting the intensity of the total noise do not
materially influence the pitches of the two parts of the
noise.

In studying the problem of effectively silencing the
exhaust noise it is necessary to consider the means to
deal with each of the two main frequency bands.
Unfortunately any silencing system designed to
suppress or damp down the high-frequency band will
not reduce the low-frequency notes appreciably.

Thus it has been ascertained that a silencer of the
absorption type consisting of a straight-through per-
orated tube in an outer casing filled with absorbent
material, such as glass, silk or bundled fine wires, will
damp down and absorb most of the high-frequency
notes; in this case the peaks of the high-frequency
pressure waves pass out through the perforations into
the absorbent material and are thereby reduced
in magnitude, but may return after some delay out of
phase with other peaks; in effect, these waves are
smoothed down so as to give a more or less continuous
pressure condition associated with low intensity noise
at exit.

This type of silencer is, however, ineffective for
dealing with the lower frequency band, for which the
best type of silencer is the *capacity type*, i.e., a plain
silencer of relatively large capacity and big changes
of sections; its action is based upon the absorption of
sound waves in the turbulent areas of each change of
section.

By combining the two types of silencer it is possible
to deal with both the higher and lower frequency
bands and thus to obtain the best silencing of the complete exhaust noise. Fig. 99 illustrates, diagrammatically, how this can be effected, whilst the practical
interpretation of these principles to composite
silencers is indicated below in Fig. 100.

One type of silencer is designed so that the exhaust
gases are divided into two parts, each following a path
through the silencer of a different length. It is so

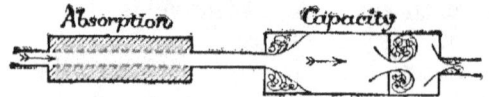

Fig. 99.—Principle of Ideal Silencer.

arranged that the two paths differ in length by half a wave-length of the sound vibration. Thus one wave will arrive at its maximum positive vibration whilst the other will be at its negative maximum, so that the two cancel out and, theoretically, the vibrations should be silenced out.

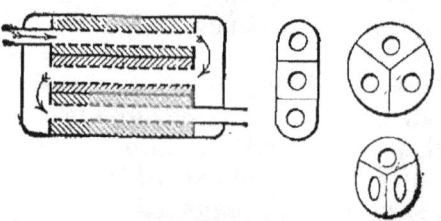

Fig. 100.—Typical Composite Silencers.

It is, of course, only possible to arrange for this cancellation of sound waves at one particular note frequency, viz., that of the normal engine speed; at most other speeds only partial cancellation occurs.

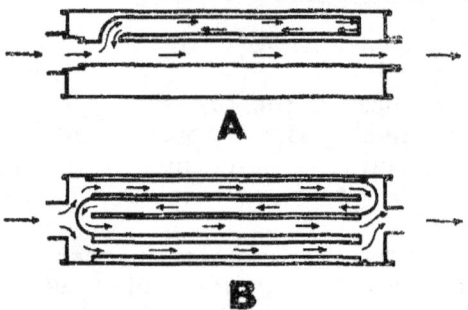

Fig. 101.—The Sound Wave Cancellation Type of Silencer.

Fig. 101 (A) and (B) show the Quincke and Herschell silencers respectively. In the former instance there is an enclosed tube of one-quarter the wave-length to be cancelled; as the sound wave in this part is reflected at the end, it has a half-wave effect when meeting the direct wave. The latter diagram shows a divided path in which the upper one is half a wave-length longer than the straight-through lower one.

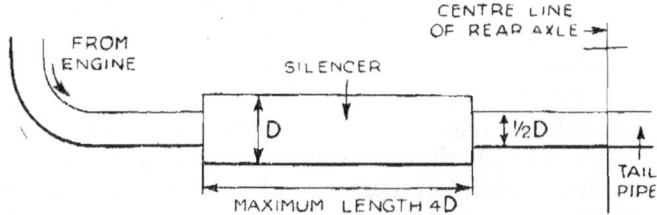

Fig. 102.—The Brooklands Silencer.

A type of silencer used on the racing track, known as the Brooklands model, is illustrated in Fig. 102; its use is compulsory on the British tracks. In this case it is laid down by the R.A.C. that the exhaust gases must be led into an expansion chamber having a capacity of not less than six times the volume swept by one cylinder of the engine, the diameter D (Fig. 102), if circular, or the equivalent dimensions, if of any other form, being not less than one-fourth of the length, the tail pipe shall have an internal diameter of not more than half the equivalent diameter of the silencer, and shall be so arranged that the gases from the pipe cannot impinge upon the road, and the pipe shall extend beyond the rear axle. Exhaust pipes may be carried outside the body.

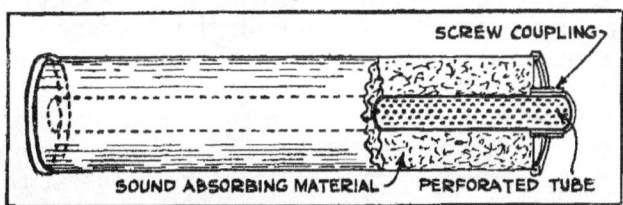

Fig. 103.—The Burgess Absorption Type Silencer.

The Burgess Silencer.—This design of absorption silencer represents a scientific attempt to silence the exhaust by storing some of the gas at high pressure and returning it at low pressure in a more or less continuous stream. The silencer consists of a central perforated tube, which is unobstructed throughout its whole length. It is surrounded by another cylinder of sheet metal, the intervening space being filled with a special sound absorbing material. The theory of this silencer is that during the high pressure fluctuations the gases pass through the perforations where they are stored in the sound absorbing material; in effect the high pressures are reduced by the damping action of the latter material. The fluctuations of pressure in the exhaust gases, which are the cause of the exhaust noise are appreciably reduced in the gases flowing out of the central tube, so that the noise must also be reduced in intensity.

The Ignition System.*—Reference has already been made to the fact that the explosive charge is ignited at the correct moment by a high tension spark. It is not possible in the present limited space to give more than a very brief outline of the principles of the ignition systems used on car engines.

Hitherto, the most popular method of obtaining the high tension spark was by means of the magneto. This device consists of a low-tension current generator, operating upon a similar principle to the dynamo. This generator, in addition to its low-tension or *primary circuit* also embodies a *secondary circuit* consisting of a very large number of fine wire windings around the primary.

The primary circuit has a circuit-interrupting device, known as a *contact breaker.* Now, it is a well-known fact that if a current is flowing in the primary circuit of an electrical transformer, and if this current is suddenly interrupted by breaking the circuit, it will induce a high voltage current in the secondary circuit.

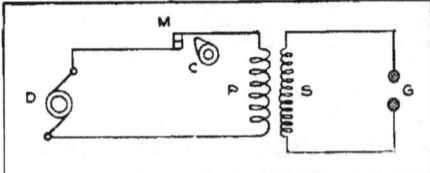

Fig. 104.—Showing Principle of Magneto.

Thus, if the latter circuit has several thousands turns of wire and the primary only a few turns, the voltage of the current induced in the secondary circuit will be of the order of thousands of volts if that in the primary be very low, viz., from 4 to 8 volts.

* A fuller account is given in Volume iv of this series.

Referring to Fig. 104, which illustrates the principle of the magneto. *D* is a dynamo, or current generating device, current being taken from it to circulate around a primary circuit *P* having a primary coil, consisting of a few turns of thick insulated wire, and a circuit make-and-break device *M*; the latter has two contacts *M*, one of which is fixed and the other opened by the rotating cam *C*. As the latter rotates, it breaks the circuit once every revolution, a spring returning the contact after the cam has passed its opening position.

The secondary circuit *S* has several thousand turns of fine insulated wire wound around a central soft iron core, the primary usually being wound over the secondary. The latter's circuit contains a spark gap *G*, such that when the contacts *M* are separated the high tension current in the secondary circuit leaps across this gap.

The magneto, which operates on this principle, uses permanent magnets of horse-shoe form to create the required magnetic field in which the primary coil (armature) rotates. The armature, in the case of a single cylinder engine runs at one-half engine speed and, therefore, gives one spark every two revolutions.

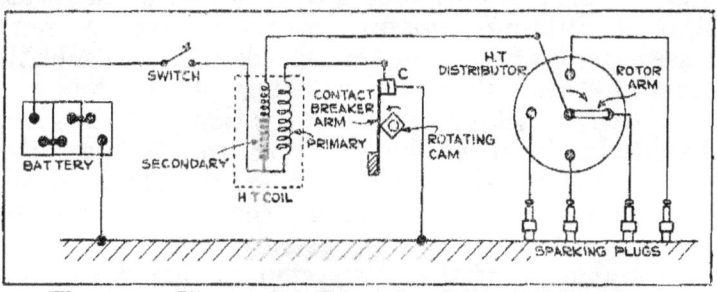

Fig. 105.—Illustrating Principle of the Coil Ignition System.

The four-cylinder magneto also runs at engine speed, but it has a *high tension distributor*, giving two sparks per revolution of the engine; the six-cylinder magneto has a six-contact distributor with rotor arm running at one-half engine speed, so as to give three sparks per revolution of the engine.

Coil Ignition.—Instead of employing a separate electrical generator, *viz.*, the permanent magnet and rotating armature device described, it is possible to use the current direct from the lighting battery, so that the primary circuit (Fig. 105) now contains the battery, a switch for the purpose of cutting out the ignition when stopping the engine, the primary circuit and contact breaker.

The secondary circuit has the secondary coil, a distributor arm which rotates around a number of equally-spaced contacts and the sparking points, or plugs.

It will be noticed that, instead of employing separate insulated circuits, as in Fig. 104, a good deal of wiring has been dispensed with by connecting one side of the battery and primary, one end of the secondary coil and one side of the sparking plugs to a common electrical conductor, known as the " earth " of the system; this is usually the metal frame of the chassis and the engine. The system in question is termed the *Single Pole* ignition one. In modern systems the positive pole of the battery is earthed.

Advantages of Coil Ignition.—Apart from being cheaper than the magneto, coil-ignition gives a practically uniform intensity of spark, whereas the usual design of rotating armature magneto gives a more intense spark at the higher speeds; unless suitable precautions are taken the sparking plug points are apt to wear quickly. The coil-ignition system gives a better spark at engine starting speeds than the magneto, but there is a tendency for the spark to fall off in intensity at high engine speeds, so that for racing engines magnetos are usually employed.

One drawback of the coil system is that it depends upon the state of the battery, so that if the car has been standing for a period of several weeks and the battery has run down, it is usually impossible to start the engine—even by hand-cranking.

Some Coil Ignition Details.—With the coil ignition system, the contact breaker and distributor form a single unit mounted on a convenient part of the engine.

The cam and rotor arm of the distributor are driven by means of a vertical or inclined shaft—usually from the valve camshaft, so that it runs at one-half engine speed.

The contact-breaker formerly used platinum or platinum-iridium contacts, but nowadays tungsten—a cheaper metal—is employed, as it is equally effective in resisting corrosion.

The contact-breaker points are only opened by a very small amount, viz., from $\frac{18}{1000}$ to $\frac{20}{1000}$ inch by the cam; the latter has as many lobes as the engine has cylinders, when it is driven at one-half engine speed, i.e., off the engine's camshaft.

In order to prevent excessive sparking across the contacts, when the contacts open a small electrical *condenser* of about 1 microfarad capacity is connected in parallel (shunt) across the two points.

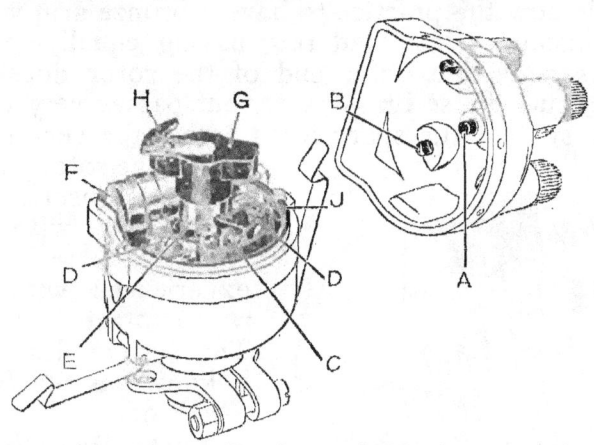

Fig. 106.—Contact Breaker and Distributor of Coil Ignition System. (Lucas.)

A—Carbon brush.   B—Electrode.   C—Contacts.   D—Screws securing contact breaker plate. E—Rotating cam. F—Condenser. G—Rotating distributor arm. H—Metal electrode. J—Contact breaker pivot.

The *high tension coil* is a separate unit bolted to any convenient part of the engine. It contains only the primary and secondary coil and a laminated iron core; the primary is generally arranged inside the secondary coil. The high tension coil is cylindrical in shape, and the central terminal is the high tension one.

The primary coil usually has a few hundred turns of enamelled copper wire of about 1 to 2 ohms total resistance; the current is from 1 to 2 amperes and voltage from 6 to 10.

The secondary coil has from 10,000 to 15,000 turns of very fine insulated wire—usually about $\frac{5}{1000}$ inch diameter and enamelled. It gives a spark of several thousands volts but of low current intensity. The layers of wire are also insulated with varnished silk.

In regard to the *H.T. Distributor*, this is simply a device for delivering H.T. current at the correct intervals to the sparking plugs. Its high tension contacts are therefore made when the pistons are practically at the ends of their compression strokes.

Instead of allowing a carbon brush to rub over an insulated disc having brass contacts inserted at equal intervals around the periphery as was previously the case, it is now the practice to have a bronze arm which rotates inside an insulated ring having equally-spaced brass inserts. The outer end of the rotor does not actually touch these brass parts, but passes very close to them, so that the spark leaps across the very small gap as the rotor passes each brass insert. This does away with the frictional resistance experienced with the wipe-contact method. The system described is known as the *Jump Spark* one.

Other Magnetos.— The earlier magnetos, such as the Lucas and Bosch, had fixed permanent magnets and rotating armatures containing the primary and secondary coils as described previously.

More recent models employ stationary coils and rotating

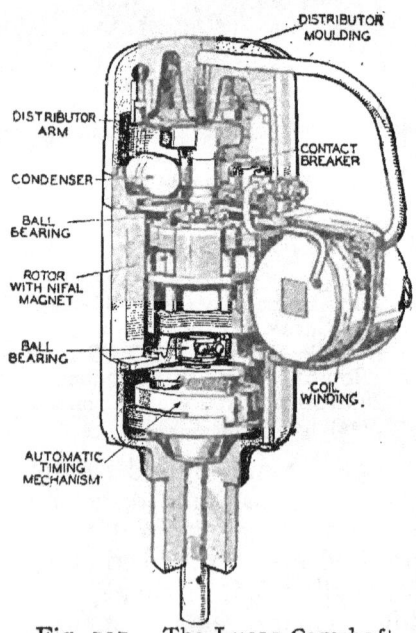

DISTRIBUTOR MOULDING

DISTRIBUTOR ARM

CONTACT BREAKER

CONDENSER

BALL BEARING

ROTOR WITH NIFAL MAGNET

BALL BEARING

COIL WINDING.

AUTOMATIC TIMING MECHANISM

Fig. 107.—The Lucas Camshaft Magneto with Stationary Coils.

magnets of special shape. The contact-breaker is operated by a rotating cam, as in coil-ignition systems. The
rotating magnet magneto is less liable to failure, since
the coils do not rotate and can be kept free from oil.

The camshaft magneto has a rotating magnet, and it
is claimed possesses the advantages of both the coil and
magneto. Examples of camshaft magnetos are the
Lucas, Bosch and Scintilla (Vertex).

One definite advantage of the modern vertical
magneto is that it is possible to move both the contact-breaker and armature together for engine timing
purposes, and thus to obtain the maximum spark at
both the fully advanced and retarded positions; this
cannot be done with the rotating armature type
magneto.

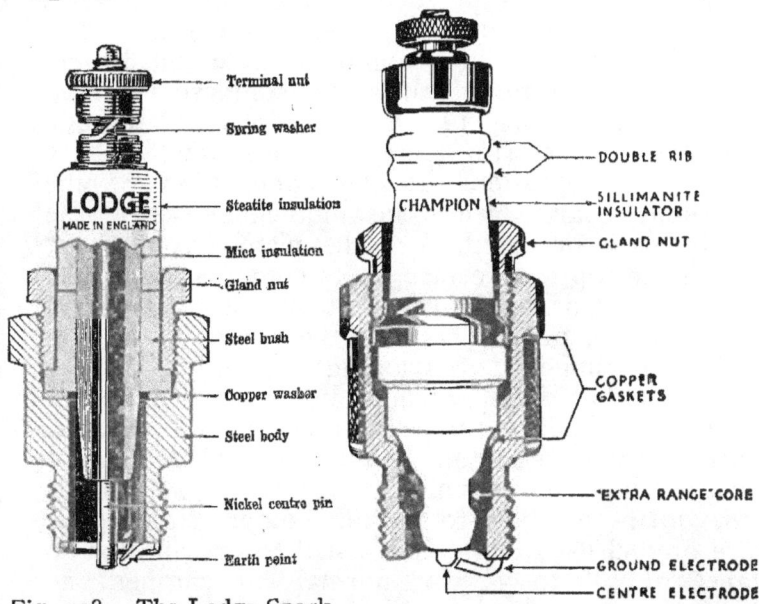

Fig. 108.—The Lodge Sparking Plug (Mica and Steatite
Insulation).

Fig. 109.—Another Type of Sparking Plug (Sillimanite Insulation).

Sparking Plugs.—The high voltage current from the
ignition apparatus is conveyed to the inside of the
cylinder by means of the sparking plug. This consists
of an outer metal body or " shell " which is screwed
into the wall of the combustion chamber—usually

F

although not always, over the cooler inlet valve—and an inner or central electrode of nickel. This electrode is insulated from the outer shell by means of mica, porcelain or steatite. The high tension cable is attached to a terminal on top of the central electrode, and the spark occurs at the other end between the electrode and the metal shell across a spark gap measuring $\frac{18}{1000}$ to $\frac{18}{1000}$ inch for magneto ignition and $\frac{18}{1000}$ to $\frac{24}{1000}$ inch for coil ignition. Many more recent engines, notably American ones, now use wide gaps, namely, from $\frac{30}{1000}$ to $\frac{40}{1000}$ in. Special high voltage high tension coils are employed with this system of wide gaps.

The sparking " points " of the plug should *be level with or slightly below the combustion chamber* walls.

There are various types of sparking plugs available, including low, medium and high-speed engine types, single and multiple points, non-detachable and detachable types (for cleaning purposes), standard (18 mm. thread) and miniature (14 mm. thread), ventilated and copper-finned types (for racing engines), etc.

Timing the Ignition.—If the combustion of the compressed charge were instantaneous, when the ignition spark occurred, then the piston would have to be on the top dead centre of its compression stroke at the moment of sparking. Actually, there is always a time interval, or lag, between the point of sparking and the attainment of maximum pressure in the cylinder, so that the spark must be arranged to occur a short interval before the piston reaches its top centre.

This is termed the *angle of advance*, and its amount varies in different engines according to the engine speed, ratio of bore-to-stroke, compression ratio, nature of fuel used, etc. It is usual to provide for an advance of 25° to 35° for normal car engines and rather more (up to 45°) for high performance engines.

Previously, the driver was provided with a hand control for moving the contact-breaker in relation to the rotating cam, or the cam in relation to the rotating contact-breaker according to whether coil or magneto ignition was used. It is now the custom to embody an automatic control in the ignition unit. This, for coil ignition, consists of a centrifugal device which operates the contact-breaker and advances it as the engine speed increases. At starting, the device is always left in the fully retarded position.

The latter position is generally arranged so that when the piston is on its top dead centre (compression), the contact-breaker points are just opening and the spark is, therefore, occurring.

Most car engine flywheels are marked with the top dead centre position of the front or No. 1 cylinder, and often with the valve-timing position also.

Another automatic ignition control, made by the Lucas firm, automatically advances the ignition when the throttle is opened at lower engine speeds, independently of the speed advance device; it is operated by the suction in the inlet manifold.

CHAPTER III.

It is now proposed to consider the various types of automatic engines in use. These range from the small motor-cycle " lightweight " engine of about 1 H.P. up to the large high-power car engines of 8 and 12 cylinders and from 80 to 160 H.P. output. Automobile engines include, in addition to the poppet valve type described in the preceding chapters, the " valveless " or sleeve-valve types, two-stroke engines, and oil or Diesel engines. Special types of engines, such as the " swash-plate," link-plate, radial and rotary engines, have been suggested and in some cases actually used in automobiles; these have usually been of an experimental nature, however. More recently, the super-charged petrol-type engine, developed primarily for aircraft purposes, has been used successfully on racing motor-cars and one or two commercial cars.

The ordinary four-stroke poppet-valve engine is by far the more commonly used on automobiles, and, is available in the single, twin, four, six, eight and twelve cylinder designs, for various horse-powers within the stated range.

The Single Cylinder Engine.—This is now employed on motor-cycles only, although it was used on a few small cars and tricars. Experience has shown that for motor vehicles the maximum size of a single cylinder engine is one of from 600 to 700 c.c. Above this size, very heavy fly-wheels are necessary, and the vibratory effects due to want of balance of the reciprocating parts become unpleasant. The average capacity of the so-called 3½ H.P. engine is about 500 c.c.; a single cylinder engine of 85 mm. bore by 88 mm. stroke gives a capacity of about 517 c.c., and has an R.A.C. rating of 4·56. The well-designed high compression engine of

164

this size would give from 20 to 30 H.P. at its maximum output.   An example of an engine of this type is given in Fig. 17.

It is not possible to balance the reciprocating parts of the single cylinder engine very satisfactorily; thus the piston, gudgeon pin and the upper half of the connecting-rod have a reciprocating motion which it is only possible to balance effectively by introducing another reciprocating mass moving oppositely; in the horizontal opposed twin engine this balance arrangement exists.   The compromise usually adopted in the case of the single cylinder engine is to supply a counter-weight in the fly-wheels, on the opposite side to the

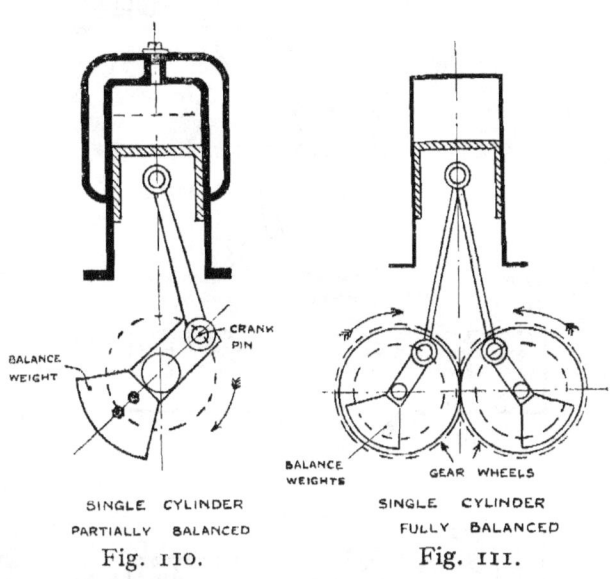

SINGLE   CYLINDER
PARTIALLY   BALANCED
Fig. 110.

SINGLE   CYLINDER
FULLY   BALANCED
Fig. 111.

crank-pin, to balance the weights of the crank-pin and the lower part of the connecting-rod (Fig. 110).   The unbalanced effect in this case is that of a rocking action (or transverse couple) which tends to vibrate the engine sideways in the fly-wheel's plane.

Another alternative is to balance one-half the weight of the reciprocating parts, assumed to be on the crank-pin, in addition to the other parts previously mentioned; this reduces the transverse rocking effect to about one-half of its previous value.   In the early

Lanchester engine almost perfect balance was obtained as illustrated in Fig. 111.

An idea of the magnitude of the unbalanced reciprocating force will be obtained when it is mentioned

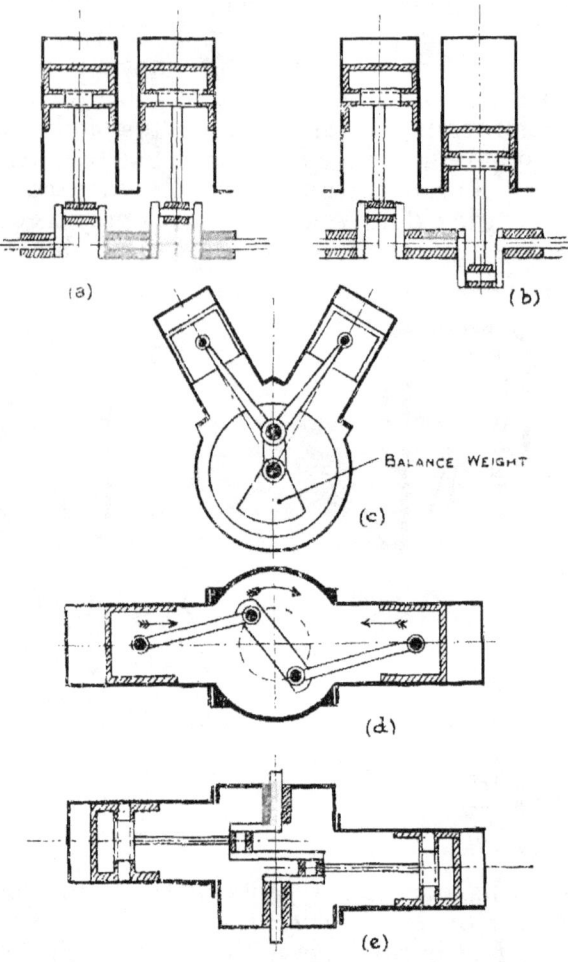

Fig. 112.—Showing the different types of Two-cylinder Engine.

that in the case of a single-cylinder engine, with a piston weighing 2 lb., the stroke being 4 inches, the greatest out-of-balance force is 907 lb. at 2,000 r.p.m., and occurs at the beginning of the stroke. Its value falls progressively to zero when near the centre of the stroke, and then increases again to a somewhat lower value, viz., 680 lbs. (usually about three-quarters of the former), and opposite in direction. In addition to this unbalanced effect, the single cylinder engine gives only one power stroke every two revolutions, so that a relatively heavy fly-wheel is necessary in order to store up sufficient energy to carry the engine over its three idle strokes.

The Two-Cylinder Engine.—The cylinders are duplicated in the case of motor-cycle and light-car engines, not only for the purpose of obtaining more power, but also in order to provide a better balance, and more even torque. Twin-cylinder engines for motor-cycles range from about 500 to 1,000 c.c. capacity, the nominal R.A.C. H.P.'s being 4½ to 9 respectively. The two-cylinder engine for three-wheelers and small cars ranges in size from about 750 up to about 1,200 c.c. In the latter case, with horizontally opposed cylinders, over 70 B.H.P. has been obtained on the test bed.

There are four possible arrangements for the two-cylinder engine as shown in outline in Fig. 112. In (a) both cranks are in line and the pistons move up and down together. The engine fires at 360° intervals, but the reciprocating parts are unbalanced; in effect this arrangement is equivalent to two single-cylinder engines side by side. The arrangement in (b) gives a better (although still only a partial) balance to the reciprocating forces, but a sideways, or transverse rocking couple is introduced due to the lines of action of the reciprocating parts not being coincident. This type of engine has unequal firing intervals, namely at 0°, 180°, 720°, 900°, etc. Neither of the arrangements (a) and (b) is now employed for automobiles.

A method of balancing the reciprocating parts of an engine, originally employed in the Gobron-Brille motor car engine as shown in Fig. 113 and more recently applied in the German Junkers compression-ignition aircraft and motor vehicle engines, is shown diagrammatically in Fig. 114.

Although both of these types have a common cylinder of approximately twice the length of the normal engine cylinder they may conveniently be regarded as two one-cylinder engines with a common combustion chamber, giving twice the power output of a single cylinder one. The chief point of difference, however, is that the number of power impulses is one-half that of a twin cylinder engine having separate cylinders, unless—as is the case with the Junkers engine the two-cycle principle is employed.

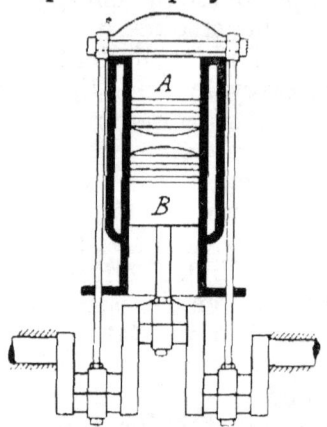

Fig. 113.—The Gobron Brille Engine. The pistons are shown at A and B.

The arrangement shown in Fig 113 consists of two pistons A and B which work in opposite directions, moving together during the compression and exhaust strokes and away during the suction and firing strokes; the combustion chamber is formed in the space between the piston crowns when in their nearest positions so that the valves would be arranged in the sides of the central portion unless use were made of ports in the cylinder walls near their outer ends as in two-cycle practice.

The upper piston A has a crosshead arrangement above carrying trunnions on which the upper ends of a pair of connecting rods rock; the lower ends of these rods work on cranks on either side of the crank for the piston B. With this arrangement, and also that shown in Fig. 114 the reciprocating forces due to the pistons are in almost perfect balance, although there are certain small unbalanced effects due to the fact of the connecting rods being of different lengths.

The disadvantage of this type of engine lies in its somewhat excessive height and greater weight, but by careful design the latter can be reduced almost to the value of a two-cylinder unit engine of conventional design.

In the Junkers engine (Fig. 114) which, as previously mentioned works on the two-cycle principle the two pistons *A* and *B* move symmetrically in a common cylinder of about twice the usual length. The combus-

tion chamber is formed in the space between the pistons
when they are closest together.   Each piston has its
own connecting rod Y and operates its own crank-
shaft; the upper shaft ECF is rotated by the upper rod
Y, and the lower shaft GDH, by the lower rod Y.

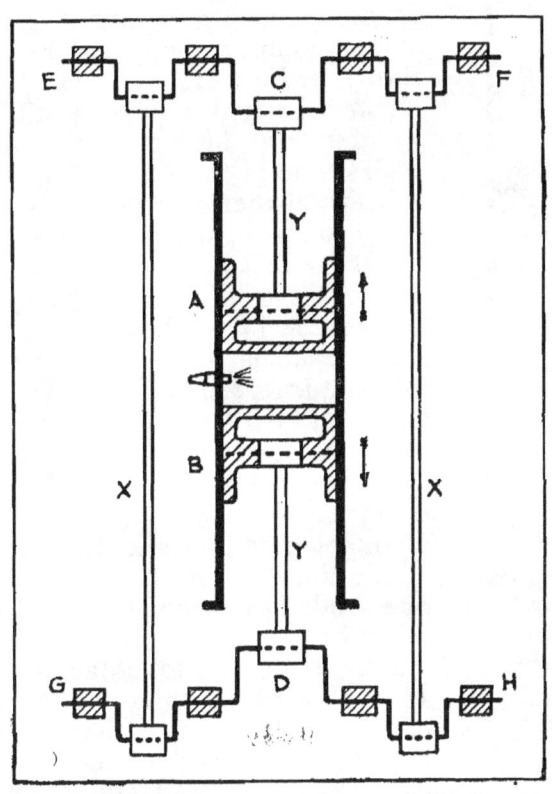

Fig. 114.—The Balanced Piston Type of Engine (Junkers).

The two shafts rotate in the same direction, and they
are coupled together by a pair of side connecting rods
X, operating on the cranks shown.

In the Junkers aircraft engine, the two crank-shafts
are connected by spur-gearing instead of rods as shown
in Fig. 115.

Another method of using opposed pistons is shown
in Fig. 116. This arrangement illustrates the Hill two-
cycle engine used for motor boats. The opposed pistons
A and B are connected by the rods Y to rocking

F*

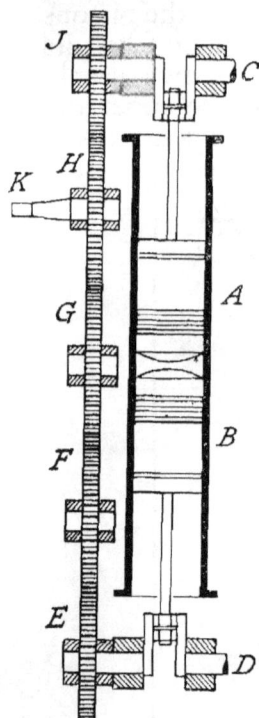

Fig. 115.--The Geared Type Junkers Engine. A and B—Opposed pistons. C and D—Upper and lower H and J—Gear crankshafts. E, F, G, wheels. K—Airscrew drive shaft.

levers M, having fixed fulcrum bearings at F. The other ends of these levers are coupled by the connecting-rods X to the crank-pins of the crankshaft C; these cranks are 180 degrees apart. This arrangement gives a more compact engine than that shown in Fig. 114.

The Vee-type engine is still fairly popular on high-powered motorcycles, and is to be found on one or two three-wheeler cars. It is economical to make, since the same type of fly-wheel and common crank-pin as in the case of the single-cylinder engine can be used, and its external shape enables it to be fitted readily into the frame or chassis. The valve gear can also be made compact, and the crank-case is practically that of the single-cylinder engine. It will be observed that the big-ends of both connecting-rods work on the same crank-pin, but in one or two cases one rod has been hinged to the other or " master " rod, near to the crank-pin, so as to obtain a larger area of bearing surface. The angle between the centre lines of the cylinders varies in the different makes, and ranges from 40° to 90° : the 60° angle is the more common.

In the case of the 40° twin engine the firing intervals reckoned from the crank position as zero at the rear cylinders are 0°, 400°, 720°, 1120°, and so on. For the 90° twin these intervals are 0°, 450°, 720°, 1170°, and so on; these intervals are less regular than in the preceding case. On the other hand the 90° twin engine can be balanced much better; thus the maximum unbalanced force is only about one-third that of the 60° twin, and the amount of vibration only about one-twelfth.

The much better balance of the 90° twin engine is usually accepted as outweighing the drawback of its

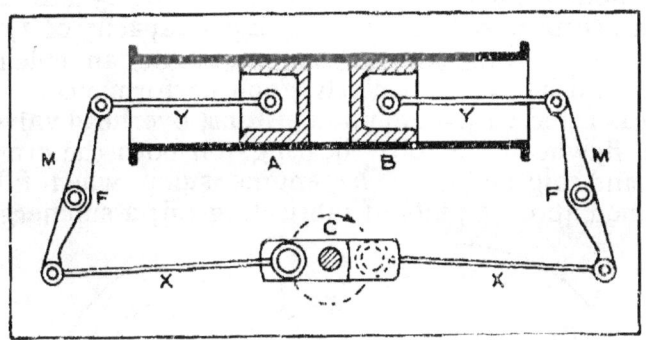

Fig. 116.—Arrangement of the Hill Engine.

less regular firing intervals compared with twin engines of smaller cylinder angles.   The Vee twin

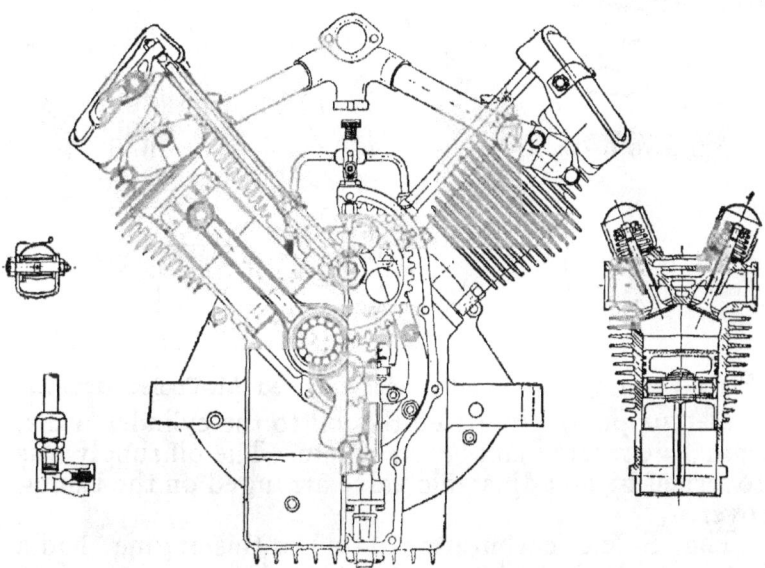

Fig. 117.—The B.S.A. Two-Cylinder Vee-Type Engine.
(Right-hand view shows Cylinder in section.)

engine is balanced, or partly so, in a similar manner to the single engine by means of a common counter-weight as shown in (c) Fig. 112.

Fig. 117 illustrates the 9 H.P. twin-cylinder Vee-type engine that was fitted to certain B.S.A. three-wheeler cars. Each cylinder has a bore and stroke of 85 mm. and 90 mm., respectively, giving a cylinder capacity of 1,021 c.c. The engine in question represents an efficient design and had a particularly good performance.

It was fitted with aluminium pistons, overhead valves, a steel flywheel and roller bearings on both the crank-shaft and big-ends. The engine sump when filled contained about 5 pints of lubricating oil; a submerged

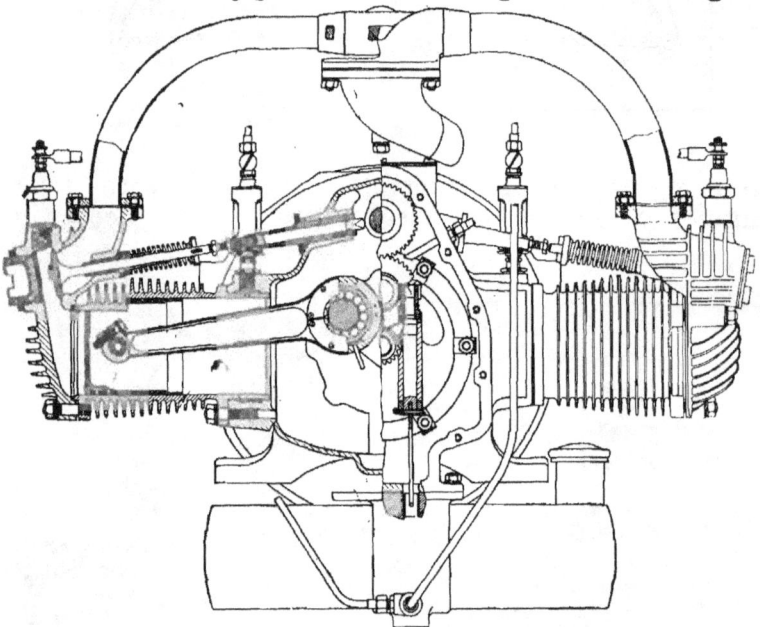

Fig 118.—A typical Two-Cylinder Opposed Air-Cooled Engine.

gear-type pump circulated this oil to the cylinder walls, timing gear and main bearings, etc. The oil supply was governed by an adjustable valve arranged on the timing cover.

The Solex carburettor fitted to this engine had a hot-spot manifold heated by the exhaust gases. The valve operating gear is, for the greater part, enclosed, the valve tappet-rods working inside fixed metal tubes.

The opposed arrangement of cylinders shown at (d) (Fig. 112) possesses marked advantages over the others. It enables the firing periods to occur at equal intervals, and the reciprocating forces to be balanced almost perfectly. It will be observed that both pistons move inwardly and outwardly together, and the connecting-rods are always in a symmetrical position so that the forces due to these masses always balance. In practice, however, it is not possible to arrange both rods in the same plane, as they would not clear or pass one another. The two cranks are therefore offset relatively to one another as shown in diagram (e) (Fig. 112). The result is to throw the lines of action of the two sets of reciprocating forces out of coincidence and therefore to introduce a rocking action; the latter can, in practice, be kept down to fairly small limits by employing narrow bearings of the roller type.

Owing to its excellent balance, and to the even firing intervals, it is possible to attain very high engine speeds with this type. As an example, the Coventry Victor Super Sports engines on standard tests attained speeds of 5,800 r.p.m.; occasional engines gave 6,000 r.p.m. and over. The air-cooled engine in question has a bore and stroke of 75 mm. and 78 mm. respectively (giving 688 c.c. capacity).

The only example of a water-cooled opposed cylinder engine used for motor cars in recent years is the Jowett one which established a good name for performance and reliability in light cars. This engine has since been developed into a "flat four" model having four cylinders arranged as two pairs of opposed ones. The bore and stroke are 2½ in. and 8⅝ in., respectively. The engine has a three-throw crankshaft with two main bearings, namely, at the ends. The centre crank pin is of twice the width of the two outer crank pins and the latter's cranks are at 180° to the former's one. The two connecting rods of the central opposed pair of cylinders were arranged side-by-side on the centre crank pin. This arrangement gives good torque and balance. A single down-draught Zenith carburettor supplies mixture to all of the cylinders in turn through a water-heated inlet pipe instead of the two carburettors previously used. Aluminium pistons are used and the engine is mounted at three points on flexible supports with hydraulic dampers.

It may here be mentioned that the opposed cylinder engine consisting of one, two or three pairs of opposed cylinders is much favoured for light aircraft engines and also for " flat " or "pancake " automobile engines —in the six, eight and twelve cylinder class.

Four-Cylinder Engines.—Although the balance and torque of the opposed two-cylinder engine are excellent, the overall length and the cylinder dimensions limit it to power units of about 8 to 10 H.P. (rated) for car use. It is also not so good as regards slow running and low speed torque as the four-cylinder

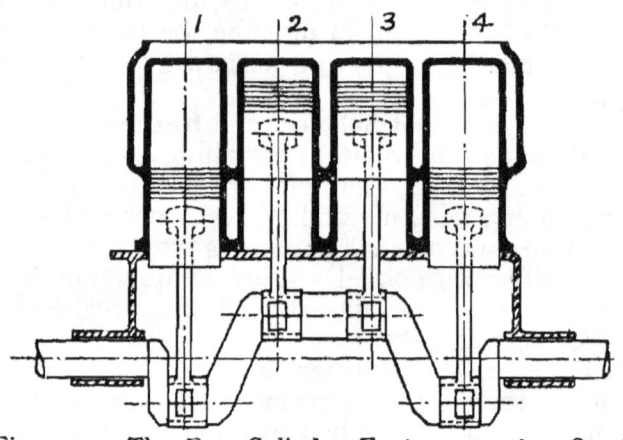

Fig. 119.—The Four-Cylinder Engine, showing Crank Arrangement.

engine. The overall length, when the engine is set laterally in the chassis, often interferes with the design of the bonnet and the general lines of the car; in the smaller light cars the cylinder heads often projected through the sides of the bonnet, air-scoops for cooling the cylinder heads being arranged in the projections.

The four-cylinder engine is by far the most popular light and medium car engine in present-day use.

The four-cylinder engine arrangement universally adopted is shown in Fig. 119. The two outer and the two inner cranks, respectively, are parallel, so that each pair of pistons is always in the same relative position. This arrangement enables a firing stroke to be obtained twice every revolution, so that the torque is very uniform as compared with that of the single-cylinder engine. Apart from the more uniform torque, the balance of the four-cylinder engine is also very good. Although not equal to that of the two-cylinder opposed, it represents an excellent compromise. It will be observed that the pairs of pistons always move in opposite directions so that the reciprocating part forces tend to balance. They do not quite balance each other, however, for, owing to the inclination or obliquity of the connecting-rods, the pistons move rather faster during the first half of the crank-shaft revolution, and slower during the latter. Thus when the crank-shaft has moved through 90° from its top centre position, the piston moves

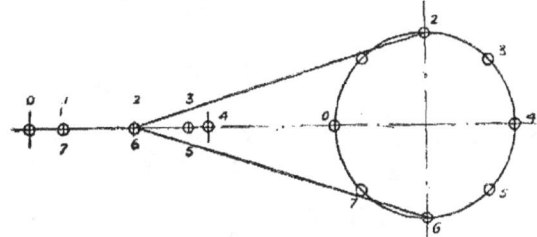

Fig. 120.—Illustrating Unsymmetrical Travel of Piston in Four-Cylinder Engines.

through more than one-half of its stroke, as shown in Fig. 120. Hence the two ascending pistons (from the bottom centre) will at first travel slower than the two descending ones, and the forces due to the masses of the reciprocating parts will not balance exactly. If very long connecting-rods could be used these forces would balance; in practice, however, design considerations limit the length of the connecting-rod.

The general result of this want of symmetry in the piston positions is that there is an out-of-balance force which causes vibrations in an up-and-down direction, at a frequency equal to twice that of the engine speed. By the employment of light connecting-rods and pistons, the magnitude of this unbalanced force can be reduced to small dimensions.

In the case of one or two of the more expensive and larger engined cars, this *Secondary Force*, as it is termed, is automatically balanced by introducing mechanically another equal and opposite reciprocating force at twice engine speed. In certain engines, this is accomplished by driving off the crank-shaft, by

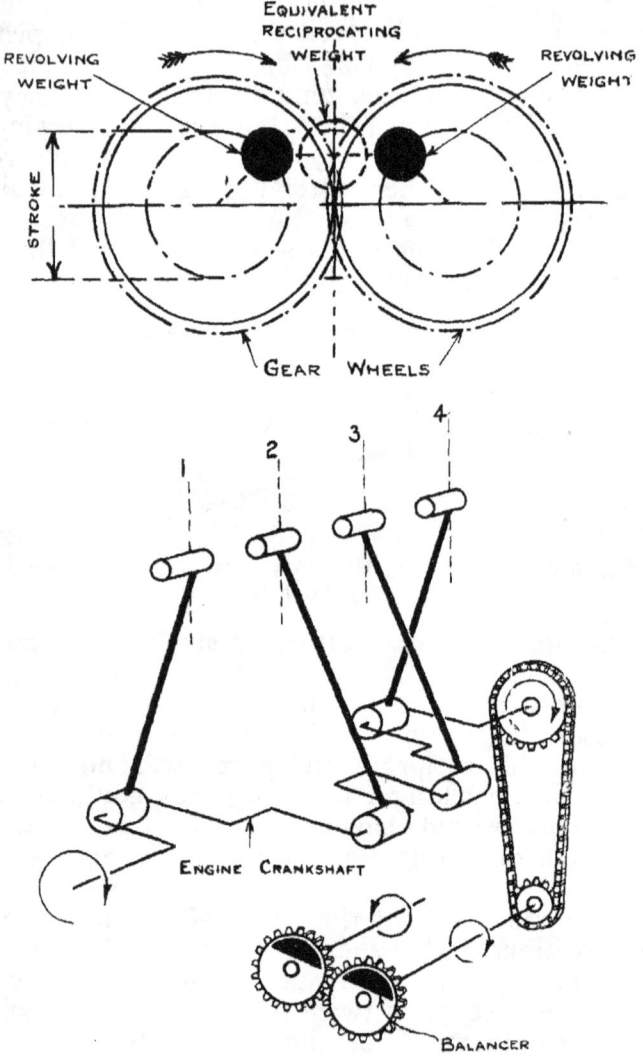

Fig. 121.—Showing how a Reciprocating Force can be Balanced by Two Equal Revolving Masses moving in Opposite Directions.

means of suitable gearing, a pair of weights (as shown in Fig. 121) whose common centre of gravity moves up and down at twice engine speed.

Ricardo has employed a direct method of balancing the reciprocating forces in the case of a single crank, two-cylinder vertical engine, by operating a pair of

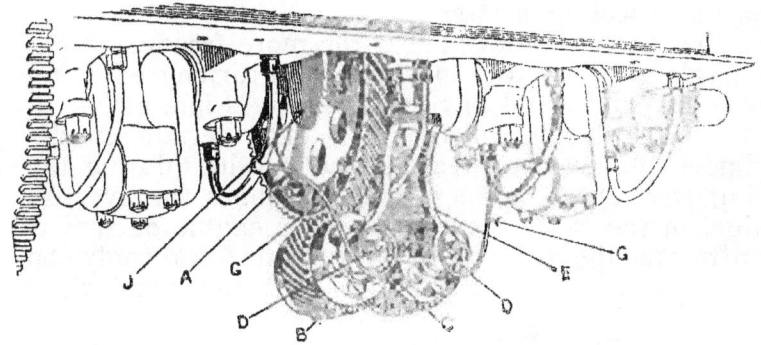

Fig. 122.—The Lanchester Balance Gear.
A—Helical gear on crank-shaft. B and C—Balance wheels moving in opposite directions. E—Bracket. G—Oil pressure lubrication. J—Crank-shaft web.

balance weights situated in the sump, by means of eccentrics and universally jointed straps. The eccentrics are mounted on the crank-shaft, and the balance weight inner members are guided up and down steel rods fixed to the crank-case. The balance weights are cylindrical

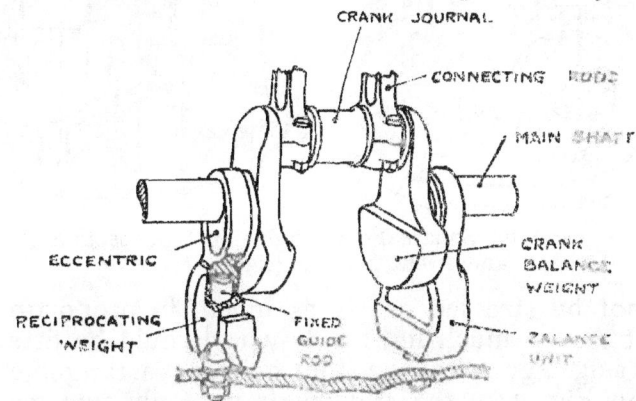

Fig. 123.—The Ricardo Reciprocating Force Balance Gear.

inside and can rock on separate cylindrical discs, which in turn slide up and down the steel guides. This allows for the sideways tilting movement of the eccentrics. Fig. 123 illustrates the principle and constructional details of this arrangement.

In ordinary unbalanced four-cylinder engines, the secondary force is usually about one-eighth to one-twelfth that due to one of the piston and connecting-rod reciprocating masses.

It has been mentioned that the more frequent firing intervals give rise to a more uniform turning effort on the crank-shaft, and in this respect it is interesting to note that whereas in the case of a certain single-cylinder engine, the greatest torque value (during the firing stroke) was no less than 8·0 times the average value, in the case of a four-cylinder engine of similar dimensions the maximum torque value was only 2·0 times the mean. This means that the crank-shaft

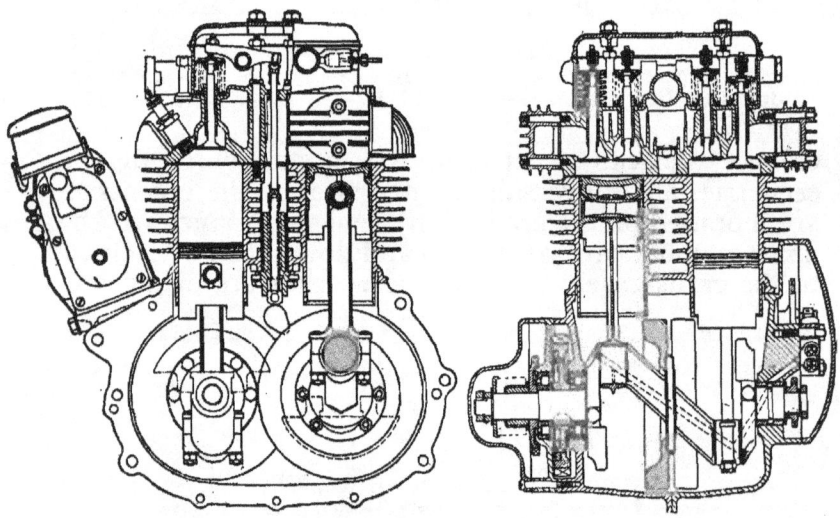

Fig. 124.—The Ariel Square Four Engine in Side Section (left) and Front Section (right).

would not be stressed nearly as much, in proportion, and that a very much lighter fly-wheel could be fitted.

Referring again to the four-cylinder arrangement shown in Fig. 119, the crankshaft has only two main bearings, namely, one at each end. This method was used in the earlier engines and in the later Austin Seven engine. It has now been superseded by the three-bearing crankshaft—with an additional central bearing —in order to overcome earlier troubles due to "whip" effects.

Square Four Engines.—Instead of arranging the four cylinders in a line as in ordinary car engine practice, they can be arranged in square formation as seen in plan view. This arrangement is equivalent to a pair of twin cylinder engines connected together.

There are two interesting examples of such engines employed in motor-cycle practice, viz., the Ariel and the Matchless Silver Hawk.

Fig. 125.—Showing Pistons and Crank-shafts of one model Ariel Square Four Engine.

The former type is shown in Fig. 124. It will be noticed that it is virtually equivalent to two vertical twin engines, each set having its own crank-shaft, the two being geared together. The engine shown has a bore of 2·56 in. and stroke of 2·96 in. giving a cubic capacity of 61 cu. in. (997 c.c.). The four cylinders are in a single casting secured to the top of a vertically split crankcase.

Each two-throw cranshaft is supported in a bronze back babbitt-lined bearing at the timing end and in a large-diameter roller bearing at the driving end, where the coupling gears are housed in a separate compartment. The rear shaft, from which the primary chain drive to the gearbox is taken, is furnished with an additional roller bearing outside the gear wheel. There is one flywheel on each of the crankshafts, the two flywheels being bolted to opposite sides of centrally located flanges on the crankshafts, thus permitting their rims to overlap. Crankpins and journals are hardened, and connecting rods are made of the R.R. aluminium alloy.

The single inlet manifold is cast integral with the cylinder head casting. Detachable finned exhaust manifolds, each serving two cylinders, are bolted to opposite sides of the head casting. Duralumin pushrods are used. To prevent interference with the free flow of air over the cylinder fins, the combined generator and ignition magneto is mounted at an angle. A compression ratio of $5.8:1$ is used, and the engine is stated to be capable of running in excess of 6,000 r.p.m.

The other example of four-cylinder engine, the principle of which is shown in Fig. 129, consists of two sets of Vee-twin engines arranged side-by-side with a common two-throw crankshaft.

Four Cylinder Firing Order.—Numbering the cylinders from left to right in the order 1, 2, 3 and 4, the order in which the spark occurs in the cylinders is 1, 3, 4, 2. There is, however, an alternative order which is occasionally used, namely, 1, 2, 4, 3. If there is any doubt in the case of a particular engine, the firing order can readily be ascertained by turning the crank-shaft by hand, slowly, in the direction of normal rotation, and noting down in turn the numbers of the cylinders as their pistons reach the top of the compression strokes. It is easy to distinguish the compression from the exhaust stroke, for both inlet and exhaust valves are closed during the former, and the inlet valve only during the latter—the exhaust valve being lifted. This method applies to any type of engine.

Typical Four-Cylinder Engines.—The four-cylinder engine was at one time popular in a fairly wide range of horse powers for cars up to about 16 H.P. rating, but more recently its place has been taken by the six-cylinder type so that, with the exception of tractor and similar industrial units it is now used only on the smallest models of cars, namely, in the 8 to 10 H.P. rated class; there are, however, one or two examples of 12 and 14 H.P. engines using four cylinders.

Fig. 126.—The Austin Ten Engine and Gear Box Unit.

Fig. 126 illustrates the complete engine and gear box unit of the 10 H.P. Austin Car. It has a bore and stroke of 63·5 mm. (2·499 in.) and 89 mm. (3·5 in.) respectively, giving a cubical capacity of 1,125 c.c. (68·7 cu. in.); the original engine developed 21·1 B.H.P. at 3,400 R.P.M., but a later model gave 27 B.H.P. at 4,000 R.P.M.

The engine is of the side-valve type, the valves being operated in the usual manner by means of the

tappets shown in the upper right hand part of Fig. 126
The camshaft, which has three white-metal-lined bearings is driven from the crankshaft by means of a double
roller chain (seen on the right in Fig. 126). The crank-
shaft is carried on three main bearings.   In the original
seven engine the crankshaft has a white-metal cen-

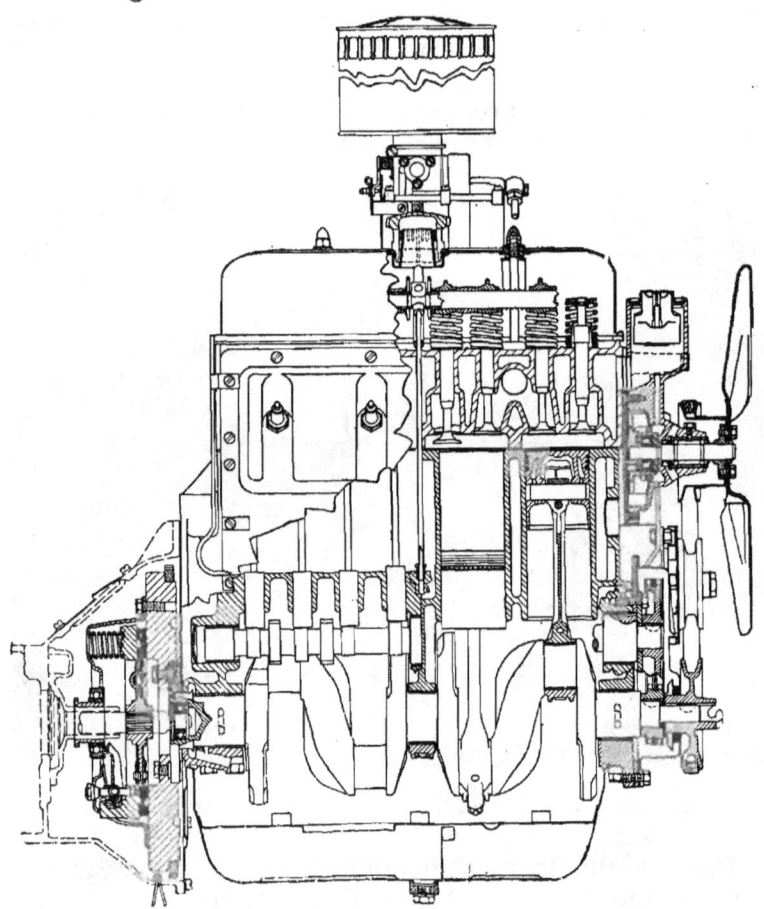

Fig. 127.—The Vauxhall Four-Cylinder Engine.

tral bearing, a roller bearing at the flywheel end and a
combination deep groove ball and roller bearing at the
front end.   It is now considered essential in four-
cylinder engines to employ three-bearing crankshafts
in order to obviate the deflection, or "whip," pre-

viously experienced with two-bearing crankshafts. The valves on the Austin Ten engine are arranged on the near side for ready accessibility. The cylinder head is detachable. Aluminium alloy pistons, each with two compression and one scraper ring are used; the pistons are oxide coated, as previously explained, to reduce initial wear.

The main supply of oil is carried in the pressed sheet metal sump below and there is a wide area oil filter through which all the used oil that drains from the inside of the engine after lubricating the working surfaces must pass before it can be circulated again by the gear-wheel pump seen to the left of the pump in Fig. 126. The oil pipes supplying oil from the pump to the main bearings can be seen in this illustration, whilst the part sectional portions of the crankshaft show how the oil is conveyed from the main bearings to the big-end ones through drilled passages.

The 10 H.P. and 12 H.P. Vauxhall four-cylinder engines are similar in general design, a part sectional view being given in Fig. 127.*   The engines have the same stroke, namely 3¾ in., and bores of 2½ in. and 2¾ in. giving cubical capacities of 73 cu. in (1,195 c.c.) and 88 cu. in. (1,440 c.c.), respectively. The 10 H.P. engine develops 34·5 B.H.P. at 3,800 R.P.M. and a B.M.E.P. of 119 lb. per sq. in. at 2,200 R.P.M. The maximum engine torque is 58 lb. ft.

The 12 H.P. engine gives 40 B.H.P. at 3,800 R.P.M.; a B.M.E.P. of 123 lb. per sq. in. at 2,200 R.P.M., and maximum torque of 72 lb. ft. The compression ratios of the 10 and 12 H.P. engines are 6·5 : 1 and 6·8 : 1, respectively.

Overhead push-rod and rocker arm valve gear is employed. The inlet valves of both models are $\frac{13}{16}$ in. diameter and the exhaust valves $1\frac{5}{32}$ in.; Silchrome heat-resisting steel is used for the latter valves.

The pistons are of aluminium alloy, each being fitted with two individually cast compression rings and one oil control ring. The gudgeon pin is clamped to the small end of the connecting rod. Water circulation is by impeller pump on the fan shaft with

* Automotive Industries.

temperature control by thermostatic by-pass. The engines use Zenith down-draught carburettors, the inlet manifolds having thermostatically-controlled exhaust heating. A feature of these engines is the use of sparking plugs with gaps of 0·037 to 0·040 in.

Ignition timing is by combined vacuum and centrifugal automatic control. Another feature is the oil-spray feed from the big ends to the cylinder walls, for lubrication purposes, referred to in Chapter VII.

Four Cylinder Vee-Type Engines.—An alternative arrangement for the four cylinders of an engine is that of two sets of vee-cylinders side-by-side, usually

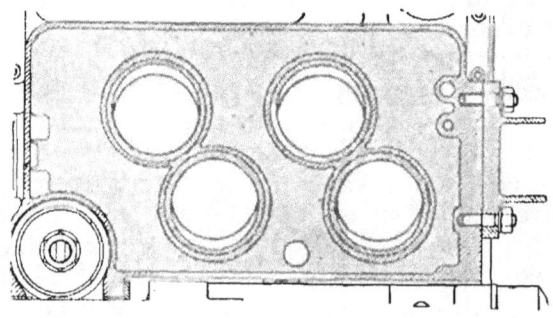

Fig. 128.—Showing Arrangement of Cylinders in the Lancia Four-Cylinder Vee Type Engine.

with the axes of the cylinders inclined at a relatively small angle to one another. In order to obtain the conventional arrangement of cranks as in the case of the four cylinder vertical engine it is usual to stagger the opposite cylinders by an amount equal to about one-half of the cylinder diameter (Fig. 128). The angle between the cylinder axes varies appreciably in the different commercial engines, but is found to lie between 14° and 20°. This results in a cylinder block which is not very much wider than that of the vertical type of four-cylinder engine. The layout provides for a very compact engine, much shorter fore and aft in comparison with the four-cylinder in line engine. A stiffer crankshaft can therefore be employed and side-by-side connecting rods arranged for.

An example of this type of engine is the Matchless Silver Hawk, illustrated in Figs. 129 and 130, in which the cylinder axes are inclined at 18°. A well-finned air-cooled monobloc cylinder casting is employed. The actual firing angle is 26° as each piston has an inclination of 4° to its connecting rod. Each pair of connecting rods works on the same crank pin, the two crank pins

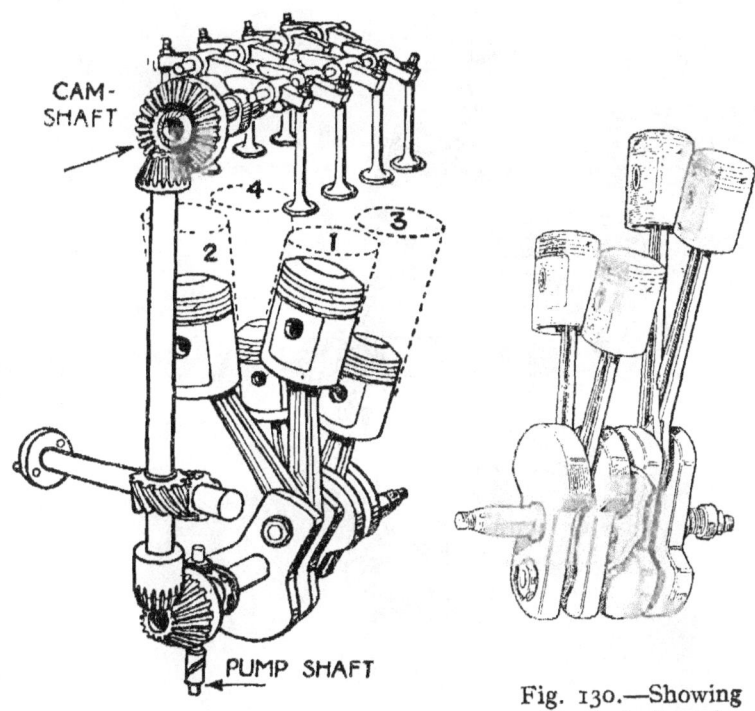

Fig. 129.—Illustrating the arrangement of the Matchless Four-Cylinder Engine. The cylinders are shown at 1, 2, 3 and 4.

Fig. 130.—Showing Crank-shaft and Pistons of Matchless Four-Cylinder Engine.

being arranged at 180° apart. The cylinders have each a bore and stroke of 50·8mm. and 73·02 mm., respectively, corresponding to a capacity of 593 c.c. The cylinder block is fitted with a one-piece head and, as shown in Fig. 129, an overhead camshaft for operating the valves; the camshaft is driven through two pairs of level gears with an intermediate vertical drive shaft.

An example of a water-cooled car engine is that of the Lancia, illustrated in Fig. 131. This has a bore and stroke of 72 mm. and 83 mm. respectively, giving a

cylinder capacity of 1,350 c.c. (82·5 cu. in.).   The compression   ratio   is   5.73:1.   The   engine   develops
47 B.H.P. at 4,000 R.P.M.   Special features of this
engine include domed combustion chambers and five-ring aluminium pistons; a special arrangement of overhead valve operating mechanism whereby a single

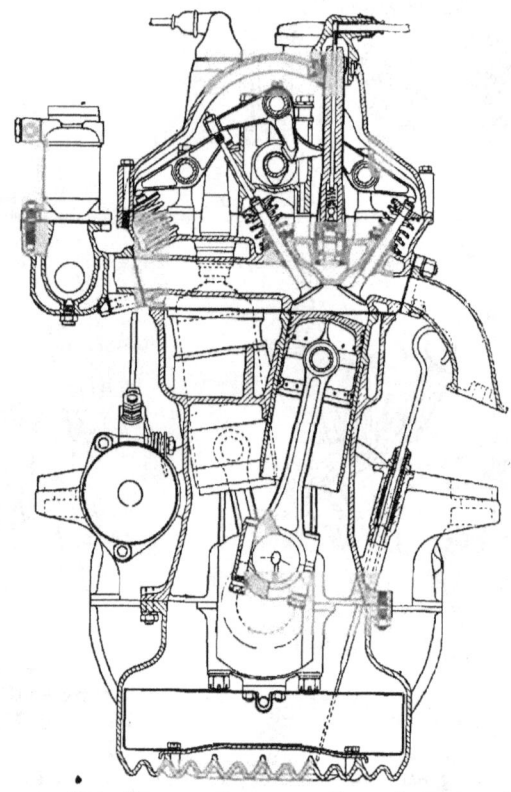

Fig. 131.—The Lancia Vee-type Four-Cylinder Engine.

roller-chain driven camshaft operates all of the inclined
valves through rocker arms of two different designs;
separate cylinder wet-type liners; three-bearing crankshaft with four crankpins arranged in a similar manner
to those of a four-cylinder in line engine; water
impeller at base of radiator and large diameter belt-driven radiator cooling fan.   The engine is of very
compact design.

The Six-Cylinder Engine.—A better balance and a more uniform torque can be obtained by using six instead of four cylinders. This latter arrangement is now employed upon the higher powered and also upon the more recent cars of moderate powers. A much smoother running, more flexible and quieter engine is thus obtained, but at the expense of extra cost and complication. The six-cylinder engine will run more slowly " on top gear " than the four, and will accelerate more rapidly when the throttle is opened; the engine vibration will be less. As regards torque, the ratio of the maximum to the mean torque in the case of the six-cylinder engine in the example considered in Fig. 81 is only 1·4, as against 2·0 for the four-cylinder, and 8·0 for the single-cylinder engine.

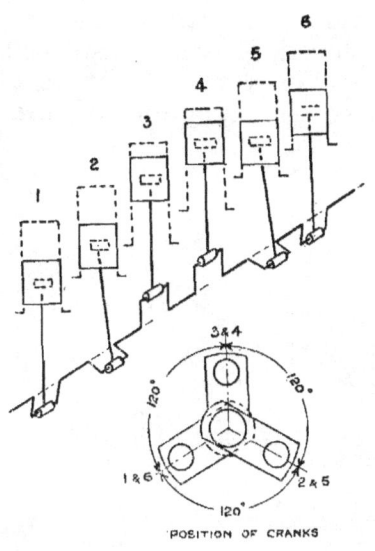

Fig. 132.—The Six-Cylinder Engine Crankshaft Arrangement.

The six-cylinder engine therefore requires only a very light fly-wheel.

The arrangement of the cranks in the six-cylinder engine is shown in Fig. 132. It will be observed that the two centre cranks are parallel, the second and fifth are also parallel and at 120° to the centre ones, and the two outer ones parallel and at 240° (reckoned in the same direction) to the middle pair. Every 120° (or one-third of a revolution) a pair of pistons will be at the top of their stroke, one on the compression, and the other the exhaust stroke. There are therefore three firing strokes per one revolution of the crankshaft. The usual firing order (Fig. 132) of the six-cylinder engine is: 1, 4, 2, 6, 3, 5, the sequence being alternatively from front to back. Another firing order, used on certain engines, is: 1, 5, 3, 6, 2, 4.

The balance of the six-cylinder engine is almost perfect, there being no "secondary" forces unbalanced; there is only a very small unbalanced force, known as the *Sixth Harmonic*, which causes a slight vibration at six times engine-speed frequency.

In the more recent car engines of 12 H.P. (rated power) and upwards the six-cylinder engine is widely employed, having displaced the four-cylinder type, except in a few instances of cars of 12 to 14 H.P. where the adoption of the four-cylinder model has

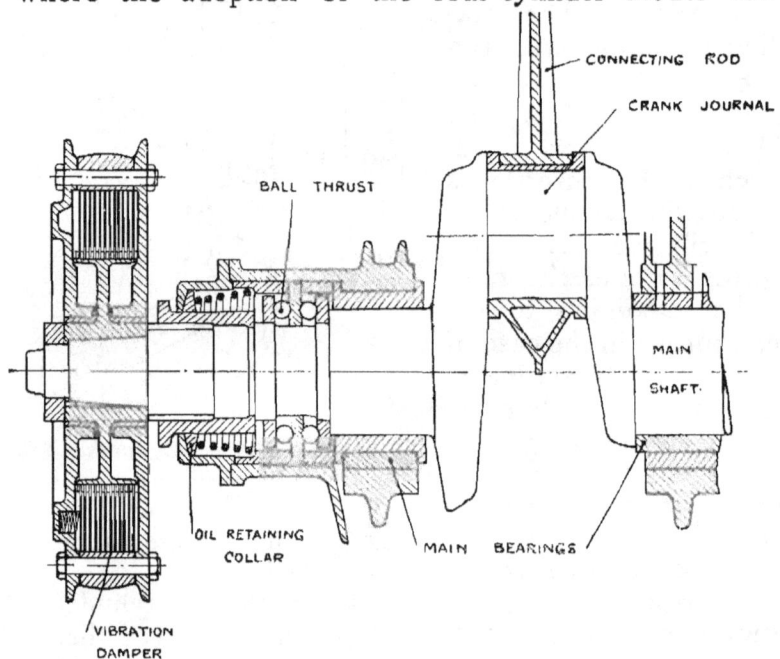

Fig. 133.—The Lanchester Vibration Damper, employed to damp out the vibrations due to the variable engine torque on the crank-shaft. The damper contains a series of plates in oil.

enabled cost to be reduced, whilst the better balancing of the engine in conjunction with the advantages of the flexible rubber engine mounting has resulted in a relatively smooth-running engine of good performance.

For small and medium powers, it is the practice to cast the cylinders in monobloc; this lessens the manufacturing costs, and enables a clean and compact design to be realized. For larger powers (e.g., about 30 H.P.

and above) it is usual to employ two sets of cylinder blocks of three cylinders each, otherwise the monobloc system would be too heavy and cumbersome.

Typical Six-Cylinder Engines.—With the exception of the six-throw crankshaft, twelve-cam camshaft and different inlet and exhaust manifolds, six-cylinder type ignition apparatus and, in general, a rather longer

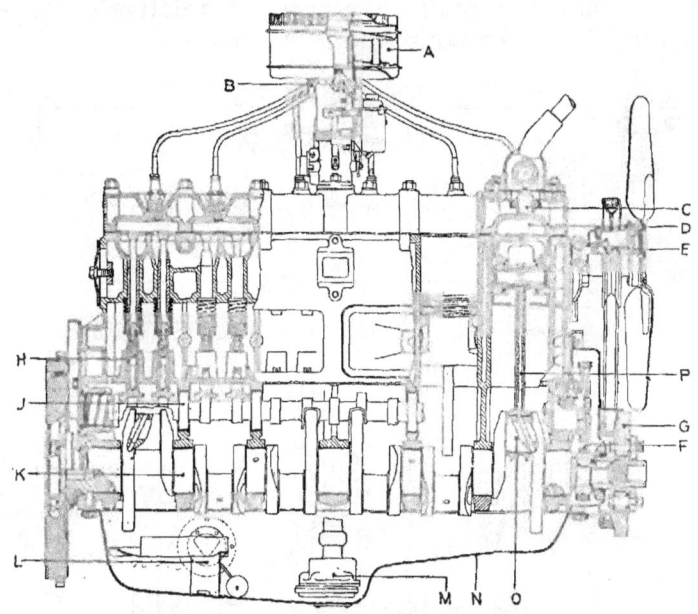

Fig. 134.—Typical Six-Cylinder Engine.
A—Air silencer and cleaner. B—Coil ignition unit. C—Thermostat.   D—Combustion chamber. E—Fan bearing. F—Timing roller chain sprocket. G—Vibration damper and fan belt pulley. H—Valve tappet. J—Cam-shaft. K—Crank-shaft main bearing. L—Oil level gauge.   M—Submerged oil pump.   N—Sump. O—Big-end bearing. P—Drilled connecting rod.

engine arrangement there is no marked difference between the general design of the six- and four-cylinder engines.   In the less expensive engines the crankshaft has only four main bearings, so that there are two crank-pins between each main bearing.

A typical example of a six-cylinder engine is given in Fig. 134, which shows a side-valve automobile engine of modern design in part side sectional view. The valves of Nos. 5 and 6 cylinders, including the tappets and cams, are shown in detail. The aluminium alloy piston, the combustion chamber and connecting rod of No. 1 cylinder are also shown in sectional view. The piston has two compression rings and one scraper ring above the hollow gudgeon pin.

The crankshaft has a torsional vibration damper at G combined with the Vee-pulley for the fan belt. The camshaft has six main bearings of relatively large diameter; these are white-metal lined.

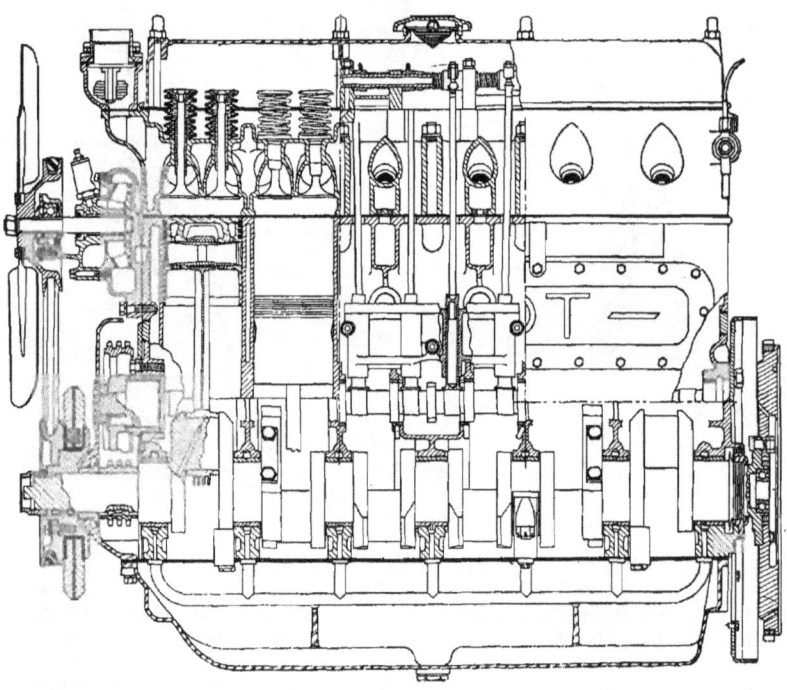

Fig. 135.—The Talbot-Darracq Six-Cylinder Engine.

Other interesting features of this design include the down-draught carburettor with large air cleaner and silencer above and pressed sheet metal oil sump.

The Talbot-Darracq four-litre engine shown in part sectional view in Fig. 135* is a good example of a well-designed overhead valve engine of the six-cylinder class. It has a bore and stroke of 90 mm. and 104·5 mm., giving a cylinder capacity of 3,996 c.c.

* Automotive Industries.

(244 cu. in.). The compression ratio is 6·3 : 1 and the power output is 105 B.H.P. at 4,000 R.P.M. This engine can also be fitted with a special head and valve-operating gear to enable the inlet and exhaust valves to be inclined symmetrically in relation to the cylinder axis. The head in question is of aluminium alloy and it provides for a hemispherical combustion chamber. The compression ratio is then raised to 7·2 : 1. All of the valves are operated from a single camshaft on one side of the cylinder axis, the inclined push-rods being in

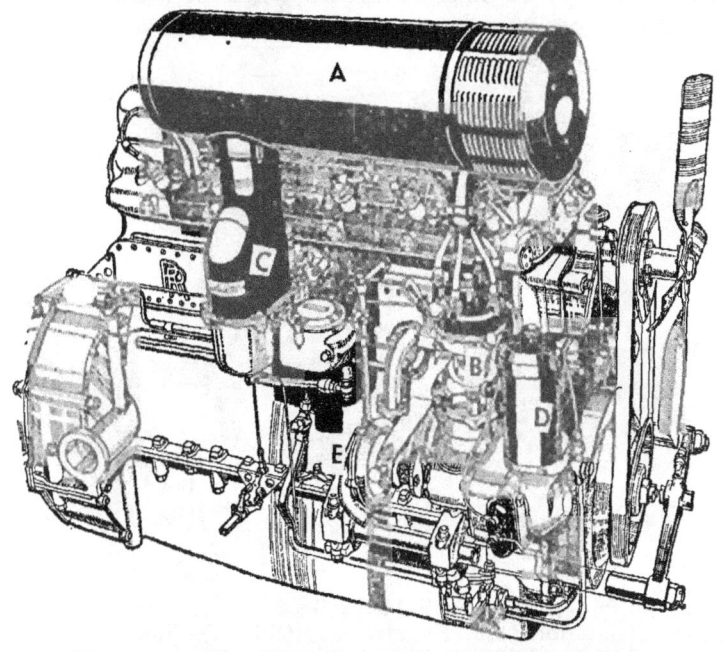

Fig. 136.—The Rolls Royce Six-Cylinder Engine.
A—Air cleaner and silencer. B—Distributor Unit. C—Carburettor.
D—H.T. Coil. E—Cooling Water Pump.

line with one another and special shapes of rocker arms used to operate the inclined valves. The engine, which has three Stromberg carburettors, develops 160 B.H.P. at 4,200 R.P.M. Special features of this engine include the use of nitralloy cylinder sleeves; Debard type R.R. alloy four-ring pistons; seven-bearing cam-shaft carried in steel-backed white-metal bearings; connecting rods having white-metal spun direct into the rod, with finned cap for cooling purposes; inclined shaft-driven gear-wheel oil pump submerged in sump below; torsional vibration damper integral with Vee-pulley for the fan drive; fan-shaft driven water impeller for circulating the cooling water and thermostat in cylinder head water outlet to radiator.

The Rolls Royce six-cylinder engine is made in two models of $3\frac{1}{4}$in. bore by $4\frac{1}{2}$in. stroke and $3\frac{1}{2}$in. bore by $4\frac{1}{2}$in. stroke, the rated H.P.'s being 25·3 and 49·4, respectively. These engines have overhead push-rod operated valves, detachable cylinder heads, water pump

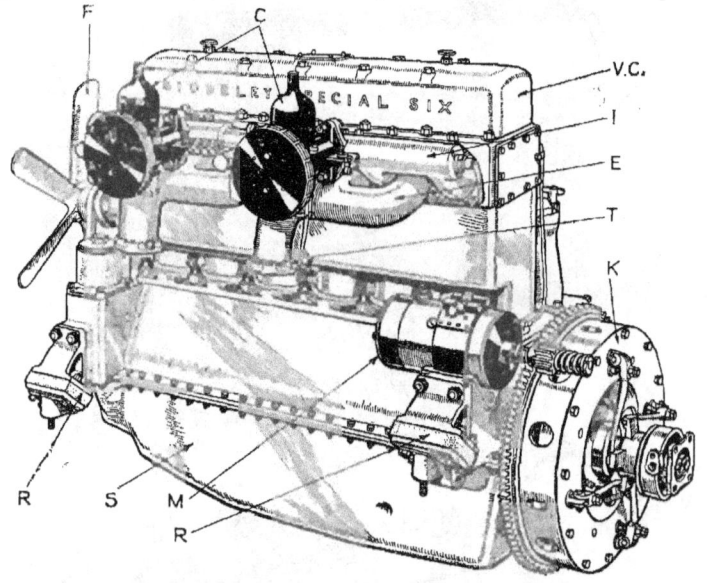

Fig. 137.—Armstrong Siddeley Six-Cylinder Engine.
C—Twin carburettors.   E—Exhaust manifold.   F—Fan.   I—Inlet manifold.   K—Clutch.   M—Starting motor.   R—Rubber mountings.   S—Oil Sump.   T—Carburettor-coupled throttles. VC—Overhead valve cover.

circulation, automatic shutters to radiator, pressure lubrication systems and battery ignition.

The six-cylinder Siddeley Special engine, shown externally in Fig. 137, has a bore of 88·9 mm. and stroke of 133·4 mm. (4,960 c.c.) and is rated at 30 H.P. The cylinders are made of cast Hiduminium or R.R aluminium alloy, and are fitted with cast-iron liners for the barrels. The cylinder head is made of the same alloy and has aluminium-bronze inserted valve seatings and sparking plug bosses.   The pistons are of the aluminium-Invar strut pattern with three compression and one scraper ring; floating gudgeon pins are used. The connecting rods are of Hiduminium also.   Over-head valves, operated by hydraulic tappets, through push-rods and rocker arms are employed.   Duplex valve springs are used for the valves, as in aero-engine practice.   The crankcase and oil sump are of light aluminium alloy.   A seven-bearing crankshaft with torsion vibration damper is employed.   The usual high pressure lubrication system, with submerged gear-type oil pump, is adopted.   The cooling system comprises a centrifugal pump, belt-driven aero-type six-bladed fan, mounted on ball - bearings, and radiator with thermostatically-controlled shutters.   The engine is mounted at four points to the frame, on rubber buffers.

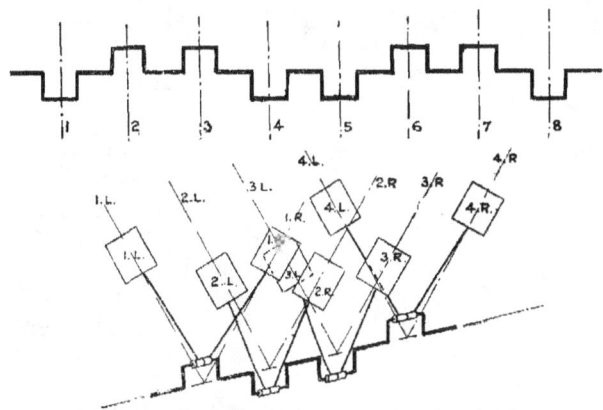

Fig. 138.—Illustrating Two Arrangements of Eight-Cylinder Engine Crank-shafts.
Top—The " Straight Eight." Bottom—The Vee-type.

The Eight-Cylinder Engine.—This type possesses an advantage over the six-cylinder engine in the matter of more uniform torque and rather better acceleration, but its balance is not, generally speaking, so good, except in one " straight eight " crank arrangement.

The eight cylinders may be arranged either in one long line, or in two rows of four cylinders inclined to one another (i.e., similarly to four sets of Vee-twin engines, one behind the other).

G

The former arrangement gives a rather long engine and therefore long and expensive crank- and camshafts. The latter arrangement is more compact, but the balance is not quite so good.

One popular arrangement of cylinders for the straight eight type is that shown in outline in Fig. 138 (top), and consisting of two symmetrical pairs of four-cylinder engines. Evidently in this case the secondary up-and-down forces, which in the four-cylinder engine are unbalanced, are in this type balanced, although there is a small rocking action due to these forces not acting in the same line.

There are several alternative arrangements of crank-shafts for straight eight engines which give equal

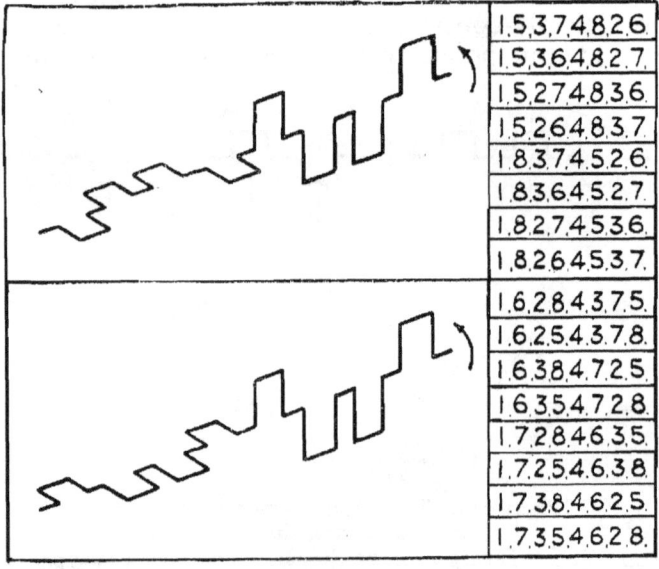

| 1,5,3,7,4,8,2,6. |
| 1,5,3,6,4,8,2,7. |
| 1,5,2,7,4,8,3,6. |
| 1,5,2,6,4,8,3,7. |
| 1,8,3,7,4,5,2,6. |
| 1,8,3,6,4,5,2,7. |
| 1,8,2,7,4,5,3,6. |
| 1,8,2,6,4,5,3,7. |
| 1,6,2,8,4,3,7,5. |
| 1,6,2,5,4,3,7,8. |
| 1,6,3,8,4,7,2,5. |
| 1,6,3,5,4,7,2,8. |
| 1,7,2,8,4,6,3,5. |
| 1,7,2,5,4,6,3,8. |
| 1,7,3,8,4,6,2,5. |
| 1,7,3,5,4,6,2,8. |

Fig. 139.—Two Alternative Crank Arrangements for Straight Eight Engines (with Alternative Firing Orders).

firing intervals, namely, four per revolution. It will be apparent that some of the cranks will have to be at right-angles to others; the particular arrangement adopted will depend also upon considerations of engine balance.

Fig. 139 illustrates two alternative crank arrangements that are used in motor-car engines. It will be observed that these may be regarded as two four-cylinder crankshafts at right-angles.

In the tables on the right-hand side of Fig. 139 are given the alternative firing orders for the cylinders that are possible. One favoured firing order at present is as follows: 1, 6, 2, 5, 8, 3, 7, 4.

The straight-eight engine was previously used to a greater extent than the Vee-eight, in spite of its greater overall length and the difficulty in damping torsional vibrations. It gives better accessibility and more ample bearing surfaces for the connecting-rod and main crank-shaft, and, generally speaking, a much cleaner design.

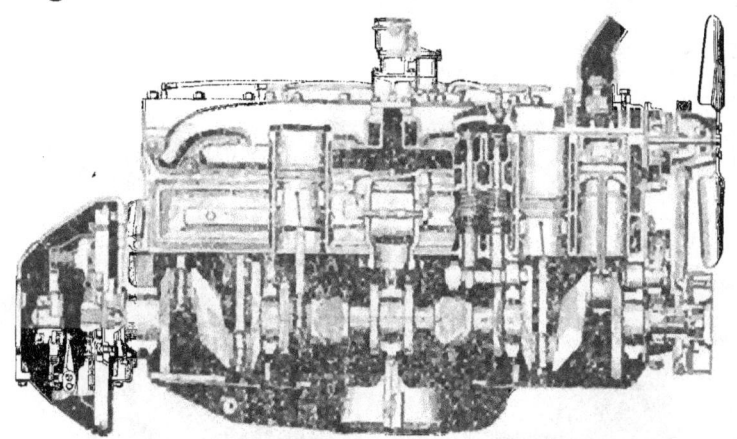

Fig. 140.—The Packard Straight Eight Engine.

An example of a high-powered straight eight engine is the Packard one shown in Fig. 140. This has a bore and stroke of $3\frac{1}{4}$ in. and $4\frac{1}{4}$ in., respectively, corresponding to a cylinder capacity of 4,620 c.c. (282 cu. in.). The compression ratio is 6·43 : 1, but a special cylinder head is available to give a ratio of 7·0 : 1 for use with high octane fuel. The engine develops 120 B.H.P. at 3,800 R.P.M., with the lower compression ratio aluminium head. The valves are of the side pattern, the long camshaft having bearings between each pair of cams. The crankshaft has five main bearings so that there is a pair of cranks between each bearing. The arrangement of the cranks, as shown in Fig. 140, consists of the two outside pairs as in a four-cylinder vertical engine and the four inside pairs, also arranged in a similar manner to the four cylinder engine, but at right angles to the outer pairs. The four cranks of the left-hand cylinders are therefore looking-glass images or reflections of those of the right-hand ones; this arrangement gives very good engine balance and torque characteristics.

The camshaft, crankshaft and big-end bearings are all of the shimless steel-backed, white-metal lined pattern, which cannot be adjusted by letting together and scraping; new bearings must therefore be used when wear occurs. The firing order of the Packard engine is as follows:—1, 6, 2, 5, 8, 3, 7, 4.

Fig. 141.—The Ford 30 H.P. Vee-Eight Engine.

The Ford Vee-Eight Engine.—The Ford V-8 Engine is a good example of a compact eight-cylinder vee-type one. This engine is made in two models, rated at 22 and 30 H.P., respectively. The former has a bore and stroke of 66 mm. and 81·5 mm. (2,230 c.c.), whilst for the latter they are 77·78 mm. and 92·25 mm.

(3,621 c.c), respectively. The latter engine develops
88·5 B.H.P. at 3,700 r.p.m.

The two cylinder blocks are cast integral with the
upper half of the crankcase; detachable aluminium
alloy cylinder heads are fitted. The engine employs
side valves of the expanded or mushroom end pattern,
similar to the 8 H.P. and 10 H.P. models. With these

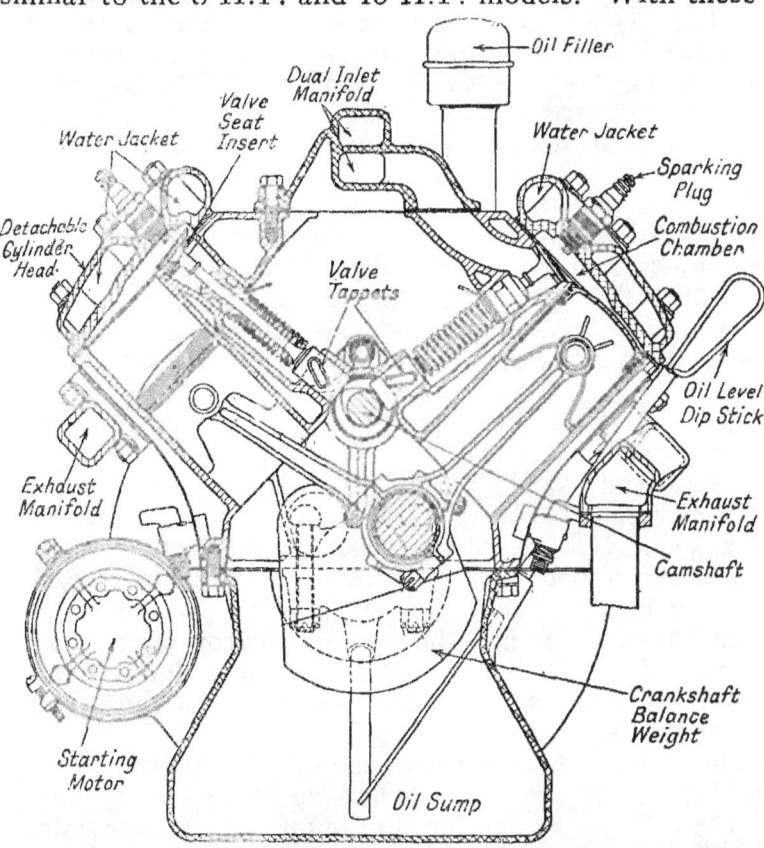

Fig. 142.—Sectional view of Ford 30 H.P. Vee-Eight Engine.

valves it is necessary to use split valve guides for
assembly and removal of the " fixed clearance " valves.
A unique feature is the provision made for removing
the complete valve, spring, collar and cotter unit,
merely by the prior removal of a cotter held in place
by the valve spring. The valves are of silicon-
chromium nickel steel.

A single camshaft, situated in the Vee of the cylinders, above the crankshaft, is employed; it is driven from the crankshaft by helical gears, the larger reduction wheel being of Bakelised fabric in order to give silent operation.

A three-bearing crankshaft is used, for the engine is relatively short. The cranks are arranged as shown in Fig 138 (lower illustration). Two pairs of connecting-rods operate on the same pin, the respective pairs being 1—5; 2—6: 3—7 and 4—8. The big ends of each pair are arranged side-by-side on their crank pin. The bearings are white-metal lined and the thrust is taken on the rear bearing. The crankshaft, which is of alloy cast iron, is very accurately balanced on an electrically-operated Olsen balancing machine, so as to be both in static and dynamic balance. The gudgeon pins are fully floating in the piston; they are each of ¾-inch diameter and hollow. Aluminium pistons are used, two compression and one scraper ring being fitted.

Lubrication is by gear-driven pump, feeding oil under pressure to the crankshaft main and big-end bearings and camshaft bearings.

The cylinder walls and gudgeon pins are lubricated by splash. Cooling of the engine is by means of two belt-driven impeller pumps located at the front of each cylinder head. These pumps draw the heated water from the engine into the upper radiator tank, from which it flows downwards through the radiator tubes into the lower radiator tank and back to the water jackets of the engine.

The Ford carburettor is of the down-draught pattern and has two separate outlets leading to the two cylinder blocks. It has an acceleration pump and bleeder valve choke and is automatic in action, all adjustments being made at the Ford works before the car is sold. There is, however, an idling adjustment which can be regulated after the car has been run in. It is important to note that it is unnecessary to pull out the throttle control when starting the engine as the throttle automatically opens by the correct amount for starting when the choke control is pulled out.

A silencer and cleaner unit is fitted to the intake side of the carburettor.

The fuel supply to the carburettor is delivered by means of an engine-operated diaphragm-type pump located on the top of the engine behind the carburettor. The pump provides a trap for any sediment or water, a drain plug being fitted for removing the latter products. An electric petrol gauge is provided for indicating the amount of petrol in the main fuel tank.

A clever system of crankcase ventilation is employed. The air from the cooling fan is forced through an open

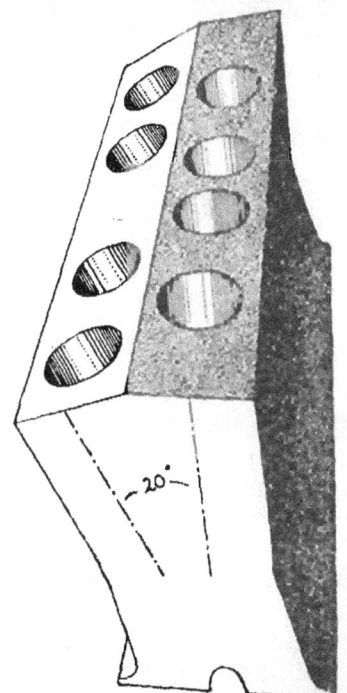

Fig. 143.—The Vee-Eight Engine Cylinder Block.

duct at the rear of engine into the closed-in valve chamber and crank-case, thus serving the dual purpose of cooling and extracting fumes, including any unburnt petrol vapour; crankcase oil dilution is thereby obviated.

Ignition is by battery and coil, the contact-breaker unit being fitted with an automatic centrifugal-type advance and retard mechanism to govern the ignition timing in accordance with the engine speed. In addition there is a vacuum-operated device, for retarding the ignition temporarily when quick acceleration is required, or when opening the throttle at low engine speeds.

The firing order of the cylinders is 1, 5, 4, 8, 6, 3, 7, 2.

Narrow Angle Vee-Eight Engine.—In order to obtain as compact an engine as possible, with a single monobloc cylinder instead of two separate blocks, the arrangement shown in Fig. 143* can be employed. In this case the cylinders are inclined at a small angle,

* Automotive Industries.

namely, 20°, in order to keep the overall width of the cylinder block as small as possible. Further, the opposite cylinders are staggered in relation to one another so that a conventional, but much shorter design of crankshaft can be employed.

It is possible with this type of engine, using an eight-throw crankshaft suitably counter-balanced, to obtain proper balancing of the primary and secondary reciprocating inertia forces. If the engine is viewed from the forward end and the cylinders in the left bank are numbered 1, 3, 6 and 8, whilst those of the right bank are 2, 4, 5 and 7, then the arrangement of the cranks which gives correct engine balance will be as shown in Fig. 144. The firing order of the cylinder is

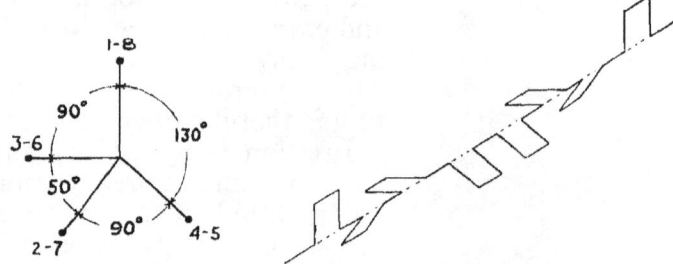

Fig. 144.—Crankshaft arrangement of Narrow Angle
Vee-Eight Engine.

then 1, 3, 2, 4, 8, 6, 7 and 5, and the intervals, in degrees, between the explosions will be 90, 70, 90, 110, 90, 70, 90 and 110, so that there is no very appreciable variation in the firing intervals.

The narrow angle engine in question, with its shorter crankshaft, is less liable to torsional vibration effects. Another advantage is the increased stiffness of the whole engine resulting from the use of a single cylinder block. On the other hand the valve operating mechanism for the inclined cylinders is usually more complicated than for the straight eight type.

Twelve-Cylinder Engines.—Evolved primarily for aeronautical use, this type has been used in some makes of car, notably in the Rolls Royce, Packard, Lincoln "Zephyr," and Daimler "Double Six." Although superior in the matter of torque, the balance is not perfect, whilst the extra complication hardly justifies its use in automobiles when such good results can be obtained from the smaller number of cylinder engines. In some cases, aeroplane engines, such as the Napier, Sunbeam and Liberty, have been fitted in racing cars in order to obtain very high powers.

The common arrangement for the twelve-cylinder engine consists of two sets of six cylinders inclined Vee-fashion at an angle of 60° as a rule. The crankshaft has the same crank disposition as in the six-cylinder engine, and forked and plain connecting-rod pairs are used. For the best results two magnetos,

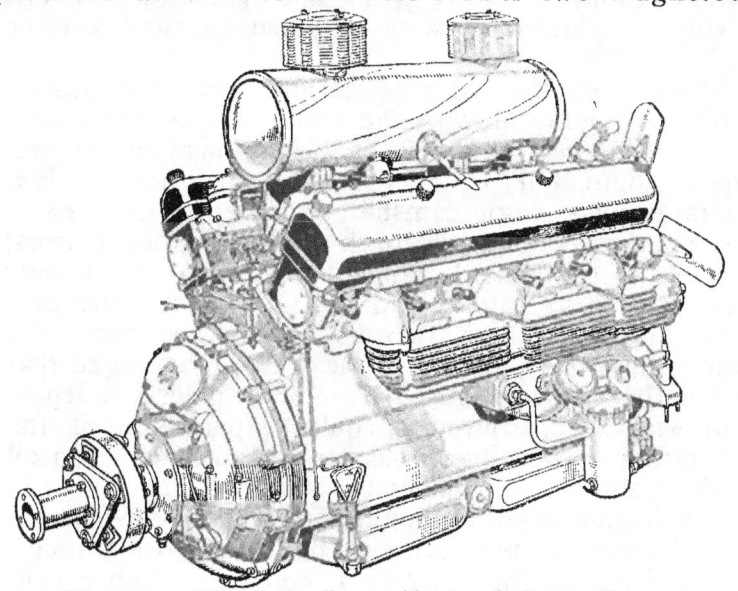

Fig. 145.—The Rolls Royce Twelve-Cylinder Engine.

two circulating pumps, and two carburettors are employed.

Two good examples of British 12-cylinder engines are the Daimler Double-Six and the Rolls Royce Phantom III types.

The former is a V-type engine of 81·5 mm. bore and 104 mm. stroke, giving a cylinder capacity of 6,500 c.c. It is rated at 49·4 H.P. This engine (1936) represents a reversion from sleeve to poppet valve design. Special features of this engine include aluminium-alloy crankcase; pent-roof combustion chamber; seven-bearing crankshaft, with enclosed

G*

automatically-lubricated torsion damper; overhead valves, operated by push-rods and rockers from two camshafts mounted in the crankcase; special cam contours permitting large tappet clearance with freedom from valve seat wear; four-point rubber mounting of engine; dual-type carburettor with air cleaner and silencer, and initial lubrication to the pistons and tappets.

The Rolls Royce 12-cylinder engine (Fig. 145) has two banks of six cylinders inclined to each other at 60°. The bore and stroke are $3\frac{1}{4}$ ins. and $4\frac{1}{2}$ ins. respectively, giving a cylinder capacity of 7,340 c.c. and a rated H.P. of 50'7.

Special features of this engine include a cylinder block cast integral with the upper half of the crankcase; cylinders fitted with cast-iron liners of the wet type; aluminium cylinder heads; overhead valves operated by a single camshaft mounted in the Vee of the crankcase, through push-rods and rocker arms; automatic (hydraulic) tappet clearance adjustment; dual-ignition by battery and coil through two independent contact breakers, two distributors and two coils; four carburettors, placed in the Vee, so arranged that two carburettors feed each row of cylinders; independent starting carburettor, quite separate from the others; large capacity air cleaner and silencer; extra oil to the walls of the cylinder for starting, etc.

Lubrication is by an engine-driven gear-type pump, the oil from the pressure side of which goes through an oil filter before entering a cooling chamber consisting of a honey-comb matrix, the temperature of which is controlled by the water discharged from the circulating pump on its way to the cylinders. The oil then enters a relief valve which determines the three different pressures used, by means of spring-loaded release valves arranged in series. The crankshaft and connecting-rod bearings are fed at the full pressure of 50 lb. per sq. in.; the overhead rocker shafts at 10 lb. per sq. in., and the timing wheels at $1\frac{3}{4}$ lb. per sq. in.

Engine cooling is by centrifugal water pump and fan-cooled radiator; the latter has thermostatically-controlled shutters to maintain the water at constant temperature.

The engine is attached to the main frame by a mounting which is torsionally insulated by rubber from the chassis frame.

Sixteen-Cylinder Engines.—Three commercial model cars have used two sets of " straight eight " cylinders inclined at an angle or vee, to give a still more even torque than that of the twelve-cylinder engine, and a practically top-gear performance. This type of engine was made by firms specialising in the " straight eight." Thus the General Motors, Ltd., made an eight-cylinder Buick and sixteen-cylinder Cadillac.

A front sectional view of the Cadillac sixteen-cylinder engine is shown in Fig. 146.* The engine has a bore and

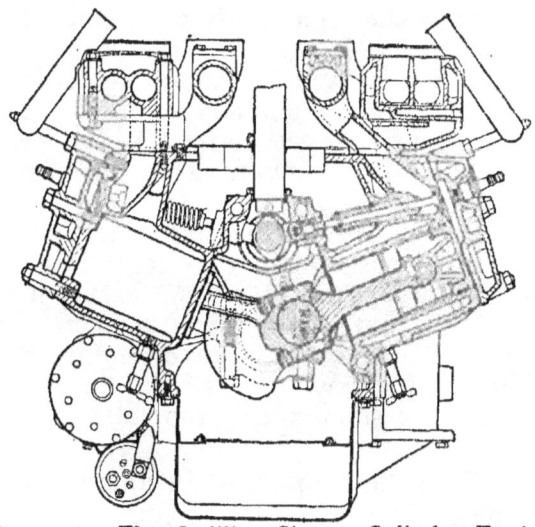

Fig. 146.—The Cadillac Sixteen-Cylinder Engine.

stroke of 88·9 mm. (3·5 in.) each, and a cylinder capacity of 7,060 c.c. (431 cu. in.). The corresponding R.A.C. rated horse power is 78·4. The engine developed 185 B.H.P. at 3,600 R.P.M. The cylinders are arranged in two banks of eight cylinders each, inclined at a relatively wide angle, so that it approached the horizontally-opposed engine arrangement and gave a wide design of engine. Both cylinder banks and the greater part of the crankcase are in a single casting. The connecting rods are shorter than in most other

* Autom. Industries.

designs using a smaller number of cylinders; this is because of the short stroke of the engine. The crankshaft has nine main bearings and two connecting rods have bearings on each crankpin, namely, side by side. A feature of the pistons and connecting-rods is the method of clamping the gudgeon pin to the small end of the rod.

Other features of this engine include the use of separate dual-pattern carburettors to each bank of cylinders; rubber type of vibration damper at the forward end of the crankshaft, with double Vee-type belt pulleys to drive both the radiator cooling fan and the generator above.

An advantage of the wide angle arrangement of the cylinders is the relatively large clear space at the top of the engine which is used for mounting most of the accessories, whilst providing ready access to the valve springs. The valve tappets are of the hydraulic compensating type for maintaining the correct clearance automatically.

An interesting feature of this engine is that practically all the accessories are in duplicate; this ensures greater reliability over long periods. Mention should here be made of the sixteen-cylinder Auto Union racing engine which had a successful career before the war of 1939. The cylinders were arranged in two banks of eight inclined at a small angle; the valve mechanism for this engine is shown in Fig. 75.

The Flat or Pancake Engine.—The amount of space occupied by the ordinary six-cylinder engine of the commercial vehicle reduces that available for passengers or goods, so that in several instances this difficulty has been overcome by placing the engine below the floorboard level. For this purpose it becomes necessary to use a "flat" type of engine such as the four, six or eight cylinder opposed ones consisting of pairs of cylinders on opposite sides of the crankshaft as shown in Fig. 112 (d). Alternatively, as is the case with the Vomag and the Büssing engine an ordinary six-cylinder or other vertical model engine can be arranged so as to work in the horizontal position. By suitable design of the chassis members the parts of the engine, e.g., the sparking plugs, valves and carburettor requiring maintenance attention can be rendered accessible without difficulty.

Several makes of compression-ignition engine for large motor vehicles and for rail cars are made in opposed-cylinder flat engines. Typical examples are the Tillings-Stevens, Vomag and Krupp.

Horizontal Six Engine.—The Hall Scott engine shown in Fig. 147 is a six-cylinder horizontal one of 5 in. bore and 6 in. stroke, giving a cylinder capacity of 11,600 c.c. (707 cu. in.). It develops 180 B.H.P. at 2,200 R.P.M. The engine is mounted underneath the floor of the motor vehicle.

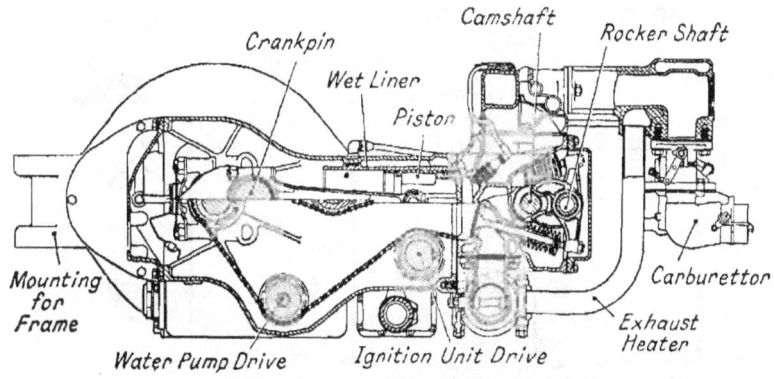

Fig. 147.—The Hall Scott Horizontal Engine.

The upper part of the illustration shows part of the cylinder and piston, and it will be noted that a wet type cylinder liner is fitted. Overhead inclined valves operated by means of overhead camshaft and curved rocker arms are used. The camshaft is driven by means of a relatively long chain which passes over four sprocket wheels; two of the latter drive the water-circulating pump and ignition cam and distributor rotor.

The carburettor is shown on the extreme right-hand side where it is mounted in an accessible position; it has an exhaust gas-heated jacket and is controlled by means of a governor.

Some Other Engine Types—Ever since the evolution of the poppet-valve engine, attempts have been made to replace the valves and their operating gear by some more simple and quieter means for attaining the same object. In some cases piston valves, resembling those of steam-engine practice, were used to govern the inlet and exhaust operations; in others rotating valves of cylindrical form, with ports cut in them, as in the Darracq rotary-valve engine, were employed. The principle of the Darracq engine is illustrated in Fig. 148, the inlet and compression operations being shown. The rotary valve consisted of a section of a cylinder rotating in a cylindrical casing having ports communicating with the cylinder, carburettor and exhaust pipe, respectively; it rotated at one-half engine speed. As shown on the left, the piston is just about to descend on its suction stroke, the

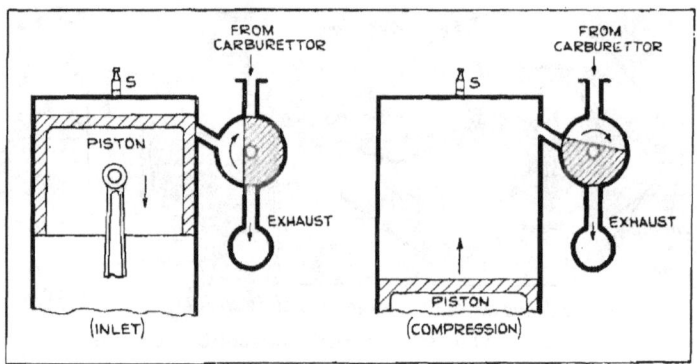

Fig. 148.—Principle of Rotary Valve Engine.

rotary valve being about to open the carburettor port. At the end of the inlet stroke, the valve shuts off both the carburettor and exhaust ports, thus sealing the cylinder. The piston is arranged to cover the port leading to the rotary valve, during the firing period so that the valve is not subjected to the maximum explosion pressure. When the piston is about three-quarters of the way down its firing (or expansion) stroke the rotary valve opens the exhaust port; the latter remains open during the next upward stroke of the piston. In the Itala engine, which was once used on cars, a vertical rotating valve, with specially designed ports, somewhat on the lines of the usual piston valve arrangement, was employed.

The cuff-valve engine arrangement, which had a short vogue, consisted of a pair of concentric slotted sleeves situated above the cylinder, the outer sleeve serving as the exhaust, and the inner sleeve as the inlet valve. These sleeves were provided with conical seatings and were operated up and down similarly to an ordinary poppet valve; in effect the arrangement was equivalent to a pair of concentric, but hollow, poppet valves of large area.

Although none of the engine types mentioned above have survived, certain new types such as the Edwards, Aspin and Cross have given promising results. The disadvantages of the early pattern engines previously mentioned were that, in the majority of cases, it was found that the effect of the hot exhaust gases in the common valve distributor reduced seriously the volumetric efficiency, and also caused lubrication and distortion troubles. Rotary and rocking valves which were exposed to the exploding and expanding gases were found to wear badly. In some cases the extra complication of the valve gears did not justify their use, and in others the explosion event tended to blow the valves off their seats.

Sleeve Valve Engines.—There is one type of engine, namely, the sleeve valve engine, which has certain advantages over the poppet valve one. Briefly speaking, this type comprises a single liner, or two cylinder liners which can slide up and down (and also rotate in one case) relatively to the piston. Ports cut in the upper portions are arranged to open and close the inlet and exhaust passages at the correct intervals. In effect, then, we have the piston working in a cylinder liner which itself can slide in another liner or in the outer cylinder casting, and controlling the inlet and exhaust operations.

The advantages claimed for the sleeve-valve engine are: (1) Quietness in operation, there being no noisy cams, tappets and valves. (2) Simplicity of construction and a considerably smaller number of working parts; this results in longer life and better reliability. (3) Higher efficiency.

Knight Double Sleeve Engine.—The principle of this engine is illustrated in Fig. 149. It consists of a pair of concentric sliding sleeves between the cylinder barrel and the piston. Each sleeve was given a small up-and-down motion by means of a crank, driven at one-half engine speed and a short connecting-rod. The upper ends of these sleeves were arranged to slide between the cylinder barrel and cylinder head, and were provided with ports to register at the appropriate moments in the cycle of operations, with the fixed inlet and exhaust valve ports in the cylinder block, as indicated diagrammatically in Fig. 149. In this diagram A represents the cylinder and B and C the two thin sleeves for valve opening and closing purposes. The sleeves have inlet ports D, and exhaust ports E, such that by a pre-determined arrangement or timing of the movements of the sleeves, the inlet ports and exhaust ports of each sleeve respectively coincide at specified times with the fixed inlet and exhaust ports. The sleeves are operated by connecting-rods F and G, which are actuated by two cranks arranged at different phase-angles from the secondary shaft H running at half engine speed.

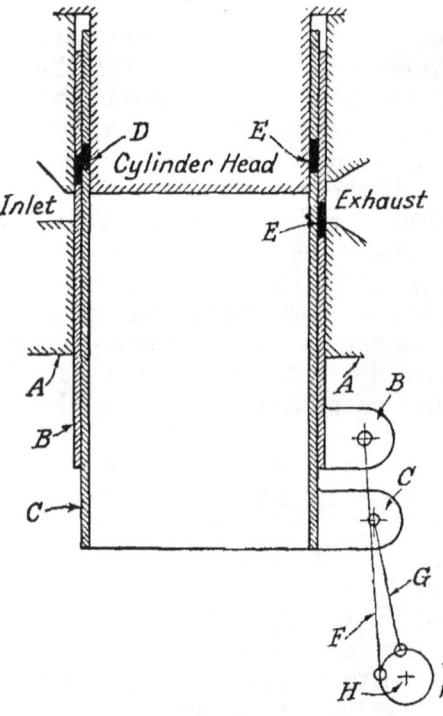

Fig. 149.—The Double Sleeve Valve Engine Principle.

The sleeves therefore have an up-and-down motion, giving zero velocity at the two ends and a maximum velocity at about mid-stroke. With this arrangement the valve-ports could be given quick opening and closing operations.

The advantages of this type of engine over the poppet

valve one were that it ran much quieter—since there
were no valve cams, tappets, valves or other impact
members—and, on account of its pocketless combustion chamber, gave a higher thermal efficiency; moreover, by suitably designing and apportioning the sleeve

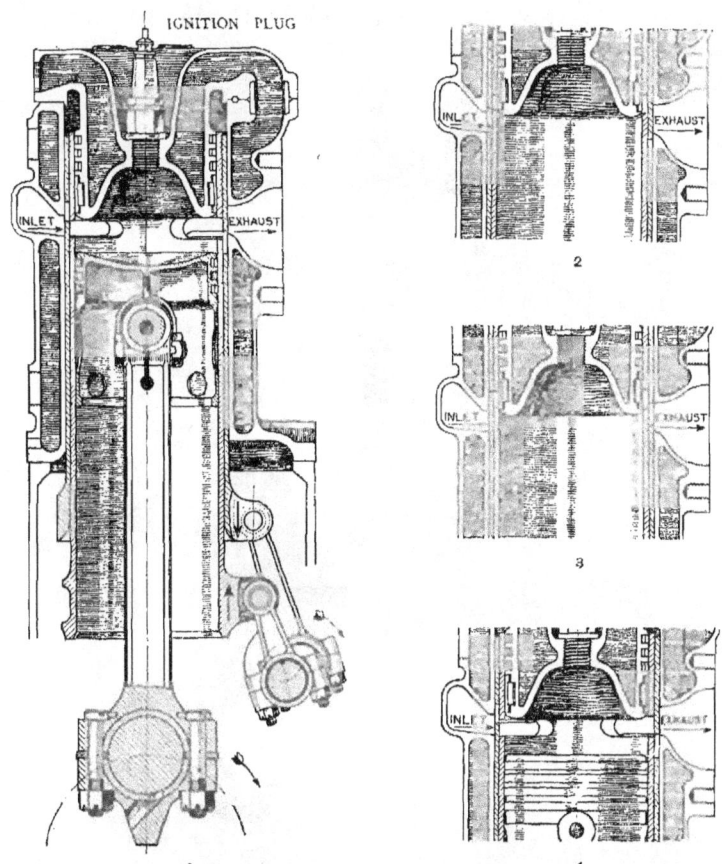

Fig. 150.—Illustrating the action of the Double Sleeve-Valve.
Engine.

1. Beginning of suction stroke.   2. End of compression stroke.
3. Firing stroke.   4. End of exhaust stroke.

and cylinder ports higher volumetric efficiencies could
be obtained than with the poppet valve type. Another
advantage is the ability of this type of engine to stand
up to long periods of operation without maintenance

attention, other than replenishing the oil for lubrication.

The principal drawback of the double-sleeve ˙engine has been its somewhat high rate of oil consumption on account of the relatively large area of sleeve surface to be lubricated. Another difficulty has been that of sleeve inertia effects which have limited the maximum engine speeds to values well below those of poppet valve engines.

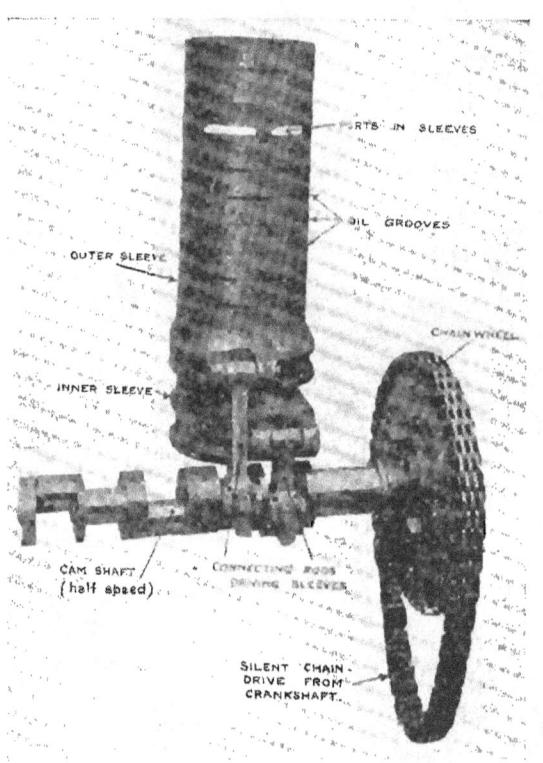

Fig. 151.—Showing Two-sliding Sleeves and Cam-shaft of the Daimler Engine.

The Knight double sleeve-valve engine was adopted by the Daimler Company in 1908, after exhaustive tests, one of which consisted of an official R.A.C. test, during which a four-cylinder engine of 124 mm. bore and 130 mm. stroke ran continuously for 5 days 14 hours, giving an average B.H.P. of 54·4, and a petrol consumption of 0·679 per B.H.P. hour. Following this it was fitted to a car and ran for over 2,150 miles on the Brooklands track at an average speed of 42·4 m.p.h. (occupying 45½ hours). It was then given a final bench test of 5¼ hours. A subsequent examination of the pistons, sleeves and other working parts revealed no perceptible wear, nor did the ports show any signs of burning or wear.

The cylinder head of the Knight engine was detachable; it had an extension, as shown in Fig. 150 (1) not unlike a stationary piston, which projected into the cylinder, and was provided with piston rings to ensure gas-tightness. The ignition plugs were situated centrally in the cylinder heads, and the combustion chamber was spherical in form. It has been shown by Ricardo that sleeve-valve engines, with their almost ideal combustion chamber shapes, are less prone to mis-firing and detonation than poppet-valve engines. The two sleeves are grooved at regular intervals, with special serrations also near the ports to ensure adequate lubrication (shown in Fig. 151).

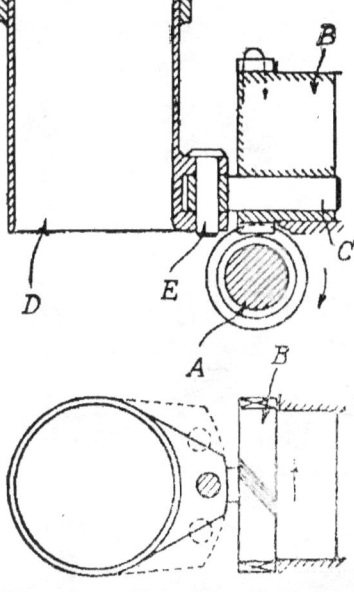

Fig. 152.—Principle of Single Sleeve Valve Engine.

The sleeve-valve principle was also embodied in the twelve-cylinder Daimler "Double Six" engine,* a Vee-type engine having two sets of six cylinders at an angle to each other.

The Single Sleeve Engine.—In this type of engine, which was built under the Burt M'Collum patents, and

* The later Daimler engines now use large clearance poppet valves

has been used by Messrs. Argylls, Barr & Stroud, and Wallace (Glasgow), a single sleeve only is employed, and the various operations are accomplished by rotating the sleeve slightly, in addition to giving it an up-and-down motion; the ports are also given a special shape.

The principle of the single sleeve valve engine is

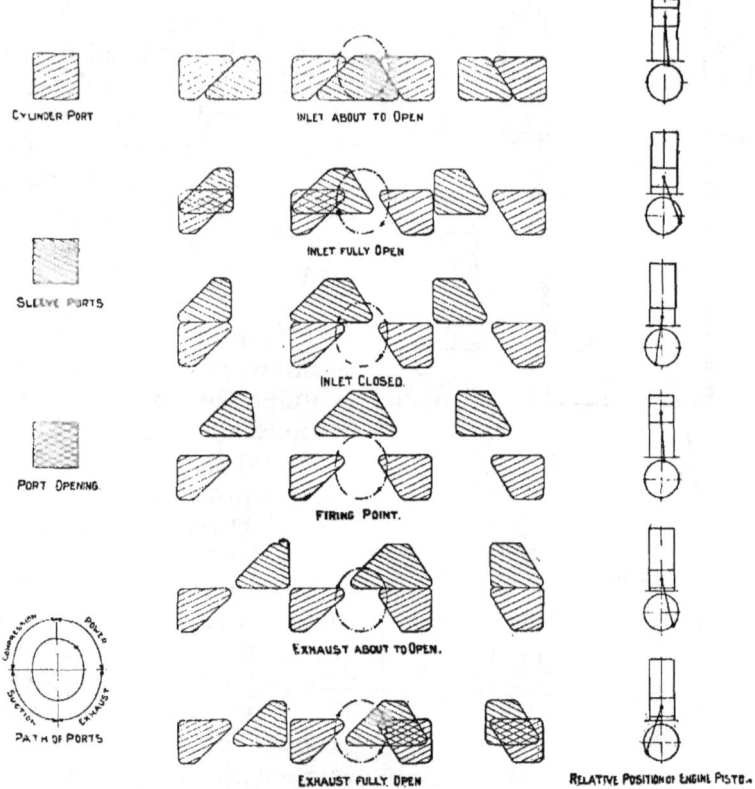

Fig. 153.—Illustrating the Method **of Working of the Single Sleeve Valve Engine.**

illustrated, schematically, in Fig. 152. This diagram shows only the single sleeve or working liner and the part of the cylinder containing the valve ports. The piston is arranged to reciprocate in this sleeve, and the latter is constrained to move both up and down and sideways, so that any point on it describes an ellipse. One method of obtaining this motion is that shown, whereby the gear-shaft A, driven at one-half engine speed by chain or gear-wheels from the crankshaft, engages with the sleeve operating crank gear B, through spiral gearing, for the axes of A and B are at right angles.   The gear B carries a tee-headed sleeve crankpin C, which is attached to the sleeve valve D by a pin E and held between suitable lugs on the sleeve.   The crankpin C is free to rotate, and can also slide endwise in a bearing contained in the gear B; it can also swing sidewise about the pin E anchored in the sleeve lugs.   This mechanism causes the sleeve to have the desired " elliptical " motion for the valve port events.   The sleeve near its upper end is provided with ports of special shape, usually varying from four to six in number.

The general arrangement of these valve ports is such that one-half of them form the inlet ports on one side, and the other half the exhaust ports on the other side of the sleeve.   There are, however, several alternative arrangements for the ports.

Fig. 153 shows a four-port cylinder with three-port sleeve arrangement in which one of the sleeve ports, namely, the larger central one, acts alternatively as inlet and exhaust port.   The key to the movements of the sleeve is indicated by the dotted elliptic diagrams in the centres and the lower left-hand diagram entitled " Path of Ports ".   The sleeve and cylinder ports are identified by the shading diagrams on the left, common over-lapping areas during the cycle of operations being indicated by the cross-hatch shading.   The corresponding piston and crank positions to the six-port diagrams are shown in the right-hand diagram.

The total movement of the sleeve is very small in comparison with that of the piston, and its elliptical movement facilitates greatly the lubrication by distributing the oil more effectively.   The sleeve ports during the firing and compression strokes pass up into the annular space between the head and cylinder wall, and are thus adequately protected.   The results of a long experimental investigation of a high-powered single sleeve engine revealed the fact that there was practically no wear between the piston and sleeve, and the sleeve and cylinder proper, and only an inappreciable amount on the eccentric driving joint. This type of engine not only yields high thermal efficiencies and power outputs, but has also been shown to wear better than the poppet-valve type. Car engines will run from 15,000 to 25,000 miles without

Fig. 154.—The Barr and Stroud Single Sleeve Valve Engine (in part cut-away view).

attention to decarbonizing, and with little falling off in power. In regard to performance it may be of interest to mention that an Argyll racing-car engine using single-sleeve valves broke a number of world's records at Brooklands in 1913. Rated at 17·3 H.P. it had four cylinders each of 83·5 mm. bore and 130 mm. stroke;

the stroke of the sleeves was 1·5 in. The compression ratio used was 5·6 : 1. The engine gave 75 B.H.P. at 3,200 R.P.M. and a B.M.E.P. of 115 lbs. per sq. in. at 2,400 R.P.M. The inlet valve ports opened on top dead centre, and closed 30 degrees after bottom dead centre. The exhaust ports opened at 60 degrees before bottom dead centre, and closed 15 degrees after top dead centre.

With the exception of a single-sleeve valve engine made by Messrs. Vauxhall Ltd., before their association with General Motors Co., there have been no commercial engines made for motor cars. The excellent performance of the single sleeve valve aircraft engines referred to later in this section indicates that this type should also give results for automobile purposes.

This principle has also been applied to motor-cycle engines, in the case of the Barr & Stroud machines, and has given excellent results. These engines were made in the 2¾, 3½ and 8 H.P. sizes, the latter being a Vee-twin engine; the corresponding cylinder capacities were 350, 500 and 1,000 c.c. In these engines five ports were cut in the sleeve, two for the inlet, and three for the exhaust. The sleeve itself was of close-grained cast iron, and was $\frac{5}{64}$ in. thick; it was ground on each surface very accurately.

Ricardo Sleeve Engine Design.—Fig. 155 illustrates a single sleeve valve engine (Ricardo) of the two-stroke fuel injection type, capable of giving a good performance. The single sleeve S is operated by the rocking lever L, and it has not only an up-

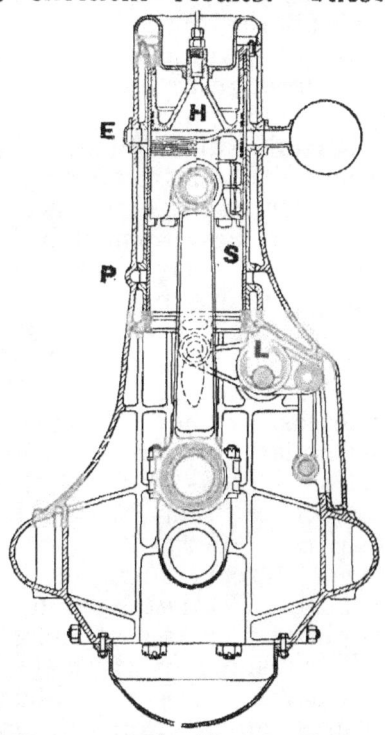

Fig. 155.—The Ricardo Sleeve Valve Engine.

and-down motion but also a small amount of rotary movement, as shown by the dotted path at the left-hand end of the lever *L*.

The sleeve has two sets of ports cut through it. The upper set when moved downwards so as to come opposite to the ports *E* are used for the exhaust. Whilst these ports are open, scavenging air is admitted through the ports *P*, and corresponding ones in the sleeve (these are shown just below the piston), so that a charge of air under pressure sweeps along the cylinder and scavenges the remaining exhaust gases through the ports *E*. The cylinder head *H* forms the turbulent combustion chamber and also acts as a guide for the upper part of the sleeve; it is provided with the two piston rings shown for this purpose. The four-cycle sleeve valve engine works in a somewhat similar manner, except that the inlet ports are at the top on the one side, with the exhaust ports also at the top and on the opposite side. A later design dispenses with the ports in the upper part of the sleeve and uses the top edge of the sleeve itself to cover and uncover the exhaust ports.

Bristol Single Sleeve Valve Engines.—The more recent Bristol aircraft engines use single sleeve valves, as shown in Fig. 156. As a result of a relatively long period of development it has been possible to obtain appreciably better performances than from poppet valve engines of similar dimensions. Thus, in the case of the " Perseus " sleeve valve engine of 5¾ in. bore and 6½ in. stroke a B.M.E.P. of 190 lbs. per sq. in. was obtained at 2,800 R.P.M. on 87-octane fuel, whereas for the " Mercury " four-poppet valve engine of similar cylinder dimensions the B.M.E.P of 165 lbs. per sq. in. was obtained at 2,650 R.P.M. When using 100-octane fuel the " Perseus " engine gave a B.M.E.P. of 220 lbs. per sq. in. at 2,800 R.P.M. and the " Mercury " engine 185 lbs. per sq. in. at 2,750 R.P.M. The latter results give the B.H.P. per sq. in. of piston area as 5·05 and 4·18, respectively.

The manner in which the Bristol sleeve valve engine has developed is well illustrated by the performance figures shown graphically in Fig. 157. It will be noted that there has been a progressive increase in the

B.M.E.P and B.H.P. together with a steady reduction in the fuel consumption over the period of years indicated.

This type of engine lends itself well to supercharging and in this connection mention may be made of the results of some tests made by the Bristol Aeroplane Company which, it is stated, could not have been obtained from an engine having poppet valves under similar conditions. With a supercharge of 14 lbs. pᵣr sq. in. and a cylinder of 168·8 cu. in. capacity, the B.M.E.P.'s obtained were 305 lbs. per sq. in. at 2,400 R.P.M. and 272 lbs. per sq. in. at 3,000 R.P.M. The

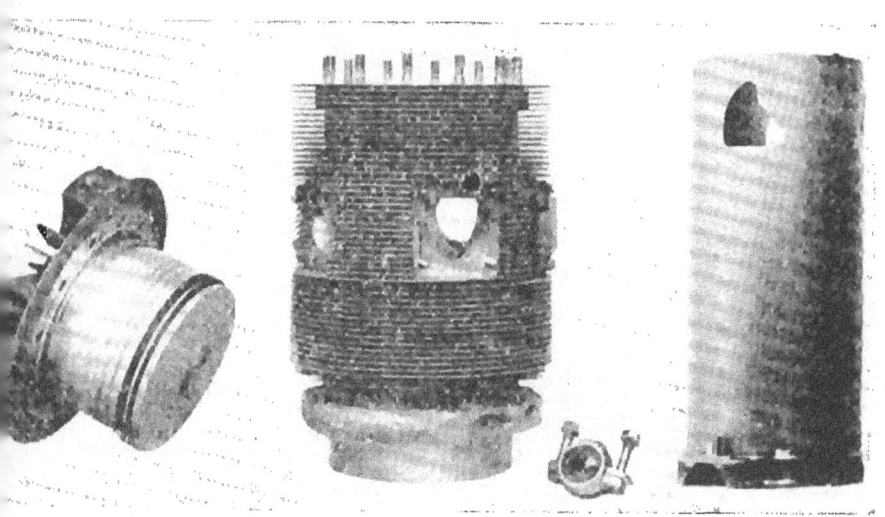

Fig. 156.—Components of Bristol Single Sleeve Valve Engines.

corresponding powers per unit piston area are 6·0 B.H.P. per sq. in. and 6·7 B.H.P. per sq. in. If the above figures are corrected for the power absorbed by the separately driven blower, the B.M.E.P's are 276 lbs. per sq. in. at 2,400 R.P.M. and 245 lbs. per sq. in. at 3,000 R.P.M., whilst the powers are 5·4 B.H.P. per sq. in. an 6·0 B.H.P. per sq. in.

Some interesting results have been obtained by Ricardo from a small sleeve valve engine as long ago as 1930, when an experimental engine ran heavily supercharged for several hundred hours on ordinary aviation

spirit at a B.M.E.P. of over 400 lbs. per sq. in. and a
specific output at least three times as great as that of
existing poppet-valve aircraft engines and that without
any signs of distress, or overheating of the engine.
With a special high octane fuel the engine would main-

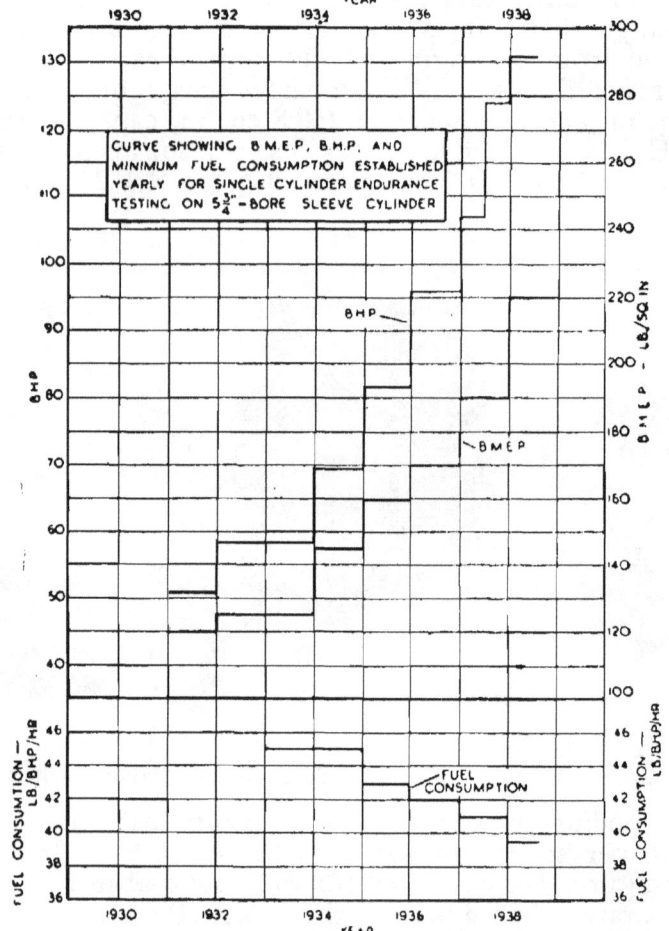

Fig. 157.—Illustrating Development of Bristol Single
Sleeve Valve Engine.

tain a B.M.E.P. of 550 lbs. per sq. in.   With these
high degrees of supercharging the exhaust is discharged from the cylinder at pressures of 300 lbs. per
sq. in. and above so that the question arises as to

whether it would not be feasible to utilize some of this pressure energy by arranging for the exhaust to operate in a low-pressure cylinder.   In this connection, the compounding of the engine is possible since the same sleeve valve can be arranged to operate as both inlet and exhaust valve to the high-pressure and as the inlet valve to the low-pressure cylinder.   In this way, greater fuel economy could be obtained and a high power-to-weight ratio.   The Ricardo engine, with a bore of $2\frac{7}{8}$ in. and stroke of $3\frac{1}{2}$ in. ran at speeds of 5,000 to 6,000 R.P.M. and gave well over 100 B.H.P. per litre; the compression ratio used was $6\cdot8:1$

Advantages of Single Sleeve Valve Engine.—The principal advantages of the single sleeve valve engine may be summarized, briefly, as follows, namely, (1) Absence of hot exhaust valve with its detrimental effect upon and limitation of the maximum power of the engine.   (2) Considerable reduction in the number of working parts owing to absence of the valves, springs, cotters, timing gear, etc.   (3) Use of higher compression ratios due to better combustion chamber shape and absence of hot exhaust valve.   (4) Greater power output per sq. in. of piston area.   (5) Lower fuel consumption per B.H.P. hour.   (6) More silent operation than for poppet valve engine owing to absence of timing gears, tappets and valves.   (7) Reduced maintenance attention due to same causes.   (8) Reduced height of cylinder owing to absence of overhead valve mechanism.   (9) Lower manufacturing costs.   (10) Greater reliability due to absence of valves and valve mechanism.

Some Constructional Features.—The Bristol engines have aluminium alloy cylinders and pistons. The sleeves are made of a special nitrogen-hardening steel, known as KE 965 steel, which has practically the same coefficient of heat expansion as the aluminium alloy.

The cylinder heads are of aluminium alloy and are well finned for cooling purposes.   The sleeve is driven from a half-speed reduction gar shaft by means of a comparatively simple mechanism as shown in Fig. 158; also in the small illustration of Fig. 156 It consists of a crank on the half-time shaft having a crankpin which engages with a spherically seated bearing, for the sleeve operating lug, which can swivel so as to remain in line with the operating pin of the crank. The motion provided by this drive is an elliptical one in which the vertical movement is about 50 per cent. greater than the horizontal one.

In regard to the upper end of the sleeve it is, of course, necessary to prevent any loss of compression or

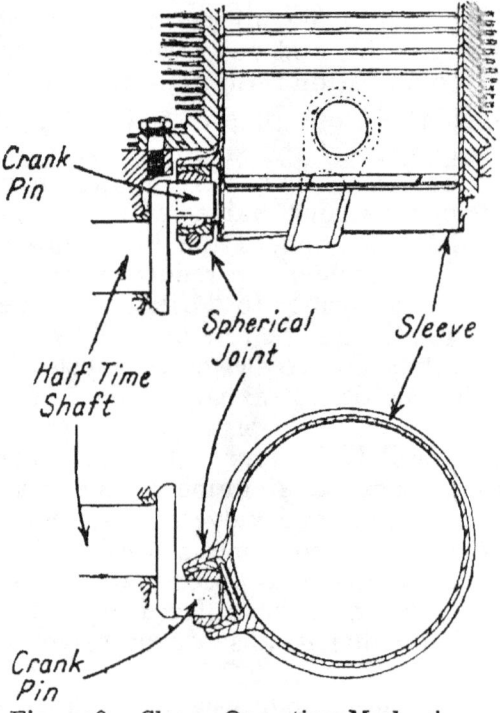

Fig. 158.—Sleeve Operating Mechanism.

combustion pressure past this end, so that a special head having a piston-like portion, provided with two piston or junk rings is employed. The inner surface of the upper end of the sleeve works over this cylindrical portion of the head, which thus seals the sleeve effectively against gas leakages. The cylinder head of the Bristol sleeve valve engine is shown on the left in Fig. 156. It is held down to the upper machined face of the cylinder by means of studs and nuts. Suitable cooling fins are arranged on the upper side of the head.

The Cross Rotary Valve Engine.—This more recent example of the rotary valve class has met with a fair measure of success as a result of its having overcome the practical difficulties connected with the maintenance of gas-tight joints and lubrication of the valve member.

Important advantages are claimed for this rotary valve design. Thus, it operates at a lower temperature than the poppet valves (the exhaust one of which is usually at a red heat), and presents a cool, polished

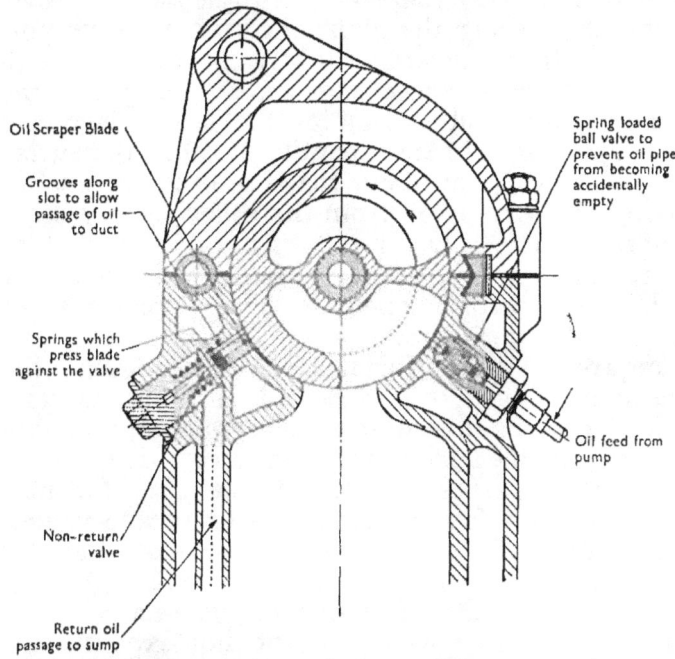

Oil Scraper Blade

Grooves along slot to allow passage of oil to duct

Springs which press blade against the valve

Non-return valve

Return oil passage to sump

Spring loaded ball valve to prevent oil pipe from becoming accidentally empty

Oil feed from pump

Fig. 159.—The Cross Rotary Valve Engine.

surface to the flame. This enables a higher compression ratio to be used. Further, it operates quietly and does not require periodic adjustment as with the tappet clearances of poppet valves. Running cooler, this valve also gives higher volumetric efficiency.

The rotary valve is arranged horizontally over the top of the cylinder (Fig. 159) and, in the single cylinder type has a diagonal rib to separate the inlet and exhaust sides. In the side of the tubular valve, ports are cut. The rotation of this tube alternatively brings the inlet and exhaust ports in communication with the cylinder; during the compression and firing strokes the plain portion of the tube seals the cylinder; the tube runs at one-half engine speed for four-cycle operation.

By making the tubular valve of larger diameter it can be made to run at quarter engine speed. The valve may be liquid-cooled for car engines, but for motorcycle ones a simple contact-cooled one is satisfactory.

The difficulty, experienced in early engines, in connection with making the tube gas-tight has been overcome by chamfering the phosphor-bronze sleeve ports and setting them inwards slightly, so that when the valve is in position these edges are sprung and thus seal the joint. Lubrication of the sleeve is carried out by supplying a liberal quantity of oil to the surface of the rotary valve at about the lateral centre line and removing the surplus oil from the other side by means of a scraper device and non-return valve so that the oil is returned to the sump. The oil supply is controlled by the throttle opening, being greater at the larger throttle openings.

In regard to the performance of this engine which has been developed over a period of about 20 years, until it is now in the commercial stage, high outputs have been obtained from small engines. Thus, in the instance of an air-cooled 247 c.c. engine (unsupercharged), using a low grade fuel of 65 octane value, a B.M.E.P. of 165 lbs. per sq. in. was developed at 5,350 R.P.M. and 176 lbs. per sq. in. at 6,000 R.P.M.

The high performance is due to the relatively cool combustion chamber which has no hot exhaust valve heads and no pockets. It is usual to employ compression ratios of 10 : 1 to 11 : 1 with normal commercial grades of petrol, since no detonation occurs with these compressions; the corresponding poppet valve engine would only be able to employ a compression ratio of about 5 : 1 with such fuels.

A special feature of the Cross engine is its relatively low fuel consumption due to the high thermal efficiency which is associated with the compression ratios previously mentioned. Thus, the 247 c.c. engine referred to gave a fuel consumption of only 0·35 lbs. per B.H.P.

per hour at 4,000 R.P.M. It is necessary to employ efficient cooling means for this type of engine so that special attention is needed in the design of the cooling fins or water jackets in order to maintain the valve unit and its housing sufficiently cool under all circumstances.

A later development of the Cross engine is the controlled valve loading system which overcomes the objection sometimes put forward against rotary valve engines, namely, that the valve is exposed to the full explosion pressure and must therefore tend to bind in

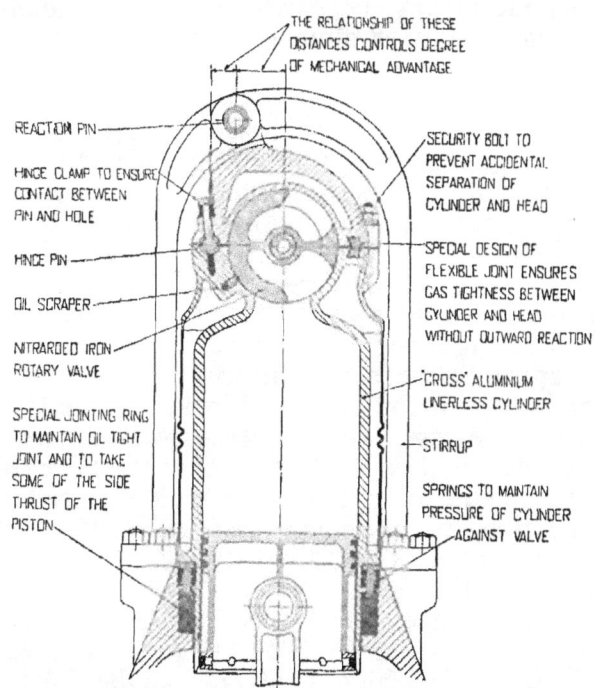

THE RELATIONSHIP OF THESE DISTANCES CONTROLS DEGREE OF MECHANICAL ADVANTAGE

REACTION PIN

HINGE CLAMP TO ENSURE CONTACT BETWEEN PIN AND HOLE

HINGE PIN

OIL SCRAPER

NITRARDED IRON ROTARY VALVE

SPECIAL JOINTING RING TO MAINTAIN OIL TIGHT JOINT AND TO TAKE SOME OF THE SIDE THRUST OF THE PISTON

SECURITY BOLT TO PREVENT ACCIDENTAL SEPARATION OF CYLINDER AND HEAD

SPECIAL DESIGN OF FLEXIBLE JOINT ENSURES GAS TIGHTNESS BETWEEN CYLINDER AND HEAD WITHOUT OUTWARD REACTION

CROSS ALUMINIUM LINERLESS CYLINDER

STIRRUP

SPRINGS TO MAINTAIN PRESSURE OF CYLINDER AGAINST VALVE

Fig. 160.—Improved Cross Engine Arrangement.

its housing or to experience excessive wear on its surface and bearings; further, differential expansion effects between the valve and its housing are eliminated.

The loading of the valve is controlled by the designer so that only sufficient pressure is exerted upon the valve to maintain its contact with the sealing port-edge lip. As the pressure rises in the cylinder it endeavours to press the valve away from the port but always the valve is given a little more pressure to seal the port than the gas can exert upon it to press it away.

The housing is divided and the two halves are always in resilient contact with the valve which merely acts as a spacing member between them. If the valve heats up more rapidly than the surroundings and therefore expands more, the housing automatically adjusts itself to accommodate the valve with no increase of pressure and by virtue of this, the expansion problem is completely eliminated.

By the same means the valve is so lightly loaded at all times that seizure is almost impossible and engines have been run dry for considerable periods without trouble.

The controlled valve loading is brought about by utilising the gas pressure in the cylinder to provide the necessary force to exert pressure on the two halves of the valve housing and controlling the amount of this force by the principle of leverage or mechanical advantage. (Fig 160). The cylinder unit is allowed to float axially and obtains the reaction to prevent it leaving the crank chamber against an overhead stirrup gear. The valve is trapped between the bottom half of the housing which forms part of the cylinder block, and the top half of the valve housing. As the pressure in the cylinder rises the greater will be the force reacting on the stirrup gear, and if uncontrolled this force would exert far too heavy a bearing load on the rotary valve, particularly in large engines where the valve is small in relation to the diameter of the bore.

The system of mechanical advantage or leverage to reduce this pressure can be brought about in a variety of ways. A pair of hinged levers can be fulcrumed on the stirrup, one end of each lever reacting on the upper half of the valve housing, the other against the lower half. A simpler method and one which works well in practice is to use the top half of the valve housing as the lever and insert a hinge pin at one of the joints between the two halves of the valve housing. The upper half of the valve housing is attached to the overhead stirrup at a position between the centre of the valve and the hinge pin. The ratio of the distances between the hinge pin and the reaction point, and the valve centre and the reaction point, is the controlling factor allotting the proportion of load which is applied to the valve and the proportion which passes directly through the hinge pin to the stirrup.

It has been found that an engine will work satisfactorily with only 1 per cent. of the explosion load reaction of the cylinder exerted on the sealing lips, although it is customary to allow between 5 per cent. and 10 per cent. to make quite sure there is no possibility of the two halves of the valve housing being separated by gas pressure. Precautions, however, are taken to prevent accidental opening of the two halves of the valve housing and adjustable stops are provided on both sides of the housing to ensure that such separation could only be of small magnitude.

It is necessary, of course, to provide some light spring pressure underneath the cylinder to maintain the upward pressure of the cylinder when there is no gas pressure in the cylinder, as, for instance, during the induction stroke.

Another important development is the adoption of the method of raised lip port-edge sealing which ensures gas-tightness over the whole life period of the engine. The lip is formed in the aluminium alloy cylinder head by removing the metal around it for the length of the valve and up to within about 10° of the split in the valve housing. The depth of this recess is only about ·004 in. The rotary valve is made of special alloy steel, nitrogen hardened, and Y-alloy is favoured for the cylinder and valve housing. The cylinder is linerless and the type of pistons and rings illustrated in Fig. 160 are employed with fully satisfactory results.

The Aspin Engine.—In this design of engine the ordinary poppet valves are replaced by a rotating conical member having ports for the inlet and exhaust operations. The engine in question has a relatively high output due to the fact that it can employ much higher compression ratios than those used in poppet valve engines without detonation effects occurring. The absence of hot exhaust valve heads and the beneficial shape of the "pocketless" combustion chamber are the chief contributory factors in this connection.

H

Referring to Fig. 161* which shows the cylinder head, rotating valve and port arrangement in detail, the rotary valve unit, shown at F is a nitralloy steel shell filled with a light alloy and coned externally to 60°. Attached to this is a cylindrical member B running on roller bearings at A and C. Above is the half-speed gear wheel E engaging at the left with the smaller gear wheel H which is driven by the vertical shaft N. Referring, again, to the left-hand sectional view the cavity in the rotating valve F constitutes the compression space, the piston crown at top dead centre coming

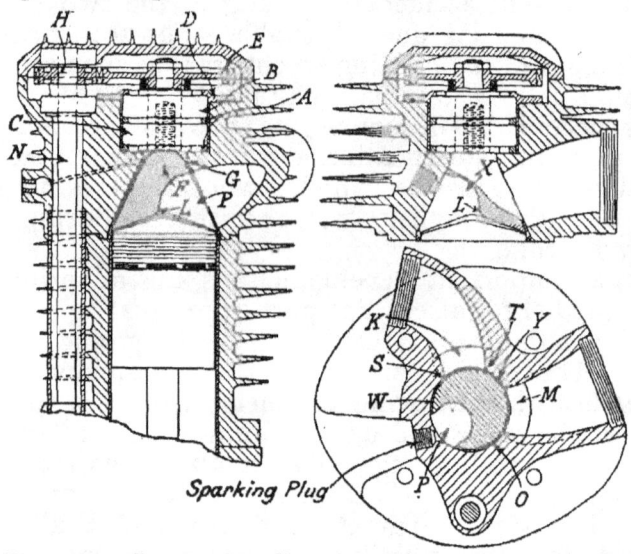

Fig. 161.—Illustrating Principle of the Aspin Engine.

close to the lower conical surface of the rotary valve; the contour of the port at L is stated to be an important item in the valve design.

The valve F rotates anti-clockwise and, as shown, is approximately in the charge ignition position and remote from any hot area. After ignition the port P rotates past the plain surface at O, during the firing stroke and thence reaches the angular position opposite M, the exhaust port, when the exhaust operation occurs. After this the inlet port K is uncovered and overlaps M to a degree which is determined by the

*The Automobile Engineer.

length of the section Y and by the location of the point T, which represents the start of the induction stroke.   The point S determines the valve closing lag, after which the port P passes over the blank surface W to give the compression operation and thence to the ignition position near the end of the compression stroke shown in the left-hand illustration.

Owng to the shape of the combustion chamber formed in the rotating valve and to the absence of the hot-spots associated with poppet valve engines, e.g., the exhaust valve heads the combustion process it is possible, in the Aspin engine, to employ higher compression ratios without detonation effects.   It is believed that after ignition there is an initial flame-front movement only, the whole of the remaining charge being then simultaneously projected radially from three sides and at varying but very high velocities upon the flame nucleus as the piston reaches its top dead centre, thus prohibiting any specifically directional movement since all avenues of frontal movement are then closed and complete combustion occurs very quickly, but progressively.

An experimental single-cylinder air-cooled engine, of 2·64 in. (67 mm.) bore and 2·78 in.(70·5 mm.) stroke, having a capacity of 15·2 cu. in. (249 c.c.) developed 5 B.H.P. at 1,800 R.P.M., 25 B.H.P. at 7,000 R.P.M. and 32 B.H.P. at 11,000 R.P.M.   It was found possible to employ a compression ratio of 14·1 : 1 with an ordinary fuel of 0·74 S.G. and heating value 18,810 B.T.U.'s per lb.   It is stated that low octane fuels, such as paraffin oil, can be used at this high compression ratio without detonation and that the engine, which was not supercharged, could be operated at a speed of 14,850 R.P.M. without failure.

The Slide Valve Engine.—The slide valve engine is a modified type of sleeve valve engine whereby a pair of sliding members, each representing a longitudinal slice of a complete sleeve, as it were, take the place of the single sleeve of the sleeve valve type engine.   Each slide has parallel sides and works in a curved recess machined in the cylinder wall; the slides and recesses are on the thrust sides of the piston.

Fig. 162 illustrates the Imperia slide-valve engine that works on this principle and shows the arrangement of the slide valves, the operating cam-shafts and rockers. This mechanism, it will be observed, is relatively simple. The arrangement illustrated gives the same valve port areas as four valves per cylinder in the ordinary poppet valve engine.

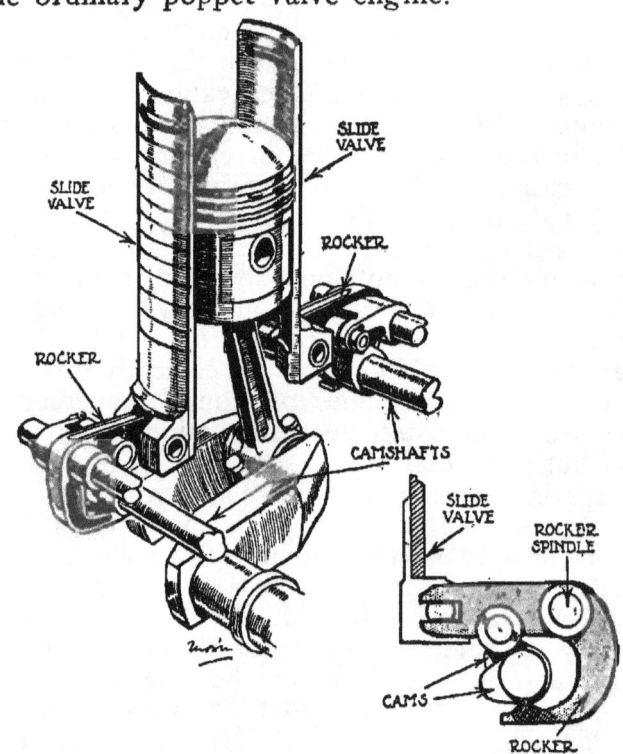

Fig. 162.—The Imperia Slide-Valve Engine.

Fig. 163.—Showing the Slide Operating Mechanism.

In connection with the design of the slides these are provided with a series of grooves for lubrication purposes, oil being supplied to these grooves and to the operating cams from a pressure pump. The movement of each slide is only 9 mm., the slides being almost stationary during the firing and compression strokes. There are two ports in each side of the cylinder wall, the top edge of the slide and the lower edge of the slide port acting as a " cut off."

The Imperia engine to which this valve gear was fitted was a four-cylinder model of 66 mm. bore and 80 mm. stroke; a six-cylinder model is also made. The former engine gave 25 B.H.P. at 2,600 r.p.m., and ran very quietly.

The Radial Engine.—In this type the cylinders are arranged in star fashion about the crankshaft's axis, so as to lie in the same plane. The connecting-rods all operate upon a single throw crank.

The two-, three- and five-cylinder air-cooled radial engines have been used, in the past, for automobile

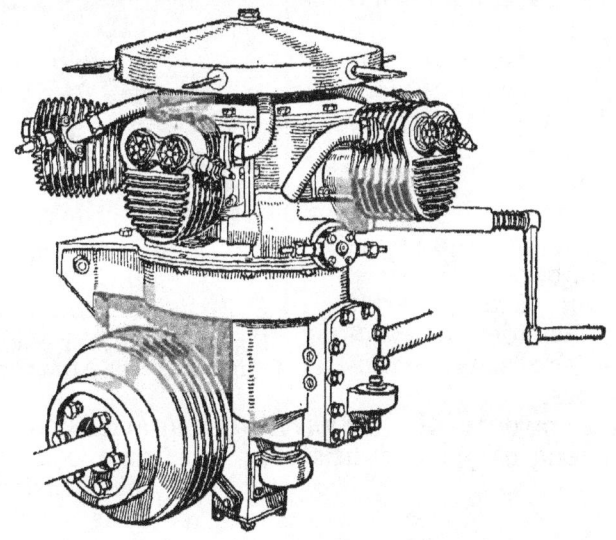

Fig. 164.—The 1,460 c.c. Five-Cylinder North Lucas Radial Engine which was used on a Motor Car. (Courtesy, *Autocar*.)

purposes, and more recently a continental racing car has been fitted with a radial engine. Engines with 7, 9, and more cylinders are used for aircraft.

The advantages of this type lie in its compactness, low weight per horse-power and accessibility. Against these merits must be offset the disadvantage of larger frontal area, more complicated exhaust pipe system and the difficulty of ensuring that all the cylinders receive equal lubrication; following successful aircraft radial engine practice, however, this difficulty can now be surmounted.

The simplicity of the single-throw crankshaft and the single cam ring for operating the valves are other attractive features of this type of engine.

The usual arrangement for radial engines is to employ an odd number of cylinders, such as 3, 5, 7 and 9 in order to obtain equal firing intervals as expressed by equal crank-angle degrees.

More recently a nine-cylinder air-cooled radial C.I. engine has been employed in military tanks made in the U.S.A. This engine, known as the Guiberson, is similar to the aircraft engine made in 1940-41 by the same manufacturers.

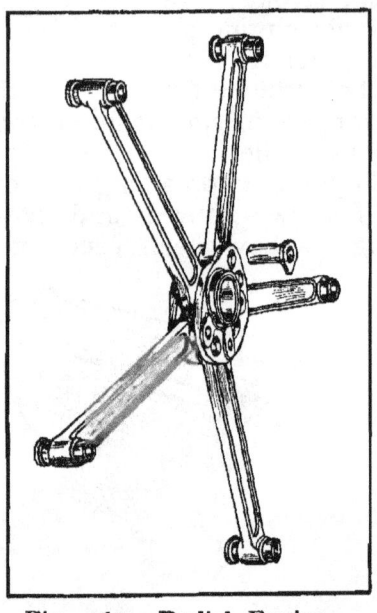

Fig. 165.—Radial Engine Connecting-rod Arrangement.

In regard to the *Connecting-rod* arrangement for radial engines this consists of one main rod, known as the *Master Rod* which is usually provided with a white-metal lined bearing running on the single crankpin. The big-end of this rod is flanged to a U-section shape and bearing pins pass through the flanges; these pins form the bearing members for the big-end bushes of the other connecting-rods, which are

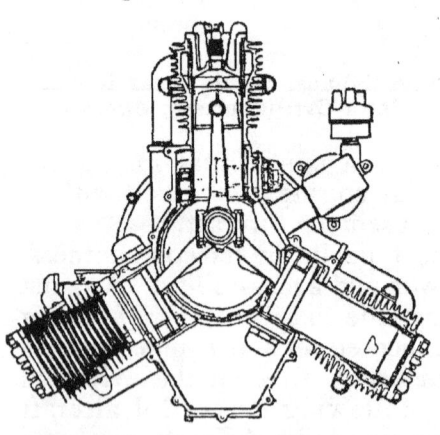

Fig. 166.—A Three-Cylinder Single Sleeve Valve Radial Engine.

known as *Link Rods*. In the illustration of a typical five-cylinder connecting-rod system the upper left hand rod is the master rod and the other four the link rods. In the Bristol radial engine arrangement (shown in Fig. 91) there is a phosphor bronze bush which floats on the crank-pin, whilst the master rod big-end bearing runs on the outside of this bush; the latter is drilled through in various places for lubrication purposes.

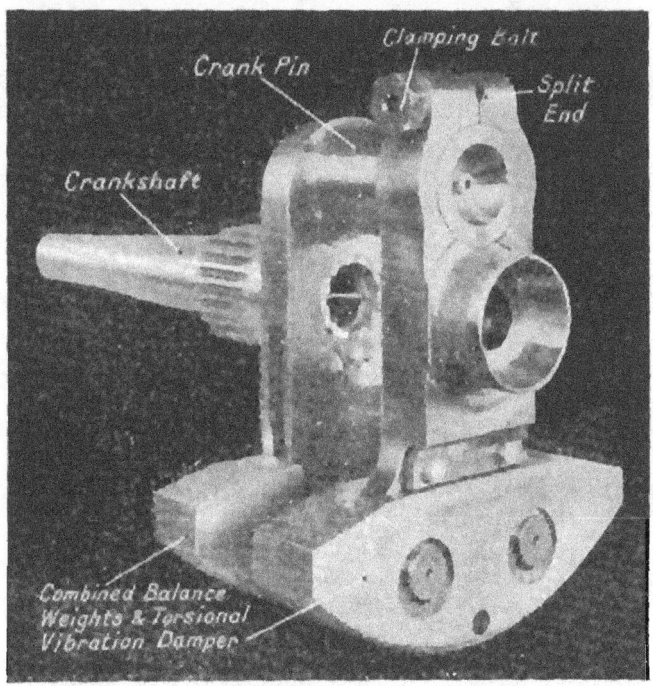

Fig. 167.—Wright Radial Engine Crankshaft.

The crankshafts of radial engines are built up, so that one crank member is detachable and the master rod can slide over the crank-pin; it is not necessary, therefore, to employ a split big-end bearing.

The crankshaft of a typical aircraft radial engine, namely, the Wright " Cyclone ", is shown in Fig. 167. The crank pin is made integral with the left hand web and balance weight member, whilst the right hand web and balance weight are keyed and clamped to the end portion of the crank pin. A special feature of this crankshaft is the provision in the balance weight unit of pendulum type torsional vibration dampers which effectively eliminate any tendency to vibration due to engine torque variations.

The Bradshaw Oil-Cooled Engine.—Although it works on the ordinary four-stroke principle the design of this engine is rather original as regards the cooling arrangements. Its special feature consists in the cooling of the cylinder barrels by the lubricating oil. The cylinders are so constructed that the long plain barrels project into the crank chamber as shown in Fig. 168. The crank chamber itself is of ample

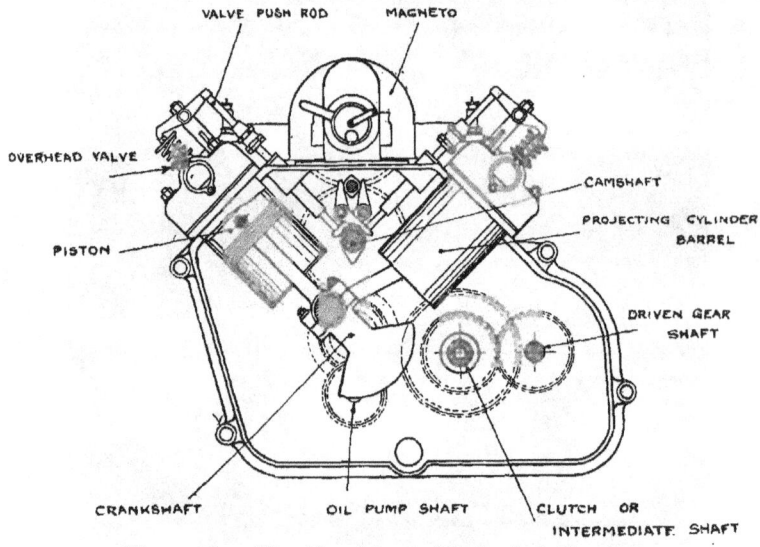

Fig. 168.—The Bradshaw Oil-Cooled Engine.

proportions, and contains in its sump about 1¼ gallons of lubricating oil. A small oil pump of the gear type circulates the oil around at the rate of about one-half gallon per minute (in the case of the 85 × 121 mm. Vee-twin engine). The oil is drawn through a filter, by the pump, and is delivered under pressure to the hollow crank-shaft, whence it passes to the big-end bearings and out again. Suitable ducts, or channels, are arranged into the big-ends and crank-pins to allow the oil to squirt out in three streams on to the cylinder barrels and to the lower surfaces of the pistons. It will be evident, therefore, that the lubricating oil, whilst lubricating the moving parts, also carries away the surplus heat. The oil is cooled by contact with the surfaces of the crank-chamber; the latter is ribbed externally. Detachable well-ribbed cast-iron cylinder heads, overhead inlet and exhaust valves, and central cam-shaft are other interesting details of this engine. The engine in question was fitted to a light car, and it was also made as a single cylinder motor-cycle unit.

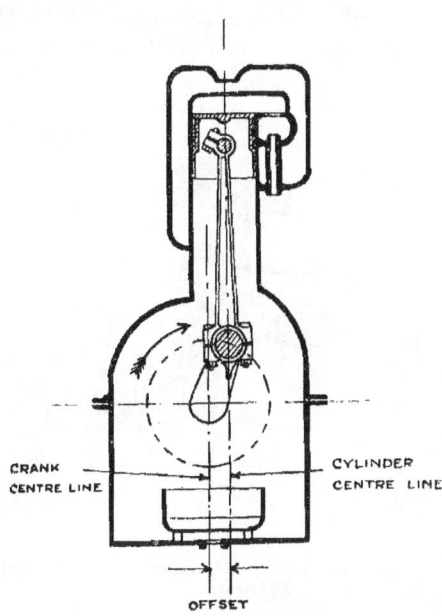

Fig. 169.—The Offset or Desaxé Cylinder Arrangement.

The Desaxe Type Engine.—In order to reduce the thrust of the piston on the cylinder walls during the firing stroke, the cylinders in some engines are set forward in the direction of rotation, so that their centre lines pass in front of the crank-shaft centre as shown in Fig. 169. This *Offset*, or *Desaxé* arrangement gives also a rather slower piston motion at top-centre and tends towards higher explosion pressures. It does not affect the engine balance appreciably.

H*

The usual amount of offset is from one-fifth to one-sixth of the cylinder diameter. The piston stroke is about 0.3 per cent. longer with this amount of offset, and with a connecting-rod to crank ratio of 4.

The practice of offsetting the engine is followed in the case of several automobile engines.

The Swash-Plate Engine.—Amongst the numerous experimental engines which have been suggested, or built, from time to time, the axial or swash-plate engine is worthy of mention, although at the time of writing it has not advanced much beyond the experimental stage. Briefly, the arrangement of this engine comprises a number of parallel cylinders, situated

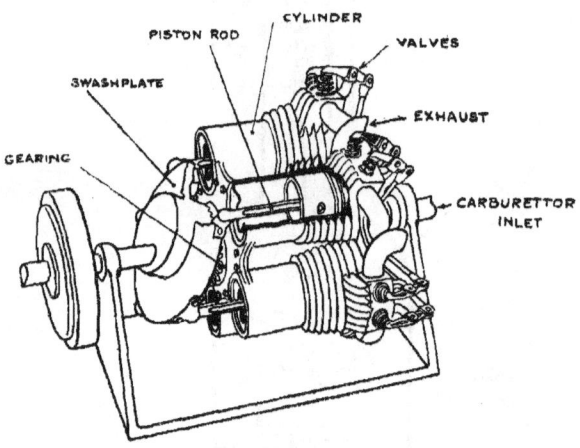

Fig. 170.—Illustrating the Revolving Swash-Plate Engine Principle.

with their centres, or axes, on a circle. Each piston has a rod provided with a ball-end which can work in a socket situated in an inclined plate known as the *Swash Plate;* the latter can rotate about its axis. The block of cylinders rotates about the symmetrical axis, and the swash-plate also rotates in the same direction and at the same speed. It will be evident, then, that as the cylinders revolve about their central axis, the pistons will move in and out, as the ball-ends of the rods are, as it were, guided around the swash-plate periphery. Each piston completes two strokes per revolution of the cylinders so that the ordinary four-stroke cycle can be followed. The length of the piston stroke is given by the distance between the perpendicu lar from the ball-ends on to the shaft axis. In this manner the piston thrusts due to the explosive loads are converted directly into turning effort, without the use of a crank-shaft.

It has always been doubtful whether this advantage, and that of the compact form of this engine, can be offset against the increased frictional losses due to the thrusts of the rods on the swash-plate.

The Bristol Axial Engine.—A particularly neat and compact axial engine of the multi-cylinder type has been built and put into experimental service by the Bristol Tramways and Carriage Company of Bristol. This engine has nine cylinders equally spaced and arranged axially about the crankshaft axis. It belongs to the engine class known as the *Wobble Plate* type, and differs from the swash plate model by the fact that the member which transmits power from the pistons to the crankshaft does not revolve. The

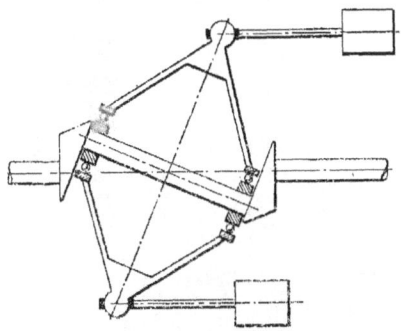

Fig. 171.—Showing Principle of Bristol Axial Engine.

wobble plate, or star member, is mounted upon a Z-shaped crankshaft (Fig. 171). The slanting crankpin is set at an angle of 22½ degrees to the line of the main shaft with the centre of the crank-pin in line with the centre of the main shaft. The star member has ball-bearings, and is mounted upon the crank-pin so that the shaft can be revolved within the star member, thus imparting a " wobble " to it. A torque member, or stabilizer, is attached to the star member in such a way as to allow universal motion but, at the same time, prevent the star member from revolving upon the shaft.

The connecting-rods and pistons of the nine cylinders are connected to the star member by spherical bushes, which encircle the ball-ended arms arranged around the periphery of the star member.

The engine is provided with a large rotary valve system. The rotary valve has four pairs of inlet and

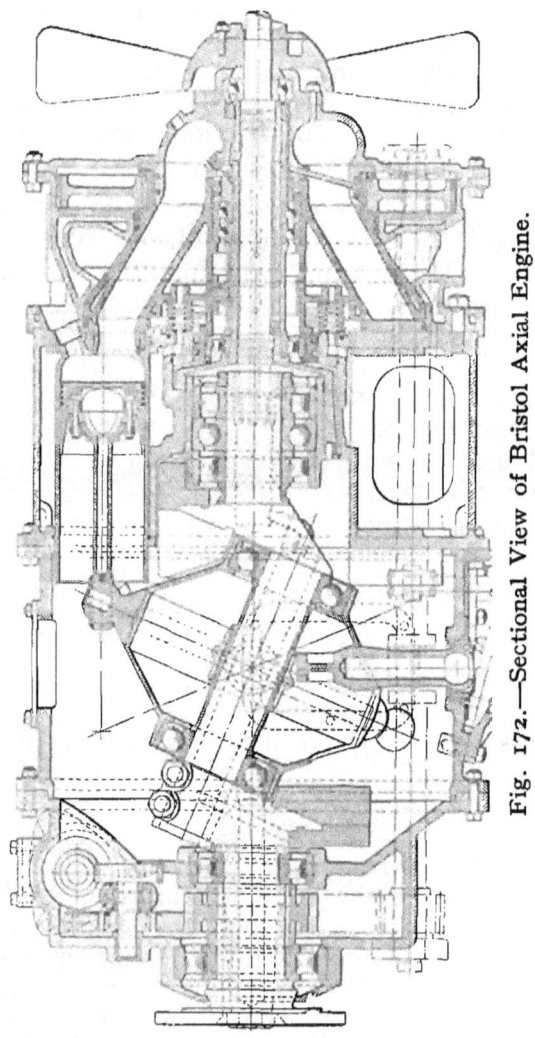

Fig. 172.—Sectional View of Bristol Axial Engine.

exhaust ports which register alternately with a single port in each cylinder end. The rotary valve rotates at one-eighth engine speed, and a perfect seal is claimed to be obtained between the rubbing face of the valve and the fixed valve plate on the cylinder ends.

The engine described was built in the 7-litre class, and was actually 5 cwt. lighter than the corresponding 7-litre vertical cylinder engine; in addition it occupied considerably less space.

The engine had a higher mechanical efficiency than the ordinary type of axial engine owing, it is believed, to the smaller number of principal bearings and absence

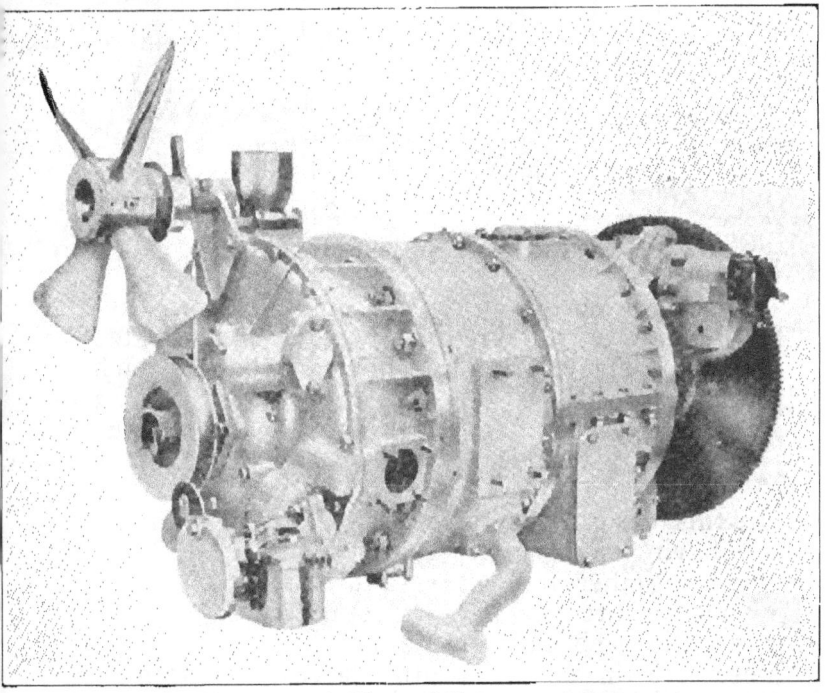

Fig. 173.—General View of Bristol Axial Engine.

of connecting rod angularity and its side thrust upon the piston.

It is stated by the makers that this engine requires only one-third of the power of an ordinary engine to motor it around.

The axial engine (7-litre) gave a maximum B.H.P. of 150 at 3,000 r.p.m. as against the 117 B H.P. at 2,600 r.p.m. of the 7-litre comparison vertical engine. The maximum engine torque was 320 lb. ft. as compared with 294 lb. ft. of the latter engine. The petrol consumption over the whole speed range was about 15 to 17 per cent. less than that of the vertical engine.

Another interesting feature of the Bristol engine is that, by the provision of two counter-balance weights to the crankshaft, it can be given a theoretically perfect balance.

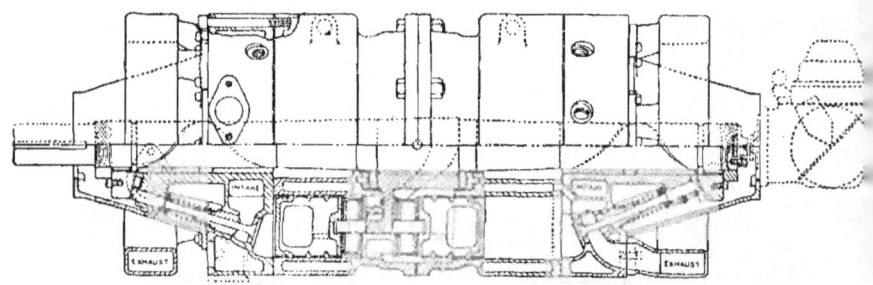

Fig. 174.—The Herrmann Barrel Engine.

A modern design of barrel engine known as the Herrmann one, of American origin is illustrated in Fig. 174. The engine has twelve cylinders each of bore and stroke of 3⅟₁₆ in. (77·6 mm.) and 3¾in. (95·3 mm.) giving a cylinder capacity of 330 cu. in. (5·407 litres); it developed 140 to 150 B.H.P. at 1,900 R.P.M.

In this design two cylinder blocks are employed, each having six parallel cylinder bores. They are joined together by flanges and bolts. The pistons are cast in pairs connected by a solid web carrying two rollers. These are in contact with the cam plate which forms two sine waves (Fig. 175). One revolution of the engine shaft corresponds to four strokes of each piston or to a complete cycle of a four-cylinder unit. All valves are located in the cylinder heads, being slightly inclined to the cylinder axes so that the stems converge towards and are in contact with bevelled cam faces on a plate mounted on the mainshaft. Each

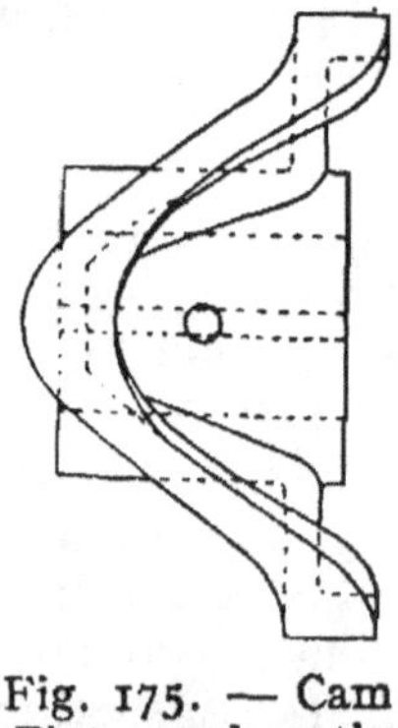

Fig. 175. — Cam Plate used on the Herrmann Barrel Engine.

plate has two cam lobes, the inner for the exhaust and the outer for the inlet valves. An annular induction manifold is formed round the main bearing having radial passages to the valve pockets, the combustion chambers being shaped to obviate any detonation tendency. The cylinders fire in direct sequence all round the block. Cylinder and valve dimensions and several auxiliaries such as the ignition unit are identical with the Ford V8 engine.

The dry weight of the engine, with cast iron cylinder blocks, without its carburettor was 384 lbs. It is estimated that with aluminium blocks and liners the engine weight could be reduced to 216 lbs. *i.e.*, about $1\frac{1}{2}$ lb. per B.H.P.

Another design of barrel engine known as the Alfaro, has been made by the Aircraft Development Inc., Boston, Mass., U.S.A. The engine has four cylinders of $2\frac{11}{16}$ in. (68·4 mm.) bore and $3\frac{3}{8}$ in. (80·6 mm.) stroke giving a displacement of 167 cu. in. (2740 c.c.). The cylinders are arranged parallel with the drive shaft and operate the latter by means of two cam plates. The two pistons of each cylinder are of the opposed type and are therefore balanced dynamically. The engine operates on the two-cycle, fuel-injection, spark ignition principle.

The pistons act on the cam plates by means of roller bearings, whilst the drive shaft rotates on single-row annular ball bearings. The inlet and exhaust are through ports in the cylinder walls, controlled by the pistons in a similar manner to the Junkers' engine. Fuel is injected at 2000 lb. per sq. in. Scavenging air is supplied from a centrifugal air compressor driven at about 10 times the drive shaft speed. The compression ratio employed is 9·5 : 1.

During tests made on this engine, using 87 octane fuel an output of 113 B.H.P. was obtained at 2,030 R.P.M., corresponding to a B.M.E.P. of 132 lb. per sq. in. The fuel consumption was 0·59 lb. per H.P. hour. Under standard atmosphere conditions 115 B.H.P. was given at 2,000 R.P.M. for a B.M.E.P. of 136 lb. per sq. in.

The engine complete with its generator and starter (weighing 29 lb. in all) had a dry weight of 269 lb., corresponding to 2.34 lb. per H.P.

The overall length of the engine is 3 ft. 8$\frac{9}{16}$ in. and the body diameter, 1 ft. 3$\frac{1}{2}$ in., excluding sparking plugs and other small projecting accessories.

This type of engine represents a compact design of low frontal area; actually, it gives a much lower frontal area than conventional car or aircraft engines of equal output.

Engines Using Gas as Fuel.* —Apart from the ordinary petrol and C.I. engines, which employ liquid fuels, there are automobile engines in use operating upon combustible gases, such as town lighting (or coal) gas and producer gas.

The use of the vehicles to which these systems are applied is limited chiefly to commercial goods carrying ores, where fuel costs are of primary importance.

Coal gas has been used on a small scale over a considerable period, more particularly for both goods vehicles and private cars during the Great War period, when petrol was both dear and limited in supply.

More recently a high-pressure gas system has been used, in which coal gas from town supply sources is first compressed to 5,000 lbs. per sq. in., and stored in strong receivers. It is used to supply commercial vehicles which are fitted for the purpose with light, high strength " Vibrac " alloy steel cylinders, with gas at 3,000 lbs. per sq. in. A typical vehicle installation would consist of six steel " bottles " connected by strong piping together and to a special pressure-reducing valve for supplying the engine with gas at a low pressure.

The engine used is the ordinary petrol type, but in place of the usual carburettor there is a gas inlet with a variable air supply or mixing valve. There is also a suction-operated device which shuts off the gas supply when the engine stops.

It has been shown that when using coal gas of 475 to 500 B.T.U.'s per cubic foot, calorific value, the

---

* A fuller account of these engines, with several illustrations. is given in " Carburettors and Fuel Systems," 4th Edition. A. W. Judge (Chapman and Hall, Ltd.).

overall running costs of an engine (including cost of compressing and storing gas) are equivalent to petrol about 8d. to 9d. per gallon, i.e., from one-half to two-thirds the petrol engine fuel costs.

Another type of vehicle.has its own gas generator in the form of a kind of " boiler " mounted on the chassis.   The fuel used is either anthracite, coke or wood.   Air is drawn through the ignited or glowing fuel in a smaller amount than necessary for complete combustion, and water vapour is also drawn into the heated fuel.   The resultant product drawn out of the " producer " is a gas containing hydrogen, methane, carbon-monoxide, nitrogen and usually a small amount of carbon-dioxide.   This gas has a high calorific value, and can be used in a similar manner to coal-gas.

A typical installation of the High-Speed Gas Company, Park Royal, London, has a vertical gas producer, supplying gas to an engine of 8 : 1 compression ratio, developing 55 B.H.P. at 2,200 r.p.m. Using a carbonaceous fuel at 30s. per ton (and incidentally, only ·06 pint of water per mile), the vehicle will cover 16 to 20 ton-miles at a speed of 20 to 25 m.p.h. for a cost of only a penny; this represents a saving of between 70 and 80 per cent. of the petrol engine fuel costs.

Petrol Injection Spark Ignition Engines.*—In this type of engine no carburettor is employed, the fuel being injected directly into the cylinder head during the compression stroke and is ignited towards the end of the latter by means of a sparking plug.   This type of engine does not operate on the compression-ignition principle and has a much lower compression ratio, namely, about 6·5 to 7·5 to 1.

Fig. 176 illustrates the cylinder head of the Allis Chambers engine, which employs a cavity type of piston.   The fuel-laden compressed air has a certain amount of turbulence and sweeps past the points of the sparking plug at the moment of ignition.

* A full account of these engines is given in " Carburettors and Fuel Systems," 4th Edition, A. W. Judge (Chapman & Hall Ltd.).

The Hesselmann engine shown in Fig. 177 also works on the same principle; it is widely used on American commercial vehicle and stationary engines.

The advantages claimed for the fuel-injection spark ignition engine are as follows:—(1) Better distribution of fuel and air (mixture) to the cylinders, since there is usually no inlet manifold since the air is drawn into each cylinder direct. (2) Higher volumetric efficiency on account of the absence of carburettor and inlet manifold; these cause a certain amount of obstruction to the mixture flow and result in the cylinders not

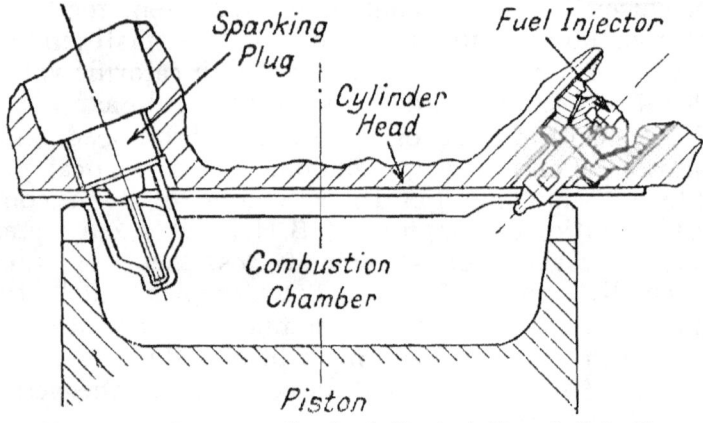

Fig. 176.—Cylinder Head of Typical Petrol Injection, Spark-Ignition Engine.

receiving their full charges. (3) Elimination of back-fires. (4) Freezing, or icing up troubles in very cold weather are avoided; in the case of aircraft engines this is a notable difficulty. (5) Better atomization of the fuel. (6) The use of alternative fuels of lower flash-point than petrol is practicable. (7) Absence of the modern complicated carburettor with its various automatic mixture compensation devices, float chamber, etc.

There is little doubt that the petrol injection system gives rather better distribution and a higher volumetric efficiency and that for certain purposes—notably in connection with the satisfactory use of lower grade fuels—it is preferable to the carburettor.

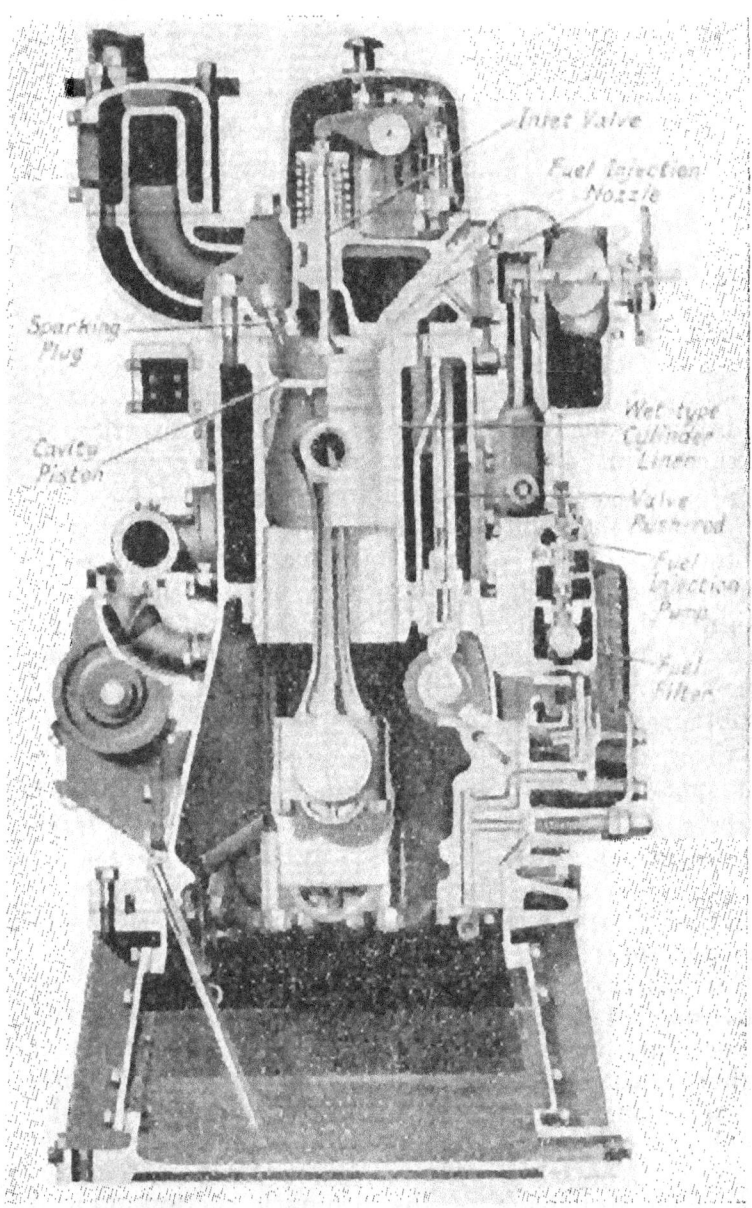

Fig. 177.—End Sectional View of the Hesselmann
Petrol·Injection, Spark-Ignition Engine.

CHAPTER IV.

Supercharged Engines.—The power output of an ordinary four-cycle petrol engine depends upon the quantity of mixture drawn into the cylinder, other conditions remaining the same. Now the quantity of mixture induced will be governed by its temperature and the difference between the pressure (about 14.7 lbs. per sq. in. absolute) outside and that within the cylinder. In the ordinary engine the pressure is atmospheric outside, so that no matter how well designed the engine is from the carburettor and valve ports point of view, there is a definite limit set to the power output for a given cylinder capacity. In order to obtain more power, it is necessary to supply air (or mixture) to the engine at higher pressure than atmospheric. The power developed increases roughly as the density of the air. A more exact relation is as follows:—

$$\text{H.P.} = k \, (\rho)^{1.1}$$

where H.P. is the horse power, ρ the air density, and k a constant.

For aircraft purposes supercharging, that is, delivering air at a greater pressure than atmospheric, is necessary in order to maintain the power output at higher altitudes, for the air density falls to about one-half its ground-level value at 20,000 ft. By means of the usual supercharging devices it is possible to obtain from 30 to 60 per cent. more power from an engine.

Another drawback of the ordinary suction-induced charge engine is that at the higher engine speeds less and less charge is drawn into the cylinders, owing to the increased frictional and other air resistance effects.

The supercharged engine, therefore, employs some form of air compressor or pump to force air at a higher pressure than atmosphere into the cylinders during the induction strokes, so that at the end of each induction stroke the cylinder has a greater weight of petrol and air charge. There is, hence, a greater amount of potential heat energy in the charge and, within certain limits, a greater amount of power is developed at the same engine speeds.

In practice the supercharging devices are arranged to give varying degrees of increased pressure to the cylinder charge (as measured just at the commencement of the compression stroke) from low supercharge pressures of 3 to 5 lbs. per sq. in. above atmospheric up to high supercharges of 25 to 30 lbs. per sq in. above atmospheric; the latter pressures are used for high output racing engines.

Boost Pressures.—It is now usual in supercharging practice to specify the degree of supercharging in terms of " boost " pressures, namely, the air pressures in the inlet manifold above atmospheric pressure. Thus a boost pressure of 5 lbs. per sq. in. indicates that the pressure in the inlet manifold or intake pipe is 5 lbs. per sq. in. above that of the outside atmosphere. At ground level the pressure in the inlet manifold would therefore be $14 \cdot 7 + 5 = 19.7$ lbs. per sq. in. Supercharging is generally employed in British aircraft engines to maintain ground power up to a specified height, known as the *Rated Altitude*; also to enable the engine to develop the extra power needed for air-craft take-off purposes.

Supercharging versus High Compression.—Of the possible methods for increasing the power output from an engine of given capacity, and fixed maximum speed there are two important ones that can be employed in practice, namely, (1) By the use of higher compression ratios. (2) By supercharging. In each case the result is to increase the brake mean effective pressure, and since the speed has been assumed constant the power output will increase directly as the mean pressure.

The disadvantage of the high compression method is that with the increase in mean pressure obtained the maximum pressure of combustion also increases. With supercharging however, as the pressure of the super-charged mixture in the cylinder at the commencement of the compression stroke increases the maximum cylinder pressures do not increase at the same rate and therefore a much higher value of B.M.E.P. can be obtained for much lower values of the maximum cylinder pressure.

Another important factor is that of cylinder temperatures which are generally lower for a given B.M.E.P. for supercharged engines than for high compression ones; further, it is not so necessary to employ fuels of such high octane values with normal types of supercharged engines.

The results of some tests made on supercharged engines have shown that in a given instance an engine having no supercharge gave a B.M.E.P. of 110 lbs. per sq. in., a maximum cylinder pressure of 495 lbs. per sq. in. and maximum combustion temperature of 4400° Fah. (abs.)

With a supercharge of 25 per cent. above atmospheric, the B.M.E.P. and maximum pressures were 132 and 600 lbs. per sq. in. respectively, with a maximum temperature of 4525° Fah. For a 50 per cent. supercharge the corresponding pressures were 162 and 735 lbs. per sq. in., respectively, with a maximum cylinder of 4590° Fah.

When the supercharge was 100 per cent. the pressures were 195 and 1000 lb. per sq. in., respectively with a maximum temperature of 4675° Fah.

It will be noted that the temperatures increase at a relatively slow rate.

An interesting comparison between the supercharged and high compression pressure types of engines giving about the same power output from a cylinder of given bore and stroke at a given speed and mixture strength has been made by A. H. R. Fedden of the Bristol Aeroplane Company. For the purposes of comparison the performance of the unsupercharged engine was also considered and the following results obtained:—

Comparison between Supercharged and High Compression Engines.

	Unsupercharged Engine	High Compression Engine	Supercharged Engine
	A	**B**	**C**
Compression Ratio	5 : 1	11 : 1	5 : 1 (Supercharged to 50 per cent.)
Indicated Mean Effective Pressure (lbs. per sq. in.)	126	154	157
Maximum Cylinder Pressure (lbs. per sq. in.)	455	1000	595

These results show clearly the advantage derived from supercharging for the given power output as represented by the I.M.E.P. is obtained for a maximum pressure of only 595 lb. per sq. in. as compared with 1000 lbs. per sq. in. for the high compression engine. The latter would have to be built much heavier in order to withstand the higher pressures.

In regard to the degree of supercharging employed in the engine in question it should be explained that the original pressure in the cylinder for the unsupercharged engine was 14·7 lbs. per sq. in. and for the 50 per cent. supercharge, 22·0 lbs. per sq. in. The effect of supercharging, however, results in a higher compression pressure, the respective values for the three engines A, B and C given in the preceding table, being approximately, 140, 265 and 210 lbs. per sq. in., respectively.

Power for Driving the Supercharger.—In most supercharged systems the air compressor is driven from the engine crankshaft by means of a gear train or other means of transmission, so that at each engine speed a certain amount of the extra power developed as a result of the supercharging is absorbed in driving the compressor and this power must therefore be deducted from the total power developed in order to give the net or useful power of the engine. On the other hand the pistons of the engine during the "suction" stroke do not have to draw air into the cylinder but are actually under the air pressure effect of the supercharger so that they perform useful work.

Gear-driven superchargers of the centrifugal and blower types absorb from 5 to 10 per cent. of the total power output to drive them so that this power must be subtracted from the total power in order to obtain the net or useful output. *

When the supercharger is of the exhaust turbine driven centrifugal compressor type the engine does not have to supply additional power to drive the compressor and the total power developed therefore requires no deduction on this account. The system in question is therefore the more efficient one. In this connection it has been shown that for aircraft engine purposes exhaust-driven superchargers—although introducing certain practical difficulties in connection with the high temperatures involved in the exhaust system—enable an aircraft fitted with a given engine to attain higher altitudes than for the gear-driven supercharged engine of similar dimensions.

Fuel Consumptions of Supercharged Engines.—Although it is necessary to use a greater quantity of fuel per unit time with supercharged engines, yet owing to the greater power outputs obtained the amount of fuel used per B.H.P. per hour is usually only slightly greater than for the same engine unsupercharged. In certain designs of gear-driven supercharger the fuel consumption per B.H.P. per hour has actually been rather less than for the unsupercharged engine. Further, with the exhaust gas turbine supercharger, owing to the utilization of some of the energy of the exhaust gases that, otherwise, would be wasted, appreciably lower fuel consumptions have been obtained than for the gear-driven and also the unsupercharged

*A full account of supercharging is given in "Aircraft Engines." Vol. 1. A. W. Judge (Chapman & Hall Ltd.) 1942.

engine of similar dimensions and nominal compression ratio.

Automobile Engine Results.—Fig. 178 illustrates the effects of supercharging the Auburn eight-cylinder passenger model car engine. The engine in question is fitted with a centrifugal air pump, gear and friction driven at six times the crankshaft speed, i.e., at speeds up to 24,000 r.p.m.

The full lines in Fig. 179 show the B.M.E.P.. B.H.P. and engine torque values of the unsupercharged engine, whilst the dotted line diagrams indicate the improvements obtained by supercharging. It will be seen that the horse power is increased over the whole speed range, and the maximum B.H.P. obtained by supercharging is about 148 as against 112 for the unsupercharged engine.

Another interesting point shown by these curves is that the average crankshaft turning effort or torque is increased over the whole range, the maximum value of the torque being about 230 lbs. ft. as against the previous value of 210 lbs. ft.; moreover, the torque curve is flatter over the speed range so that the top gear performance of the engine will be better.

It is of interest to note that the Auburn passenger cars fitted with superchargers are certified to do at least 100 miles per hour.

In some cases the supercharger is provided with a clutch so that it need not be used at low to medium engine speeds, but only to increase the power output above the medium speed values for acceleration, improved hill-climbing and top-gear performance; the clutch is

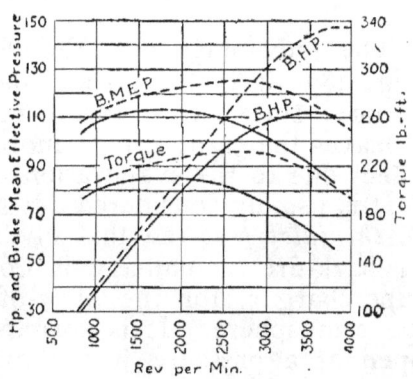

Fig. 178.—Effects of Supercharging on Performance.

then usually inter-connected with the throttle so as to come into operation after the throttle has opened a certain amount.   One model of the Mercedes car employed such an arrangement.

Position of Supercharger.—The supercharger can be fitted in two alternative positions, namely: (1) between the carburettor and the engine; (2) outside, with the carburettor between itself and the engine (Fig. 179). The former arrangement (Fig. 183) is that more generally preferred, for, in the latter case, owing to the increased air pressure on the petrol jet in the carburettor, it becomes necessary to fit a compensating device to increase the pressure in the float chamber. This is done by connecting the top of the float chamber (which is otherwise sealed) to the carburettor (compressed) air-inlet pipe.

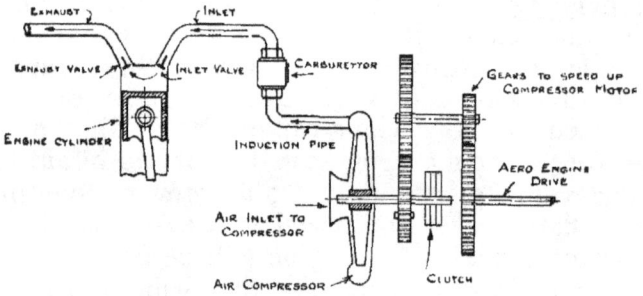

Fig. 179.—The Gear Driven Centrifugal Air Pump type of Supercharger.

The advantage of the arrangement shown in Fig. 183 is that a normal design of carburettor can be used, since the mixture is not compressed until after it leaves the carburettor; moreover, there is a cooling effect due to the evaporation of the fuel.

It is usually considered advisable to fit *a safety-valve in the inlet pipe,* so that in the event of a blow-back or back-fire no damage is done to the inlet pipe or supercharger, for the blow-off valve opens direct to the atmosphere.   It is spring-loaded and adjusted to open at a pressure just above the maximum supercharge value.   In certain instances, however, the designer omits this blow-off valve, relying upon the fact that one of the inlet valves is usually open and can release the excessive pressure into its cylinder.

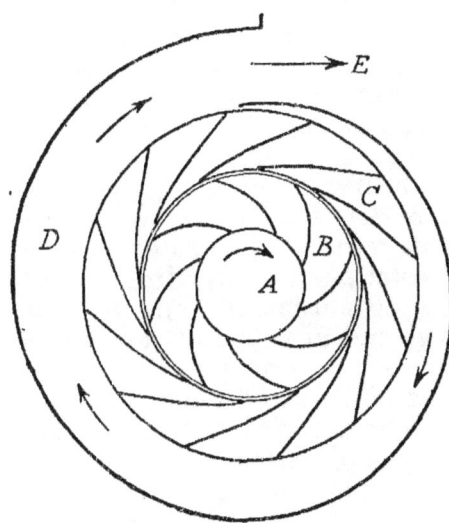

Fig. 180.—Principle of the Centrifugal Supercharger.

Another factor that is generally taken into consideration is the *increased temperature of the charge* due to its heat of compression. This surplus heat tends to reduce the quantity of charge forced into the cylinder, and it is necessary to get rid of it by cooling the charge between the supercharger and the engine. Aluminium inlet manifolds provided with air-cooling fins or intercooler units are used for this purpose.

Types of Supercharger.—There are three principal classes of air-compressor employed for supercharging engines, as follows: (1) The *Centrifugal Air Pump;* (2) the *Eccentric Drum* and *Vane Pump;* and (3) the *Rolling Drum* or *Roots' Blower.* The ordinary reciprocating air pump is not used on high-speed petrol engines on account of its size and weight.

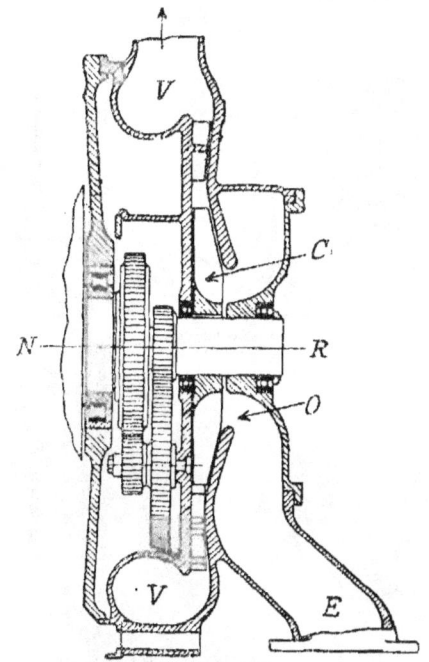

Fig. 181.—Centrifugal Compressor Gear Drive.

The principle of the centrifugal air compressor is illustrated in Fig. 180, which shows a rotor A carrying a vaned portion B; this rotates within a casing containing a set of fixed vanes C, the latter being mounted within an outer casing D of gradually increasing cross section, *i.e.*, it is of volute or spiral shape, the largest section being at the exit portion E. The air inlet to the compressor is at the centre A, the air flowing axially into this region. The rotor—which requires high speeds of rotation, namely, from about 18,000 to 25,000 R.P.M. for automobile and aircraft engines—draws air from A and in virtue of the centrifugal effect on the air forces it outwards so that at the tips of the rotor B the air is delivered under pressure tangentially to the outer circular periphery of B and then enters the set of vanes C, which are arranged tangentially and give an expanding section towards their outer periphery. These are termed " diffusers " and their purpose is to convert the velocity head of the air into pressure head, by slowing down the velocity and thus building up the pressure. The air is then delivered into the volute chamber D in increasing volume, which the gradually expanding casing of the compressor accommodates.

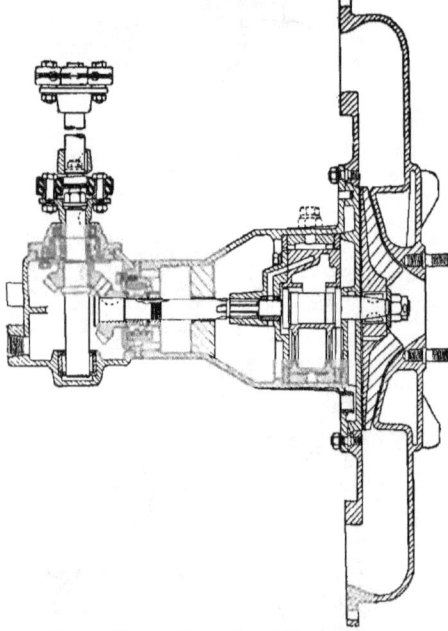

Fig. 182.—Showing the Auburn Centrifugal Supercharger and Bevel Gear Drive.

This type of compressor is usually driven by a train of gears such as that shown in Fig. 181. The engine crankshaft N carries a larger gear which meshes with the smaller one, shown immediately below, and a larger gear on the same shaft as the latter meshes in turn with a smaller one at the end of the rotor shaft R, so that a double step-up in gear ratio is thus obtained. This arrangement enables the rotor shaft axis to be in line with that of the crankshaft. The air or mixture from the carburetter below is drawn in at E and thence enters the central opening O of the rotor casing. It is discharged into the volute chamber V and finally leaves the latter at the upper end as indicated by the arrow; the compressor rotor is shown at C.

A single stage compressor of this type will give from 5 to 8 lbs. per sq. in. pressure, but if higher pressures

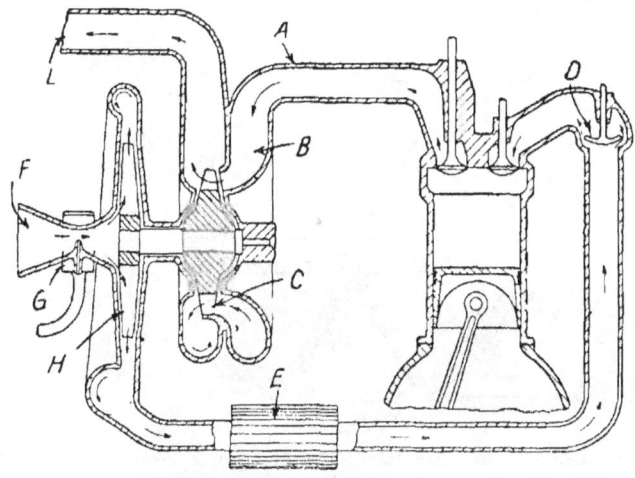

Fig. 183.—Exhaust Driven Supercharger with Outside Carburettor.
A—Exhaust gas outlet.   B—Turbine casing.   C—Turbine rotor.
D—Automatic blow-back prevention valve.   E—Mixture cooler.
F—Air entry to carburettor G.   H—Centrifugal pump rotor.
Note.—The arrows show the paths of the compressed mixture and of the exhaust gases.

are required two or more stages must be used, each rotor unit discharging into the central inlet of the next one, thus building up the pressure.

Another method employed on aircraft is to utilize the energy of the exhaust gases to drive a high-speed turbine, known as the Rateau type (Fig. 183). The rotor shaft of this turbine also forms that of the centrifugal pump so that the latter is driven at the same speed as the exhaust gas turbine, viz., at 25,000 to

30,000 r.p.m. The turbine in question consists of a narrow wheel tapering inwards from the centre to the periphery—in order to give the necessary dimensions for strength purposes—and provided with curved blades around its periphery. The high velocity exhaust gases meet and pass through this blade system, thus giving up some of their velocity energy to the turbine wheel.

There is an increased exhaust back pressure when a Rateau turbine is used, but the practical difficulties associated with this effect have been surmounted; in particular the cooling of the exhaust valves has been improved considerably.

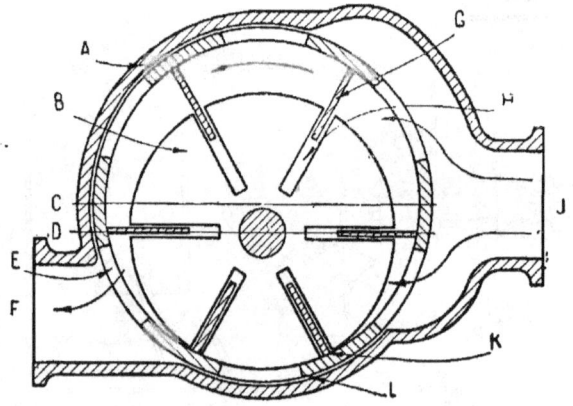

Fig. 184.—The Cozette Supercharger. A—Outer casing. B—Slotted rotor. C—Centre line of outer casing. D—Centre line of rotor. E—Perforated rotating drum. F—Delivery port. G—Sliding blade. H—Blade slot in rotor. K and L— Air tight portion of drum and rotor.

The second type of supercharger previously mentioned operates on the well-known eccentric drum and moving-vane principle. It has been used for car engines in this country, and is represented by the Cozette and Powerplus models illustrated.

The ordinary moving vane eccentric drum type of pump is not suitable for high-speed car engines, on account of the friction and wear of the blade ends. The modern eccentric drum pumps embody improvements for obviating wear of the blade ends and minimising the wear in the drum slots; ball-bearings for

the rotor, effective end seals and an adequate lubrica
tion of the moving parts are other improvements.

It is usual to mix oil with the petrol when this type
of supercharger is fitted, unless, of course, an indepen
dent lubrication device is employed for the rotor
blades. The bearings are almost invariably lubricated
from the engine system.

The Cozette model has a barrel eccentric with the
drum, and a revolving slotted casing. Six radial blades
are held against the inside of the latter by centrifugal
force and rotate with this casing, moving in and out
of their guides in the revolving drum as they rotate.

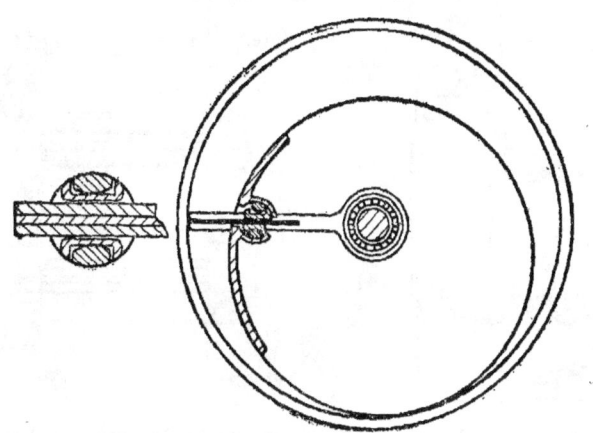

Fig. 185.—The Powerplus Supercharger, showing
one Oscillating Blade Slot.

Fig. 184 illustrates, in sectional view, this type of supercharger. It is possible to obtain increased pressures
above atmospheric up to about 10 lbs. per sq. in. with
rotory compressors of this type, when driven at engine
speed and well lubricated.

The Powerplus supercharger (Fig. 185) works on the
same principle, but utilizes blades of the oscillating slot
type to minimize slot wear; the blades are always
radial to the lay-shaft; further, there is a fixed clearance
between the blade and the casing.

The third class of air-compressor, known as the
Root's type, is a very old idea that has been modern-
ised in design for higher speed operation. In its

simplest form it consists of a pair of rotors, each of a double-loop or "figure-8" section, the shapes of these rotors and positions of their axes of rotation being so arranged that, no matter what the positions of the rotors, they are always in line contact. They are driven positively by gearing so as to rotate in opposite directions and are provided with a casing which practically touches the rotors. The rotors rotate at about 1½ times engine speed.

Thus, as the rotors revolve—the top one clockwise and the bottom one anti-clockwise—they draw in the air (or charge) from the left-hand opening (Fig. 186) and discharge it on the right-hand side.

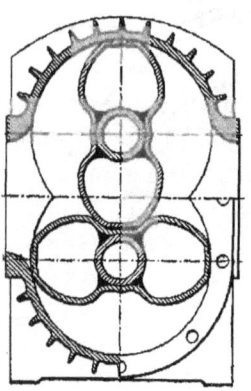
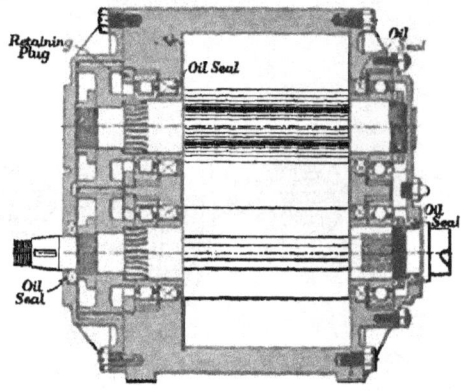

Fig. 186.—The Roots Supercharger in Cross Section.

Fig. 187.—Side sectional view of Roots Supercharger shown in Fig. 186.

The Marshall-Drew design of Root's supercharger, shown in Figs. 186 and 187, has rotors ground to involute tooth form, the tips being grooved to improve the air sealing. The rotors run in ball and roller bearings and are driven by spur gears. The outer casing is of aluminium or magnesium alloy, and provided with cooling fins.

The supercharger illustrated measures 9 × 8 × 5 inches, will deliver 60 cu. ft. of free air per minute, and give a charge pressure of 5 lbs. per sq. in. above atmospheric, at 3,000 r.p.m. This supercharger is suitable for an engine of 52 cu. in. (852 c.c.), and will increase its power by about 30 to 40 per cent.

The clearances between the rotor and casing are kept very small, being of the order of ·003 to ·004 inch. The gears are lubricated from the engine, but no lubrication is required for the rotors.

Another design of Root's compressor uses three-lobed rotors as shown in Fig. 188. This pattern gives a more uniform pressure and better sealing of the charge in its passage through the compressor.

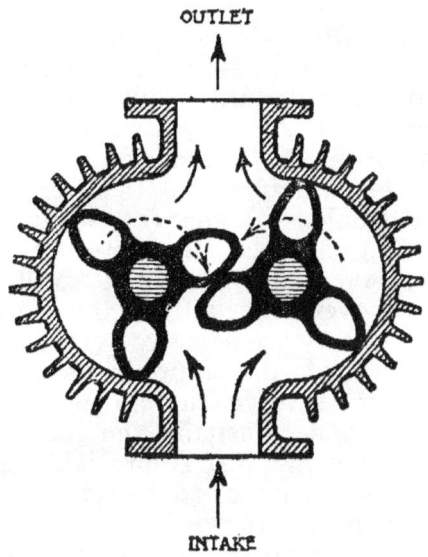

Fig. 188.—The Roots Type Supercharger, with Three Lobe Rotor.

1

CHAPTER V

Two-Cycle Engines.—This type of engine, the principle of which is described on pp. 33 to 36, has always been an attractive one to inventors, since it has a higher output for a given cylinder capacity, gives a more uniform torque, is lighter, and in general less complicated to manufacture and therefore cheaper to produce. The smaller number of working parts is also an advantage. The majority of two-stroke engines which have been produced to date do not appear to have overcome all of the disadvantages associated with this type, namely: (1) Lower m.e.p. than in the case of the four-stroke, thus giving appreciably less than twice the power output for a given cylinder capacity; usually the increase is only from 40 to 60 per cent. (2) Appreciably higher fuel consumption, namely, from 20 to 40 per cent., due usually to an escape of charge into the exhaust when both inlet and exhaust ports are open together, and to the lower compressions which are used. (3) Tendency to overheat if run on open throttle for long periods, due to almost twice as much heat being generated within the cylinder as in the case of the four-stroke. (4) Want of flexibility, the result being a much lower speed range as a two-stroke; at low speeds this engine tends to run as a four-stroke and thus loses part of its possible power.

Although a large number of different designs of two-cycle engines has been patented, most of these engines can be classified into the following categories: (A) Crankcase compression. (B) Double piston. (C) Differential piston; and (D) Separate compressor type engines.

(A) Crankcase Compression Engines.—This class includes the simplest and, incidentally, the cheapest types of engines as used on small motor-cycles and in some cases light cars. In this design the under-side of the piston is employed to draw the mixture into the crankcase—as the piston ascends—and then to compress it on the down-stroke of the piston.

Suitable ports are arranged in the cylinder walls, which are uncovered by the top edge of the piston when the latter is near the bottom of its stroke, for the exhaust escape and also for the transfer of the mixture from the crankcase to the upper side of the piston. In all cases the piston on its next ascent closes, first the transfer and then the exhaust ports,

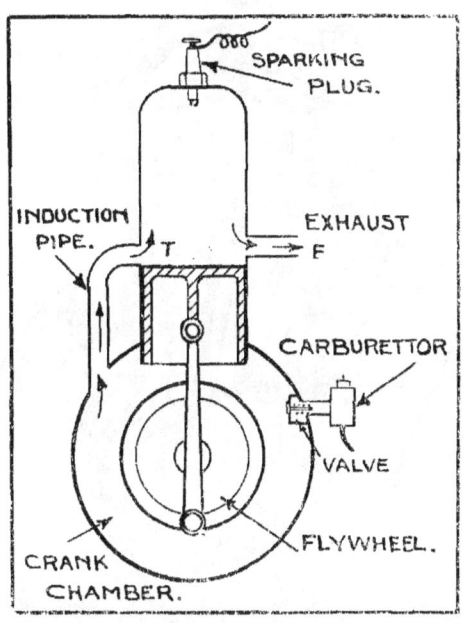

Fig. 189.—The simple Two-port Two-cycle Engine.

thus sealing the cylinder for the mixture compression stroke.

The various designs of two-cycle engine in this class differ in the arrangement, number and dispositions of their ports. They are known, respectively, as two-port, three-port, four-port and six-port engines according to the equivalent number of ports in the cylinder wall and in some cases, the piston.

The principle of the two-port engine is illustrated in Fig. 189. In this example the carburettor is connected with the crank chamber by means of a port having a non-return valve. The upward stroke of the piston draws the charge through this port into the crankcase, where it is compressed on the next down stroke and, as the piston uncovers the transfer port T, the charge flows into the cylinder, there to be compressed by the piston on its next ascending stroke. The exhaust port E is uncovered by the piston towards the end of its next descending (or firing and expansion) stroke.

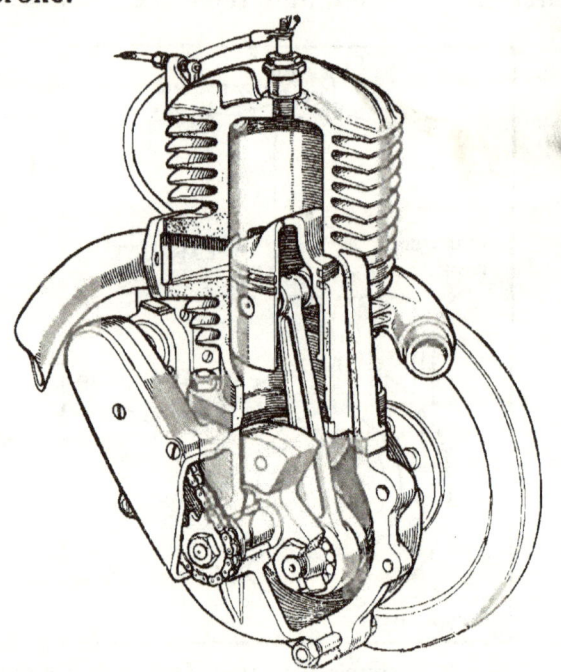

Fig. 190.—Part sectional view of Three-port Engine.
(Courtesy *Motor Cycle*.)

This type although extremely simple and cheap to build is inefficient as the cylinder cannot be given its full charge, owing to the limited period of the charging stroke and to the escape of part of the charge through the exhaust port.

The three-port engine, the operation of which is described on page 33, represents an improvement on

Fig. 191.—The Levis Six-port Two-cycle Engine.

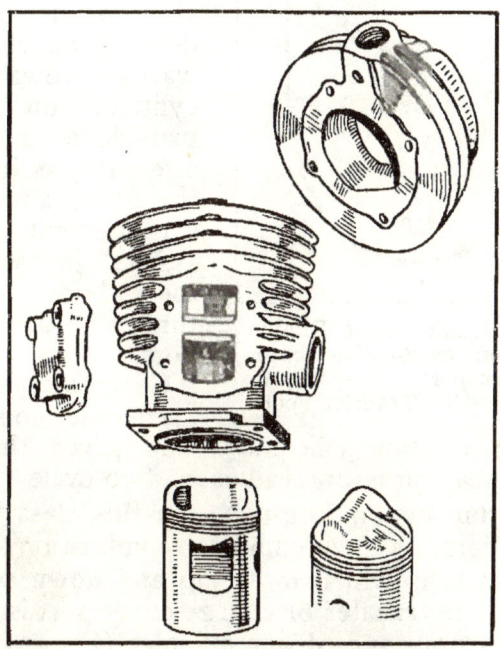

Fig 192.—Showing components of the
Levis Six-port Two-cycle Engine.

the two-port type, mechanically, but its volumetric efficiency is low.

The six-port engine (Figs. 191 and 192), of which the Levis is an example, employs a port in the piston itself to introduce the crankcase charge into the transfer port; the chief object of this is to allow the ingoing charge to cool the piston crown—as this part becomes hotter in two-cycle than in four-cycle engines. This cooling effect, however, is obtained at the expense (partly) of heating the charge, and thus reducing the volumetric efficiency.

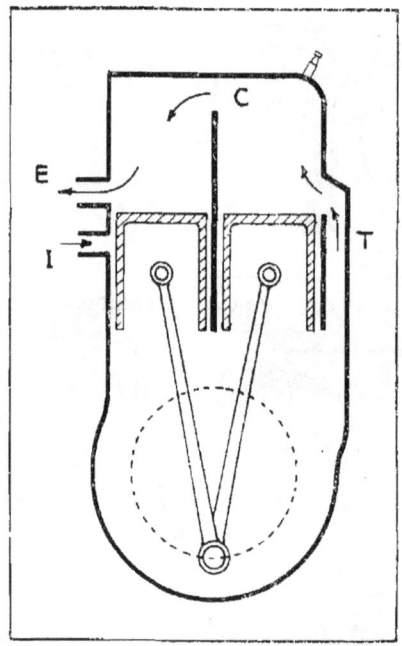

Fig. 193.—Double Piston Engine. C—Common combustion head. E—Exhaust port. I—Mixture inlet port. T—Transfer port.

Practically all engines of the class described have pistons provided with deflector crowns, the object of the latter being to deflect the charge upwards towards the cylinder top so as to provide a longer leakage path, as it were, to the exhaust ; this arrangement a l s o assists in scavenging the burnt g a s e s through the exhaust port.

Generally speaking the crankcase compression engine is of low efficiency, and gives the lowest cylinder mean pressure values of two-cycle types.

(B) Double-Piston Engines.—In this class there are two cylinders, with a common combustion chamber, and two pistons which move up and down practically together. The transfer or charge entry port is arranged near the bottom end of one cylinder (Fig. 193), whilst the exhaust port is near the bottom of the other cylinder. This arrangement it will be seen places the transfer and exhaust ports as far away as possible, so that the incoming charge sweeps right through the two

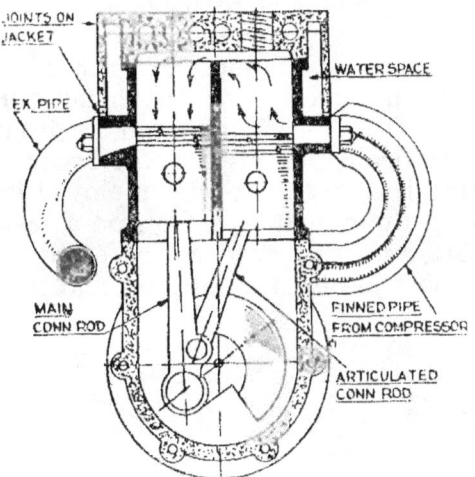

Fig. 194.—The D.K.W. Two-cycle engine.

cylinders, always in the same direction, i.e., there is a uni-directional flow of the charge.

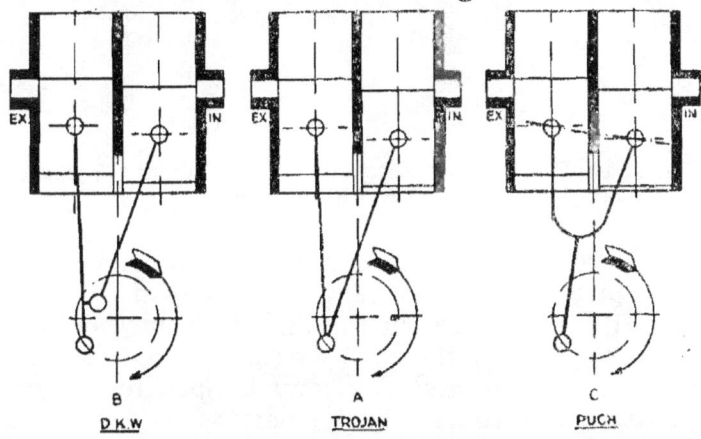

Fig. 195.—Typical examples of double piston engines.

Some typical engines of the double-piston double-cylinder pattern are the D.K.W., Trojan, Puch and Zoller. Fig. 194* illustrates the general layout of the

*English Mechanics.

D.K.W. engine as made in the form of a small working model and shows the special arrangement of the
connecting rods employed to enable one piston to be
in advance of the other for more effective scavenging
and charging; it will be noted that the right hand connecting rod is hinged to the left hand one for this
purpose. The D.K.W., Trojan and Puch engines are
shown, schematically, in Fig. 195 in order to indicate
the connecting rod arrangements employed for the purpose of giving the exhaust port piston a lead over the
charging port one; this ensures the earlier closing of
the exhaust port before the inlet or charging port.

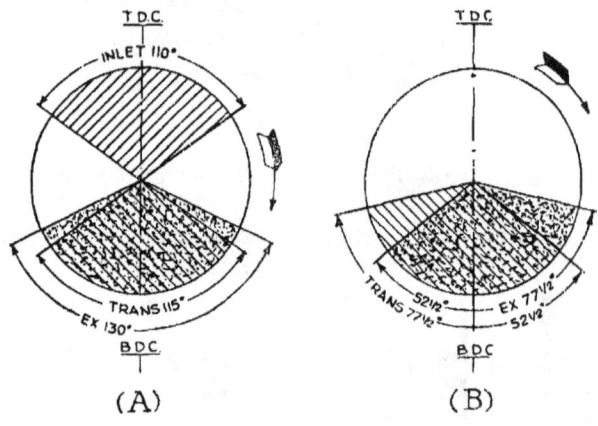

Fig. 196.—Port Timing Diagrams. (A) Ordinary Engine.
(B) Double Piston Engine.

The port timing diagrams of an ordinary and a
double piston two-cycle engine given in Fig. 196
indicate the differences in the timing arrangements of
the two types. In the later engine the exhaust port
opens at $77\frac{1}{2}°$ before B.D.C. and is open for a crank
angle of $25°$ before the transfer port opens. The latter
remains open for a total angle of $130°$ and the exhaust
port for the same period or angle. After the exhaust
port has closed the transfer port remains open for
another $25°$. This arrangement gives a longer cylinder
charging period and a lower charge loss than that of
the ordinary two-cycle engine.

Instead of having the two cylinders side-by-side, a long straight cylinder having two pistons working in opposite directions can be used, as in the Junkers and Werry engines. In this case the combustion chamber is formed by the space left between the pistons when the latter are nearest together. The inlet and exhaust ports are arranged at opposite ends, so as to obtain a uni-directional effect.

Whichever arrangement is employed, the principle is the same. The double - opposed piston scheme necessitates two crankshafts, which must be coupled together by connecting-rods or gearing.

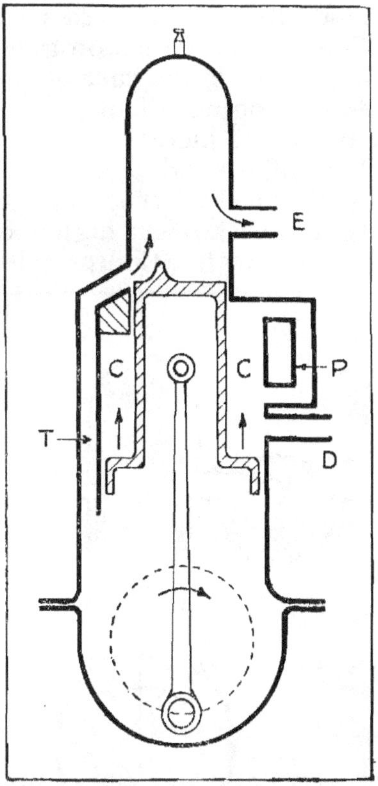

The uni-directional method can be used with crankcase compression or with a separate pump for compressing the charge.

Fig. 197.—Differential-Piston Engine. C—Compression chamber. D (on right) Mixture inlet E—Exhaust port. P—Transfer port for compressed mixture to crankcase. T—Transfer port to cylinder.

Owing to the better scavenging and to the fact that less charge is lost through the exhaust, this type of two-cycle engine is more efficient than that previously described.

(C) **The Differential Piston Engine.**—In this class of two-cycle engine the piston is made with two parts of different diameter, the smaller portion corresponding to the piston of an ordinary engine; the larger portion works in its own enlarged cylinder and is used to draw in and compress the fresh charge (Fig. 197).

I*

Usually this charge is then delivered to the crankcase, whence it is transferred to the smaller cylinder.

The differential piston gives a higher pressure to the charge than in the case of the ordinary crankcase compression engine. Thus, the Dunelt engine gives about 50 per cent. increase in charge pressure.

The differential piston engine is larger and heavier than the normal type; but it definitely gives a bigger output and shows a higher overall efficiency.

(*D*) **Separate Compression Type Engine.** — The charge, in this type of engine, is forced into the cylinder by means of a separate pump, so that crank-case compression is obviated.

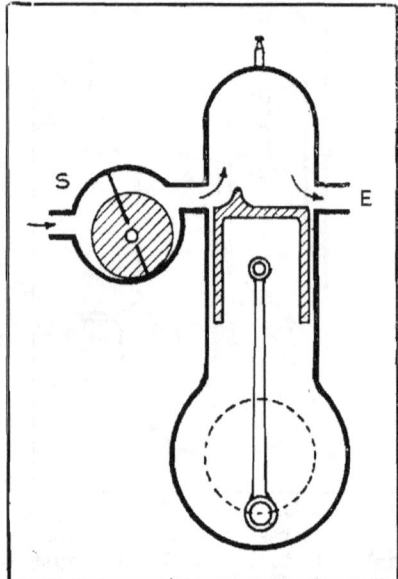

Fig. 198.—Supercharged Two-cycle Engine (Two-port)
E—Exhaust. S—Supercharger.

There are many variations of this method, although the principle is the same. For example, a reciprocating or rotary air pump can be used to force the charge either through the usual ports at the bottom of the piston's stroke, or through a mechanically-operated or simple non-return valve in the cylinder head.

Similarly the engine can be of the double-cylinder type previously described, and the air-pump can deliver the fresh mixture through the inlet port at the base of one cylinder, whilst the exhaust gases are ejected through the port at the base of the other cylinder. This method is used in the Junkers engine.

The principal advantage of the separate pump method is that the engine can be supercharged, so that a greater amount of power can be obtained from a given size engine. In order to obtain the best results, some method of scavenging the burnt gases is **necessary**; this also assists in cooling the cylinders.   This can be done by the supercharger although sometimes a separate scavenging air-pump is used, the mixture being introduced afterwards.

The small air-cooled two-stroke engines, such as the Enfield 2¼ H.P. described on p. 33 can be effectively

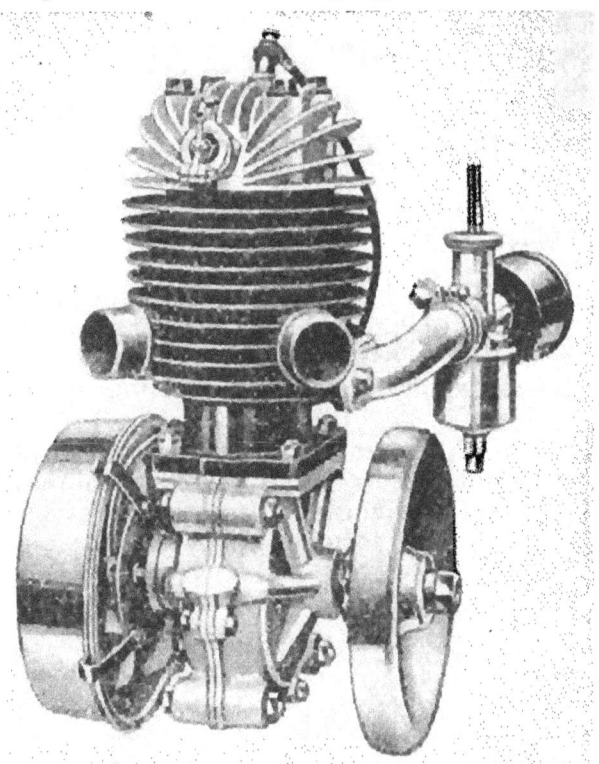

Fig. 199.—The Villiers Two-Stroke Engine, showing the aluminium cylinder head, and twin exhaust ports.

cooled, and if properly designed give quite satisfactory results, but for larger units special attention must be paid to the cooling, fuel consumption and power output if these are to compete seriously with the highly developed four-stroke types.

Fig. 199 illustrates a typical motor-cycle two-stroke engine, namely, the 3½ H.P. Villiers.   This engine has a cast iron cylinder fitted with a detachable aluminium alloy head well ribbed to ensure adequate cooling. The crank-case unit is also of aluminium alloy. The engine is fitted with a light alloy piston with floating gudgeon-pin. The connecting-rod has a full roller-bearing big-end.

The bore and stroke of the model illustrated are 70 mm. and 90 mm., respectively, giving a capacity of 346 c.c.

An interesting feature of the Villiers engine is the patent inertia ring fitted above the top piston ring to prevent the other rings from gumming up. The inertia ring does not actually touch the cylinder walls, but can rotate freely and has a slight up-and-down movement, thus preventing any film of oil forming above the other rings.

The engine is fitted with a decompressor valve (seen in the centre of cylinder head) that can be operated from the handle-bars of the motor-cycle; its object is to reduce the compression for starting or slowing down.

Lubrication of this engine is by the Petroil method, whereby the lubricating oil is mixed with the petrol and thus enters the engine through the carburettor. The petrol and oil consumptions when the engine is fitted to a motor-cycle are about 70 and 1,800 m.p.g., respectively.

Air Scavenging Considerations.—An important requirement for the satisfactory performance of the two-cycle engine concerns the elimination of the burnt gases remaining inside the cylinder when the piston is near the lower end of the firing stroke and the exhaust port or valve is opened. The fresh mixture or a separate air charge is employed to get rid of these gases, or scavenge the cylinder, but the scavenging operation must be carried out during the very limited period available for this purpose. Thus, in the case of an engine operating at 1,800 R.P.M., with a scavenging period of say 100° of crank angle—and this is a liberal allowance—the whole process of scavenging must be completed within a period of 0·009 sec. Unless the residuals are effectively scavenged the fresh air charge will be reduced in amount and the engine will not develop its full power.

It is not possible to scavenge efficiently engines of the type shown in Fig. 198, the only satisfactory method being to arrange for the air scavenge (or charging) and the exhaust ports to be at opposite ends of the cylinder, so that uniflow scavenging is obtained.

Referring to Fig. 200*, the arrangement shown at A is equivalent to that previously given in Fig. 196

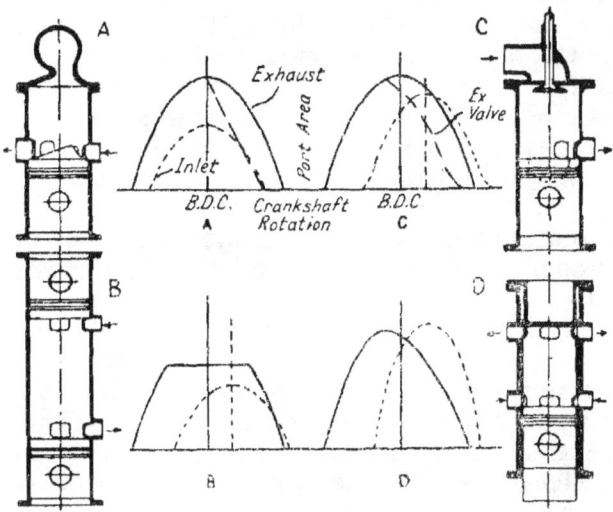

Fig. 200.—Types of Air Scavenged Two-Cycle Engines and Port Opening Diagrams.

for exhaust and scavenge ports near the bottom of the cylinder. The corresponding air inlet and exhaust port opening diagrams are shown to the right of the cylinder diagram. Although cheaper to manufacture than other designs, this type of engine suffers from the disadvantages of low volumetric efficiency and higher fuel consumption previously explained.

The most promising alternative designs are those illustrated in diagrams B, C and D, Fig. 200; in each of these uniflow scavenging is provided for. The method used in diagram B employs two opposed pistons in the same cylinder. One piston controls

*High Speed Two Stroke Oil Engines. P. E. Biggar.
Autom. Engr., April, 1930.

the exhaust ports at its end of the cylinder, whilst the other piston governs the air inlet ports at its end. This arrangement, which is used on the Junkers' motor vehicle and aircraft two-cycle C.I. engine, permits of efficient uniflow scavenging and avoids excessive loss of scavenging air since the exhaust port controlling piston is given a lead over the inlet port and closes just before the latter—as shown by the port opening diagram; the residual exhaust gases are thus reduced to a minimum.

The arrangement shown in C (Fig. 200) has the air inlet ports at the lower part of the cylinder and a poppet type exhaust valve in the cylinder head; as in the case of the opposed piston engine, the exhaust valve opens well before the inlet ports—which are, of course, piston-controlled—and it closes in advance of the inlet ports. This system provides for much better piston and exhaust port cooling than would be possible if the upper valve were the air inlet and the lower ports the exhaust ones.

The engine shown, diagrammatically, at D (Fig. 200) represents a single-sleeve valve engine, with the exhaust ports near the top of the cylinder controlled positively by the sleeve valve; the inlet ports near the lower end of the cylinder are also controlled positively by the sleeve valve. The exhaust has a lead over the inlet, as shown by the port opening diagram to the left in Fig. 200, and it is possible with this controlled exhaust and inlet port to obtain the most efficient scavenging action.

It has been shown, experimentally, that the power developed by a two-cycle engine increases with the amount of air supplied for scavenging at any given speed. There is, however, a limit to this quantity of air since the size of the compressor and the power required to drive it both increase with the amount of air delivered and with the pressure of the air supply.

For a satisfactory performance the quantity of air delivered by the compressor for scavenging is from $1\frac{1}{2}$ to 2 times the cylinder volume for an air pressure supply of 4 to 6 lb. per sq. in. above atmospheric pressure; the pressure in question seldom exceeds 7 lb. per sq. in.

The power required to supply air to the cylinders under these conditions, allowing for the unavoidable wastage of 25 to 30 per cent. through the exhaust ports, is of the order of 10 to 15 per cent. of the total or gross output of the engine. In general, for a direct-driven air compressor, about 1/60th of the gross output will be absorbed for each lb. per sq. in. of

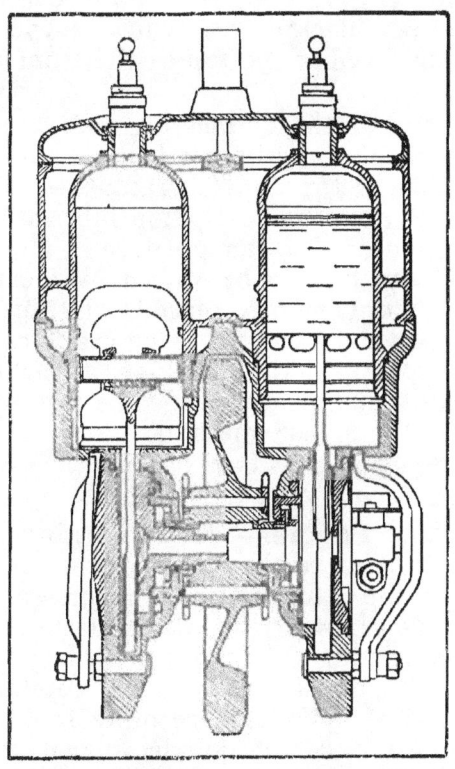

Fig. 201.—The Scott Two-Cylinder Two-Stroke Engine, with central flywheel and chain sprockets.

scavenging air pressure. The weight of the air compressor, using modern light alloys and alloy steels, would be about 0·1 to 0·25 lb. per B.H.P.

The Scott Engine.—This engine, which has been fitted to the well-known Scott motor-cycle over a long period is a three-port. two-cycle, two-cylinder inclined model with a central fly-wheel and cranks at 180° apart.

In the earlier models (prior to 1930) the chain sprocket for the rear drive was placed between the two crank-cases, as shown in Fig. 201.

In the more recent model the fly-wheel is enclosed, four bearings being provided instead of two, for the crank-shaft; the chain sprockets are now placed on the outside of the crank-case.

The Scott engine employs water-cooled cylinder jackets, a neat design of twin honeycomb-pattern radiator being fitted at the top of the front frame down tube.

The big-ends of the connecting-rods are fitted with roller bearings, the crankshaft being built up to allow of its assembly in the double crank-case. The engine employs positive mechanical pump lubrication, two oil-pumps being fitted for the purpose. The amount of oil delivered can readily be varied by means of regulating screws conveniently placed. In this system oil is delivered under pressure to each crank-case individually through ports in the crank-shaft packing glands, embodied in the main bearings. The internal parts are lubricated by the splash method.

For racing purposes an auxiliary hand-operated cylinder wall oiling system is fitted.

The Trojan Engine.—An interesting two-stroke engine is that fitted to the Trojan light car (Fig. 202); the design of the latter is quite unconventional. The engine consists of two pairs of water-cooled cylinders, each pair consisting of two parallel cylinders communicating with one another at the combustion ends. The two pairs of cylinders are quite separate, and lie side by side. Each pair of cylinders has its own pistons, but there is a common connecting-rod of Y-shape terminating in a single big end. The two cylinders act as a single one and each pair fires alternately.

The Trojan engine has a bore and stroke of 63·5 mm. and 117·5 mm. respectively, and it is designed to give its maximum power over a range of speeds, so that its power curve is a fairly flat one. Crank-case compression is employed, each pair of cylinders having a common crankcase. The mixture is drawn into the crank-case while the pistons are moving outwards (from the crank-shaft), and is pumped into the upper cylinder of the pair (the cylinders lying horizontally one above the other) as the pistons move inwards. It enters the second cylinder through a port in the combustion head. The lower piston, which controls the exhaust port, moves in advance of the upper piston, which governs the transfer port, so that exhaust occurs before the transfer port is opened. As the ports are situated in opposite cylinders of a pair, there is, as we have stated previously, less likelihood of charge loss through the exhaust. The exhaust port closes

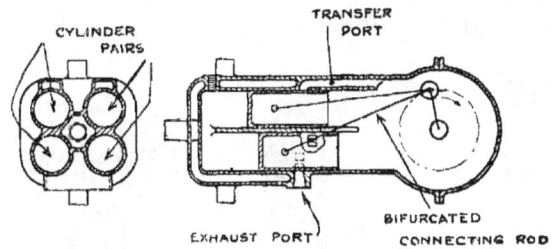

Fig. 202.—The Trojan Light Car Two-Stroke Engine.
(Four Cylinder.)

before the transfer port, and owing to their differential motion at the top of the stroke, when ignition occurs, the gases rush through the port connecting the two cylinders, thereby causing a turbulent action which is beneficial to the combustion process. It is interesting to note that the Trojan engine contains only seven moving parts.

A three double-cylinder Trojan engine has also been designed for use on light motor vehicles.

The Püch Engine.—This is a good example of the type of engine shown at C in Fig. 195. It consists of two parallel cylinders having a common combustion chamber. A forked connecting-rod is employed in order to give the pistons different relative positions near the bottoms of their strokes.

The angularity effect has no practical significance during the compression stroke.

Referring to Fig. 203, Diagram *A* shows the pistons near the completion of the firing stroke, the burnt gases flowing out through the L.H. cylinder ports, and the fresh charge from the crankcase flowing into the R.H. cylinder.

Diagram *B* shows the relative piston positions a little later, when the L.H. piston is ascending and has closed its exhaust port; the R.H. transfer port, it will be observed, is still open. Diagram *C* shows the end of the compression stroke and the ignition occurring. The fresh charge is entering the crankcase on the R.H. side.

This engine is made in single and double unit sizes, with two and four cylinders, respectively. The bore

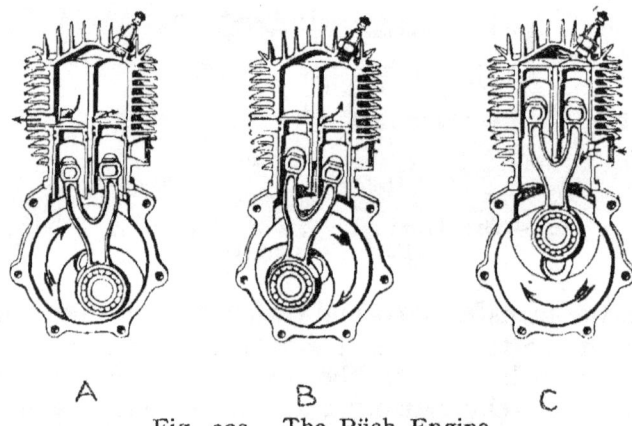

A   B   C

Fig. 203.—The Püch Engine.

and stroke are 45 mm. and 78 mm., respectively, giving cylinder capacities of 248 c.c. and 496 c.c., for the single and double unit engines. The smaller engine gives about 7 B.H.P. at 3,000 r.p.m.

The Schliha Engine.—This is an interesting example of a two-cylinder opposed engine using differential pistons with tubular extensions (Fig. 204). The system employed gives a unidirectional flow for the charge. Referring to the upper illustration, it will be observed that the two cranks are at 180°, so that the pistons move inwards and outwards together.

The mixture ·enters, through the ports *I* (upper diagram), into the crank chamber on the outward strokes of the pistons and is compressed on the inward strokes. When the two pistons reach the positions shown in the lower diagram the ports H are uncovered and the compressed charge flows from the crank chamber, through the ports H, into the annular combustion spaces. Previously, the exhaust gases had been ejected through the exhaust ports E; the direction of flow of the fresh charge and exhaust gases is indicated by the arrows.

On the next outward strokes the charge is trapped and compressed in the annular chamber, by the trunk

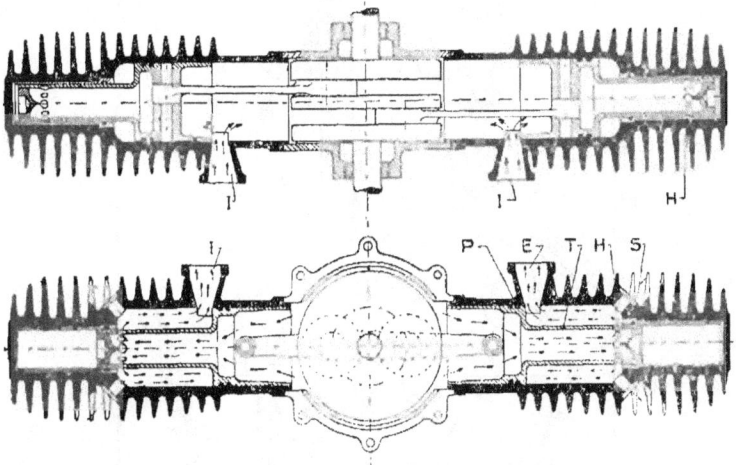

Fig. 204.—The Schliha Opposed Differential Piston Engine.

piston P. The sparking plug position is shown at S.

The advantages of this engine are that it gives excellent balance and the mixture serves to cool the hollow pistons—possibly at the expense of a diminished volumetric efficiency. The overall length of the engine may, however, prove a drawback, in some applications.

The Jamieson Engine.—This is a double slide-valve engine and was designed for high speed automobile and aircraft purposes. The engine which is shown, sectionally, in Fig. 205* has four cylinders each of 72 mm. bore, which, in conjunction with a piston stroke of 90 mm., give a capacity of $1\frac{1}{2}$ litres.

*Machinery.

The Meehanite cast-iron cylinder block is of inverted V-form, and in chests at each side of the block, actuated through links *B* by auxiliary crankshafts there are flat slide-valves *A*—two per cylinder— actuated through links *B* by auxiliary crankshafts driven from the main crankshaft. A feature of the drive to the auxiliary crankshafts is the arrangement

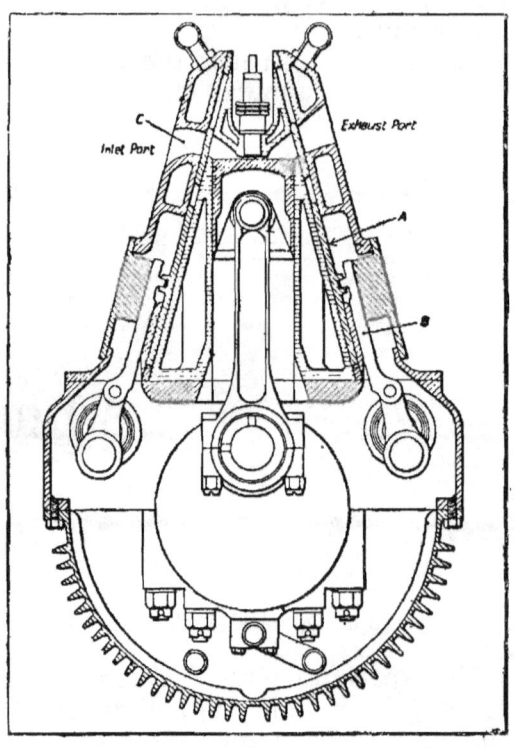

Fig. 205.—The Jamieson Engine.

whereby the timing may be varied at will, while the engine is running.

In operation, after the charge has been fired, with both ports closed, of course, the piston is forced down-wards, and some distance before the bottom of its stroke—50 deg. before bottom dead-centre—the exhaust valve opens. It then remains open while the piston moves upwards again for part of the stroke. Just before the exhaust valve closes, the inlet valve opens, and a fresh charge, under high pressure, enters the cylinder through the inlet port *C* from the supercharger. The inlet valve is open, however, for only a very short time, and when it closes, the charge, already under pressure, is further compressed during the remainder of the piston travel to top dead-centre. The two-stroke cycle is thus completed. The slide valves are adequately cooled by water jackets and they are lubricated by oil under pressure, which enters the valve chests through jets.

The Magoucy Engine.—The principal object aimed at in the design of the Magoucy engine shown in Fig. 206, is to overcome the loss of charge through the exhaust ports experienced by many existing designs of two-cycle engines.

The engine employs a shuttle-type slide valve which is automatically operated by the pressure difference on its upper and lower surfaces and controls the transfer port into the cylinder.

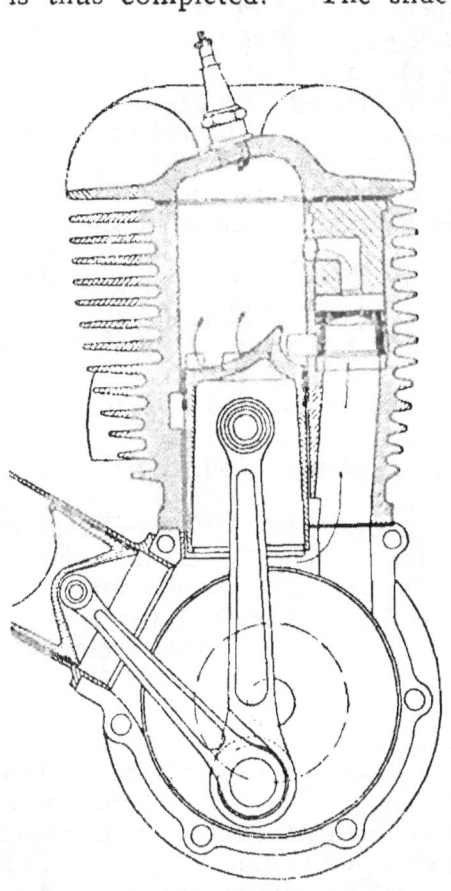

Fig 206.—The Magoucy Engine.

The engine shown is arranged for crankcase precompression which is augmented by an auxiliary cylinder with piston making an angle of about 60 deg. with the working piston. In Fig. 206 the piston in the working cylinder is shown near the end of the power stroke, exhaust having just begun. The slide valve has been forced down by the pressure in the working cylinder (with which the space above it communicates) and closes the transfer port. However, shortly after the beginning of the exhaust period the pressure in the crankcase surpasses that in the cylinder, the slide valve is moved upward by this pressure difference, and uncovers the transfer port. It will be noticed that in this engine the transfer port is located at a higher level than the exhaust port, and charging of the cylinder therefore can continue after the exhaust ports have been closed. As the piston proceeds upward on its compression stroke, the pressure in the working cylinder surpasses that in the crankcase and the slide valve is forced down again so as to close the transfer port.

The inventor has worked out a number of different designs of his shuttle valve and the one here shown, designated by the symbol R-14, is said to have given about the best results. The shuttle is made of magnesium alloy and weighs only 2 and 2.5 grams per sq. cm. of its section. In the experimental engine the parts against which the shuttle abut are made of cast iron.

Reverse Flow Two-Cycle Engines.—Instead of using a deflector on the piston, in order to guide the incoming charge, the latter can be directed in an oblique upwards direction by specially cast ports. Two inlet ports are usually employed, so that the two charge streams meet above the level of the exhaust port and, travelling upwards, tend to force the exhaust gases downwards and through the exhaust ports. A later model Villiers engine of 125 c.c employs this method. It has a flat-topped piston, two exhaust and four inlet ports. The reverse flow scavenging effect described is also arranged for in the D.K.W. engine (of German origin) shown in Fig. 207.

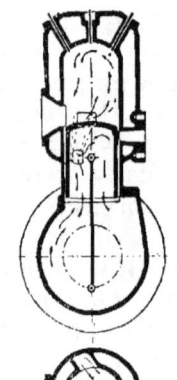

Fig. 207.
Reversed flow
D.K.W. Engine.

The beneficial effects of this method of scavenging the exhaust is indicated by the relatively high mean pressures and low fuel consumptions obtained. Thus, with the D.K.W. reversed flow engine, a brake m.e.p. of 57 lb. per sq. in. was obtained at 3,000 r.p.m., whereas the same type of engine using the conventional deflector piston gave 43 lb. per. sq. in. The fuel consumption of the reverse-flow engine was about 15 per cent. less than for the deflector type. Incidentally, a single cylinder engine of 88 mm. bore and 76 mm. stroke gave 11 B.H.P. at 3,000 r.p.m.

Direct Injection Engines.—In order to overcome the drawback of loss of fuel in the portion of the charge that escapes through the exhaust ports in many designs of two-cycle engines the engine can be designed to operate with a charge of air only, during the transfer, charging or scavenging processes. After the exhaust valve or port is closed the petrol can be injected into the charge during the compression stroke, so that in this way no fuel is lost.

The direct injection engine in question uses compression ratios of the same order as in ordinary petrol engines and employs a sparking plug to ignite the charge so that it should not be confused with the compression-ignition engine which utilizes the heat of compression of the air charge to ignite the fuel. The principle of fuel injection spark ignition has been employed successfully on four-cycle commercial motor vehicle, stationary and aircraft engines but it has, at present, only been used to a limited extent for two-cycle engines.

It was applied as long ago as 1910-12 by the late Prof. W. Watson to a converted Day two-cycle petrol engine and has since been used in experimental engines.

CHAPTER VI

KNOWN also as the high-speed oil, or Diesel engine, this type operates on an entirely different principle to the ordinary petrol engine; the latter works on the Otto cycle, whereas the former follows closely what is known as the *Constant Pressure Cycle* of operations.

The compression-ignition, or C.I. engine requires no carburettor or ignition apparatus, and it uses a heavier and cheaper type of fuel, known as Diesel oil, instead of petrol.

The earlier C.I. engine was hitherto debarred from automobile work on account of its excessive weight for a given horse-power output; moreover, it generally worked at relatively low speeds, namely, below 300 r.p.m.

As a result of intensive experimental work it was shown that the C.I. engine could not only be run at speeds comparable with those of the petrol engine, namely, 2,000 to 4,000 r.p.m., but by using the same design methods and materials as for petrol engines it could be made only a little heavier than the petrol engine; actually the petrol engine must always be the lighter type.

As a result of these developments, the C.I. engine has proved a formidable rival to the petrol engine in automobile, stationary and rail-car applications. At present the C.I. engine is used chiefly in automobiles of the passenger-carrying and goods types, for it shows a considerable saving in the matter of fuel costs as compared with the similar size petrol engine vehicle; moreover, it has other advantages in connection with lower maintenance costs, freedom from fire-risks, better engine torques, quicker starting and pulling from the cold.

More recently, some of the smaller models of C.I. engines, e.g., the Victor two-cylinder opposed,

Perkins four-cylinder, Gardner four-cylinder, Ricardo-Citroen, Dodge, Hanomag and similar engines have been fitted to (mostly experimental) motor car chassis and have established their superiority in the matter of high torque performance at relatively low road speeds and marked superiority in fuel consumption; usually the road mileages per gallon of fuel have proved from 30 to 40 per cent. greater than for the cars fitted with petrol engines of similar horse power rating.

The Citroen four-cylinder C.I. engine, of $2\frac{3}{4}$ in. bore and 4 in. stroke developing about 46 B.H.P. has been used on light delivery vans and a similar model has been offered as an alternative to the petrol engine fitted to the larger Citroen motor cars made in this country.

In the experimental cars the maximum top gear speeds with the C.I. engines were generally less than for the petrol engines of similar rating and the acceleration times were not so good as for the latter type of engine. The engines were invariably heavier although much progress has been made more recently in the direction of weight reduction. Thus, the more recent Perkins C.I. engines are not appreciably heavier than petrol engines of the same type and power rating. The present indications show that it is unlikely this type of engine will prove a serious rival to the petrol engine for motor cars although developments may occur with the high compression spark-ignition fuel injection engine which will establish it as an alternative to the petrol engine of the carburettor-fed class.

Higher Efficiency.—The C.I. engine operates at much higher compression ratios than the petrol engine. Thus, it employs compression ratios of 13.1 to 18 : 1, whereas the ordinary motor car engine uses ratios of 5.5 : 1 to 6.5 : 1.

It is well known that the efficiency of an engine becomes higher as its compression ratio is raised so that the high compression C.I. engine must be much more efficient than the lower compression petrol engine.

A direct result of this higher heat efficiency is the lower fuel consumption for a given horse-power output; the C.I. engine uses from 30 to 45 per cent. less fuel per H.P. on this account. Further, it can use a cheaper and heavier grade of fuel oil having a higher flash point (and, therefore, reduced fire risk).

How the Four-Cycle Oil Engine Works.—The C.I. engine resembles the petrol engine in its general design as it has the same kind of trunk piston, connecting-rod, inlet and exhaust valves, etc. It differs, however, in the manner of introducing the fuel into the cylinder head and in the method of igniting the fuel.

The principal difference in the two types lies in the design of the combustion head and in the fuel supply arrangements to this head. Before, however, describing the four-cycle engine let us consider, briefly, the principle of the C.I. engine.

Instead of introducing a mixture of fuel and air into the cylinder, the C.I. engine draws in pure air only. This air is then compressed by the ascending piston to a considerably higher pressure than that used in the petrol engine; thus the air is compressed to $\frac{1}{12}$ to $\frac{1}{16}$ of its original volume, giving a compression pressure of about 400 to 550 lbs. per sq. in.

As a result of its high compression the temperature of the air is raised considerably—usually to about 500° to 550° C. Now, the temperature required to " self-ignite " the Diesel oil is about 350° to 450° C. It will be evident, therefore, that the effect of injecting the Diesel oil into the combustion space containing this highly compressed and heated air charge will be to cause it to burn very rapidly. There is thus no necessity for any electric spark to ignite the oil spray and compressed air. Immediately after the given amount of oil has been injected the temperature of the combustion products is raised considerably, namely, to about 2,000° C. to 2,700° C., but the pressure does not rise appreciably above the compression pressure as in the case of the petrol engine; usually, it reaches a maximum value of 700 to 850 lbs. per. sq. in. Thereafter the gaseous products expand, forcing the piston downwards (or inwards) thus providing the energy for the power stroke.

It should here be mentioned that the fuel is injected by means of a special fuel pump, through a fine nozzle into the combustion chamber, so that it issues in the

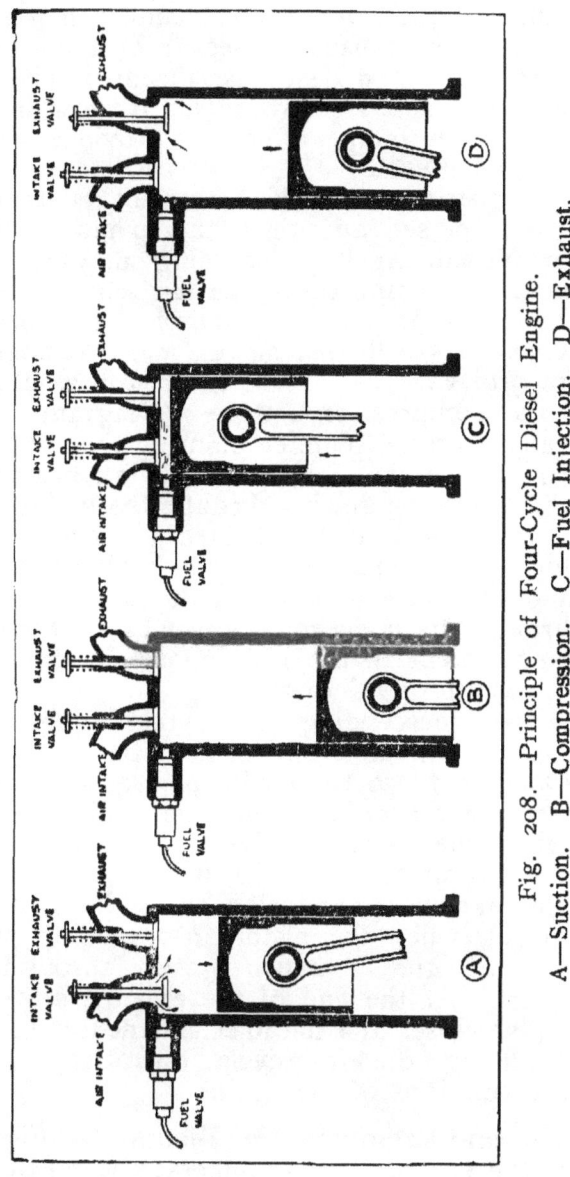

Fig. 208.—Principle of Four-Cycle Diesel Engine.
A—Suction. B—Compression. C—Fuel Injection. D—Exhaust.

form of a highly atomized spray to mix with the air
already in the combustion chamber.

Reverting now to the four-cycle C.I. engine (Fig. 208), Diagram A shows the piston near the top of its exhaust stroke, the combustion chamber being filled with the remnant exhaust gases from the previous stroke. As the piston descends, the air-intake valve is opened mechanically, allowing air to flow into the cylinder. The air intake is usually provided with an air-cleaner and silencer.

When the piston reaches the bottom of its stroke the inlet valve is closed and since the exhaust valve is also kept closed, on its succeeding upward stroke (Diagram B) it compresses the air charge, finally giving the latter—as previously mentioned—a pressure of about 300 to 550 lb. per sq. in. and a temperature of 500° to 550° C. Just before the piston reaches the top of its compression stroke (Diagram C), a mechanically-operated plunger pump is timed to force the fuel oil for combustion, under a high pressure of 1,000 to 8,000 lbs. per sq. in., through the fuel injector, where it emerges into the heated air charge as a conical highly " atomized " spray. It almost at once ignites and continues to burn all the time the fuel is forced through the injector; this usually continues for a period equivalent to the rotation of the crankshaft through an angle of 15° to 30°.

On the next descending stroke the highly heated exhaust gases expand from their initial combustion pressure of about 650 to 900 lb. per sq. in. down to atmospheric pressure (14·7 lb. per sq. in.) which occurs when the exhaust valve opens, towards the end of expansion stroke. The inlet valve is closed during the expansion and also the exhaust strokes (Diagram D), when the piston in ascending sweeps out most of the exhaust gases through the exhaust port. At the end of the exhaust stroke the exhaust valve closes and the inlet commences to open, thus completing the four-cycle operation, in two complete revolutions of the engine.

Pressure and Temperature Diagrams.—Fig. 209 illustrates the pressures and temperatures which occur within the combustion chamber of a typical C.I. engine during the major portions of the compression and expansion strokes. The compression ratio used, namely, about 15 : 1, would give a compression pressure of about 450 lb. per sq. in, but fuel injection commences at about 14° before T.D.C. when the compression pressure is about 300 lb. per sq. in and, as shown by the air temperature curve the temperature of the compressed air is about 560° C. The curve of ignition temperatures for the fuel employed shows that when the injection commences the temperature required to ignite the fuel spray is about 260° C., so that as the air charge temperature is considerably higher, namely,

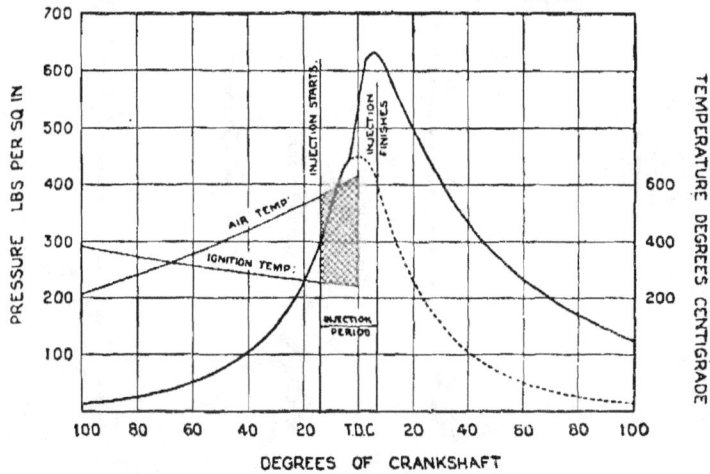

Fig. 209.—Pressure and Temperature Diagrams for C.I. Engine.

560° C., the fuel will at once ignite. The injection continues until about 7° past T.D.C. crank position. During this period, owing to the combustion of the fuel the pressure rises until it attains the value of about 630 lb. per sq. in. at the finish of the injection period. Thereafter the hot gases of combustion expand progressively, doing useful work on the piston, as shown by the upper right hand line. The dotted line below is the expansion line for the compressed air charge when no injection takes place.

The Two-Cycle C.I. Engine.—This type operates in a similar manner to the petrol engine types, the only differences being as follows:—

(1) The C.I. engine employs pure air for charging the cylinder, instead of a mixture of petrol and air. The air charge is compressed and introduced by exactly the same means as in the various two-cycle petrol engines.

(2) The C.I. engine uses a much higher compression ratio, namely, from 13:1 to 18:1.

The two-cycle C.I. engine is relatively more efficient than the petrol type, since there is no loss of fuel through the exhaust ports. On the other hand it is not yet so efficient, in automobile sizes, as the four-cycle C.I. engine.

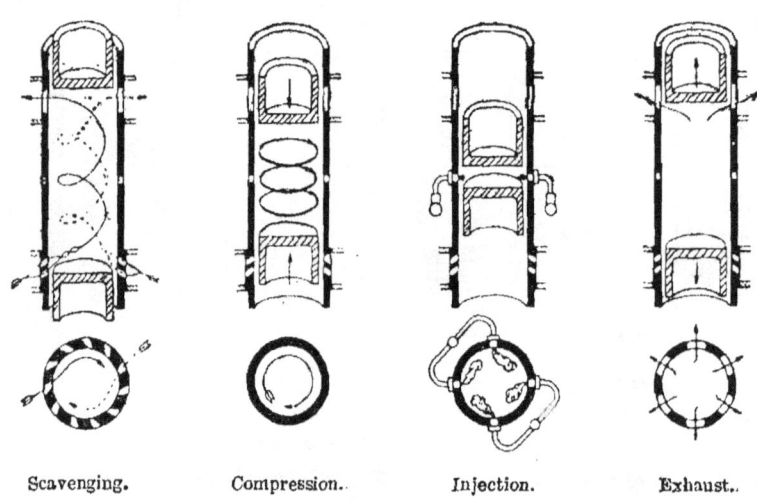

Scavenging. Compression. Injection. Exhaust.

Fig. 210.—The Junkers Two-Cycle Arrangement, using Opposed Pistons.

The best-known example of a two-cycle C.I. engine is the German Junkers aircraft engine, the principle of which is illustrated in Fig. 210). This engine has opposed pistons, which move in and out together. They are connected, by means of their connecting rods to their own crankshafts, one at the top and the other at the bottom of the cylinder block. The two shafts are connected together by means of gearing in aircraft models; the original automobile Junkers engines had long connecting rods to connect the two crankshafts to enable them to rotate together.

The long common cylinder has inlet ports for the fresh-air charge at the lower end and exhaust ports at the upper end. Referring to the right-hand diagram in Fig. 210 the pistons have reached the ends of their expansion strokes, the exhaust gases having escaped through the upper ports. The air-charge flows into the cylinder through the tangential ports shown, and is thereby given a spiral path; it effectively scavenges the exhaust gases.

The still swirling air is next compressed by the inwardly-moving pistons, as indicated by the " Compression " diagram. When fully compressed into the cylindrical space formed between the pistons when nearest together, the Diesel oil is injected tangentially through four equi-distant nozzles, in the form of fine sprays. These mix intimately with the swirling air and ignite, owing to the high temperature of compression (see " Injection " diagram). The pistons then move outwards until the upper one uncovers the exhaust (see " Exhaust " diagram) when the two-cycle process is completed.

The Junkers aircraft engine weighs only about 2 lb. per H.P.

Types of C.I. Engine. — There are numerous examples of high-speed C.I. engines used for automobiles. Whilst these all operate on the compression ignition principle, they differ widely in the designs of their combustion chambers and in the fuel injection systems employed.

It is only possible in an elementary book of the present nature to give a very brief outline of the principal types, but those seeking fuller information are referred to the author's book* on the subject.

The various designs of C.I. engines may be grouped into three broad classes as follows: (1) *The Direct Injection;* (2) *The Pre-Combustion;* and (3) *The Turbulent Head types.* There are, however, certain other designs that utilize the advantages of two or more of these types.

* High Speed Diesel Engines. A. W. Judge, (Chapman & Hall Ltd.),

(1) **The Direct Injection Engine.**—This is so called from the fact that the fuel is sprayed directly into the cylinder itself, there being no separate combustion head. (Fig. 211.)

The piston R is shown nearly at the top of its compression stroke, both the inlet valve I and exhaust valve E being closed. Commencing at the fuel supply end, fuel oil is drawn through a fuel filter F on

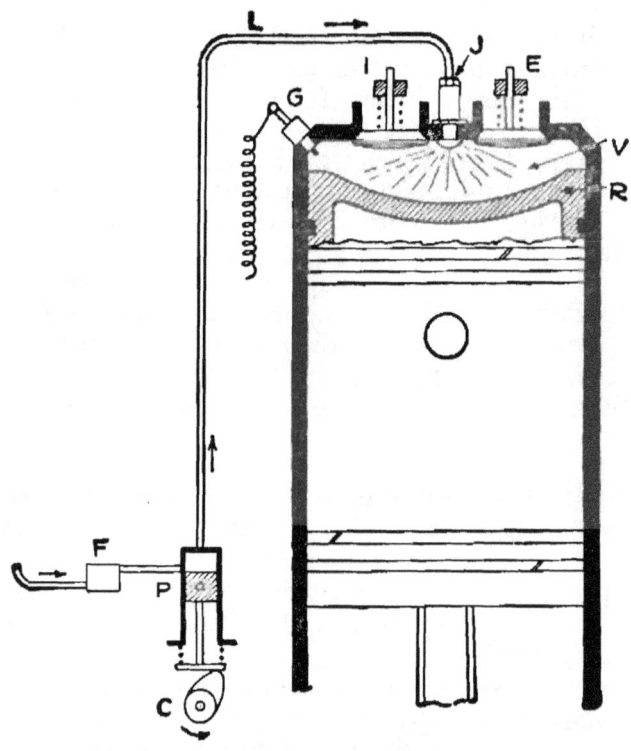

Fig. 211.—The Direct Injection Engine.

the suction stroke of the fuel pump P, in order to rid it of all solid particles; the presence of the latter would not only cause excessive wear of the moving parts of the fuel pump but would also tend to choke the fine passages in the fuel injection nozzle J.

The fuel pump accurately measures out the predetermined quantity of oil and at the correct moment delivers it, past a non-return delivery valve into the fine bore fuel pipe *L*. The pump, which is mechanic-

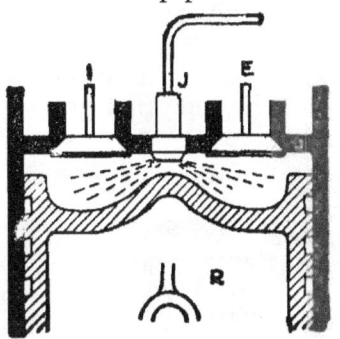

Fig. 212.—Direct Injection Engine.
with Special Piston Head.
E—Exhaust.   I—Inlet.
J—Injector.   R—Piston.

ally driven from the engine, at one-half engine speed, in the four-cycle engine, is designed to give a high pressure to the oil in order to force it through the injection valve in spray form against the compression pressure (450 to 550 lbs. per sq. in.) existing in the cylinder at the time of injection. The usual fuel pressures given by the fuel pump are about 900 to 1,500 lbs. per sq. in., although pressures up to 3,000 lbs. per sq. in. have been used on some engines.

The bore of the steel injection pipe is kept as small as possible, viz., from $1\frac{1}{2}$ to 3 mm. according to the size of engine.

The usual form of injection nozzle *J* is that of a spring-loaded plunger having a conical or specially-shaped end, normally held on a seating, located near the wall of the combustion chamber, by the spring-pressure. During the period of injection, however, the high fuel pressure given by the pump overcomes the spring pressure and lifts the valve off its seating, thus allowing the oil to enter the combustion chamber. The nozzle is designed to give the desired shape of spray, while the injection pressure is made high enough to project the spray right across to the farthest parts of the combustion chamber, in order to ensure complete burning of the fuel.

There are other shapes of piston crown and cylinder head, in direct injection engines, but all operate on the same basic principle.

The plug shown at *G* (Fig. 211) contains a small coil of resistance wire which can be heated to redness by means of a current taken from a battery.

This *Heater Plug* is only used for cold starting purposes and is switched off once the engine has started.

K

Many direct-injection engines will start from the cold without the aid of heater plugs, but in very cold weather the latter are usually essential.

The direct-injection engine uses a multiple-hole injection nozzle and requires a higher fuel pressure than most other types of C.I. engine.

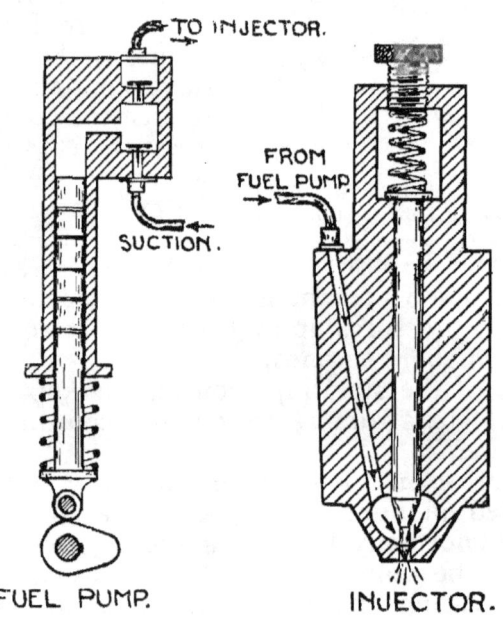

FUEL PUMP.   INJECTOR.

Fig. 213.—Illustrating Principle of Fuel Pump and
Injection Valve.

(2) **The Pre-Combustion Chamber System.**—In this type there is a small auxiliary chamber which communicates with the cylinder itself by means of a relatively narrow neck. (Fig. 214). The principle aimed at is to promote the projection of burning fuel particles from the auxiliary chamber, through the neck or throat into the cylinder head, in order to obtain a more progressive and less rapid combustion effect. At the end of the compression stroke the piston almost reaches the upper end of the cylinder and thus forces the air charge into the auxiliary chamber. At the correct moment the fuel is injected into the latter space in a partially pulverized condition. It immediately begins

to ignite and in so doing projects a mixture of burning and unburnt fuel from the auxiliary chamber through the throat into the cylinder. In some cases the throat is replaced by a perforated plate of refractory material. A relatively large fuel orifice can be used with this system. The direction of the spraying is not very important, nor is a high injection pressure essential.

The Bosch-Acro engine is an example of this class, and was at one time widely used.

The pre-combustion chamber engine has to some extent been supplanted by the turbulent-head and

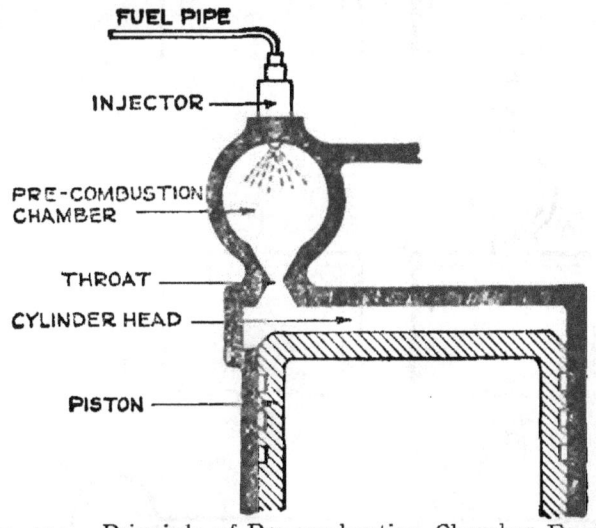

Fig. 214.—Principle of Pre-combustion Chamber Engine.

direct-injection types, for it has the disadvantages of requiring a high-compression ratio; is difficult to start, (heater plugs being necessary), and is not so efficient on account of the heat loss from the burning fuel to the throat of the combustion chamber. Moreover, it is prone to " Diesel knock " when idling and also when under maximum power conditions.

(3) **The Turbulent-Head Engine.**—Whilst it is true that there is a certain amount of movement or agitation of the compressed air in the combustion chamber just prior to the injection of the fuel, in practically all types of engine, in many cases this turbulence is more or less accidental and of relatively low degree. The

type of combustion chamber under consideration is that in which the designer has deliberately arranged to impart a moderate-to-high degree of turbulence to the

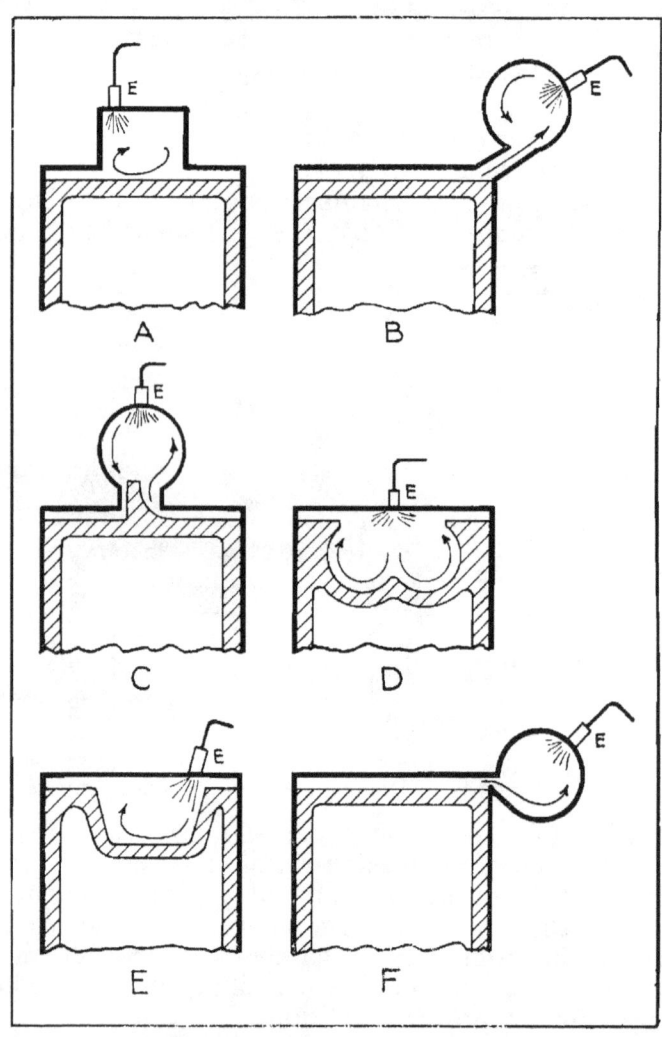

Fig. 215.—Types of Turbulent Combustion Chamber Engines. A—Ricardo Rotary Swirl. B—Ricardo "Comet." C—Clerestory. D—Saurer Dual Turbulence. E—Leyland Cavity Piston. F—Leyland Spherical Anti-chamber.

compressed charge for the purpose of securing a satisfactory admixture of the air and fuel particles.

The injector, in this case, is usually of the single hole type, having a relatively large hole. It is placed in the combustion chamber so that the turbulent air stream sweeps rapidly past the outlet hole and each particle of

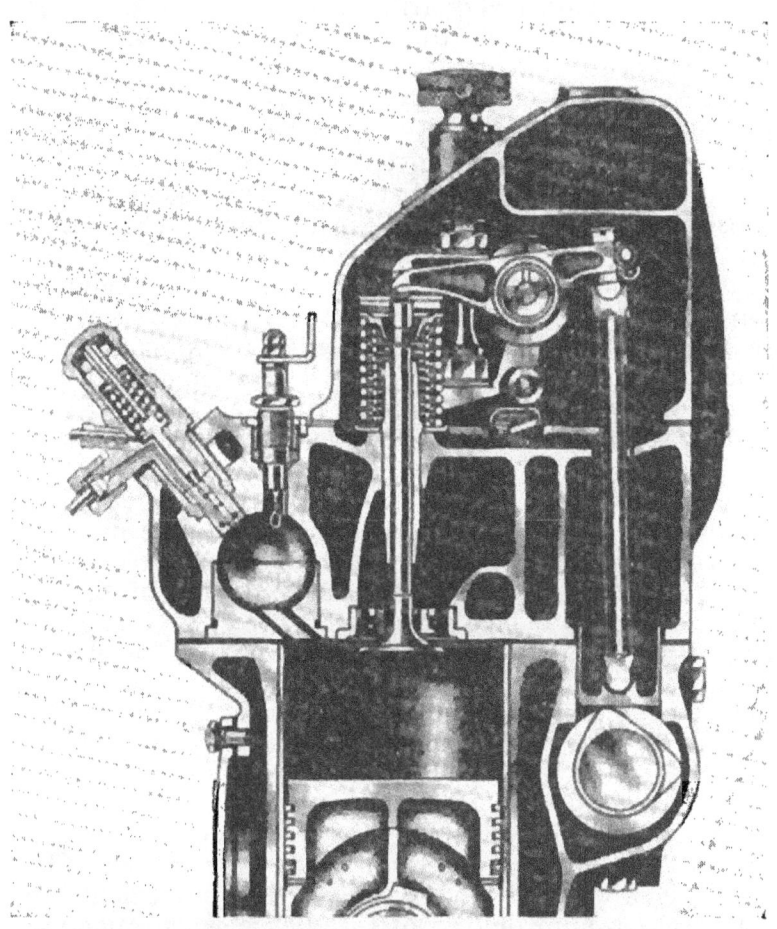

FIG. 216.—The A.E.C. Motor Bus C.I. Engine, showing Ricardo "Comet" Combustion Chamber, Fuel Injector (on left) and Heater Plug for starting purposes (on top). The high position Camshaft is also shown.

fuel is given its proper air supply for combustion soon after it emerges from the nozzle. There are various methods of producing this turbulence, one common process being that in which the piston forces the

compressed charge into an auxiliary chamber. The
piston, which may be plain or with a central projection,
is arranged almost to touch the cylinder top at the end
of its stroke, thus forcing the remaining air charge
through comparatively narrow spaces into the auxiliary
chamber; the air is thus given a high degree of
turbulence.

Some Representative Types.—The diagrams given
in Fig. 215 illustrate the principles of some of the
leading British turbulence-type C.I. engines.

The Ricardo rotational-swirl system shown at *A*,
consists in giving the air in the turbulence chamber a
rotational swirl, by means of specially inclined air-
inlet ports. The piston almost touches the cylinder
top and at the end of its compression stroke forces the
swirling air charge into the smaller cylindrical turbulence chamber, whence the fuel is projected from the
nozzle *E* in a direction across the air stream. This
system is used in the larger single sleeve valve engines.

The Comet turbulent head is shown at *B*. This
employs a small connecting throat arranged tangentially to the spherical chamber shown. The injector
E projects the fuel across the swirling air charge.
This type of head is used in a large number of British
and Continental makes of engine. The A.E.C. engines
(Fig. 216), used on London buses employ it. The clerestory head shown at *C* gives a similar turbulence effect;
the combustion chamber is cylindrical, i.e., with flat
ends in which the inlet and exhaust valves are seated.
The projection on the piston is arranged off-centre in
order to give a tangential entry to the air.

The Armstrong-Saurer dual turbulence system
shown at *D*, utilizes a special cavity type piston. The
air is given both horizontal and vertical movements so
that it has a high degree of turbulence when injection
occurs. A wide angle spray is employed, the fuel
injection nozzle having several holes for this purpose.

The Leyland cavity piston system is shown at *E*.
This is really a combination of the turbulence and
direct-injection systems, the heat losses being low and
cold starting relatively easy.

Another type of Leyland ante-chamber turbulence-
head is indicated at *F*. This belongs to the spherical

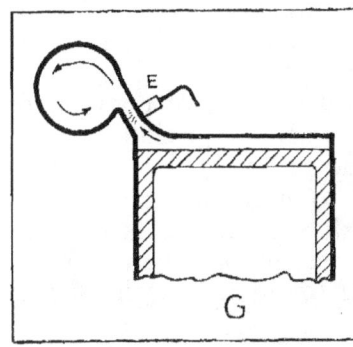

Fig. 217.—The Perkins Combustion Chamber.

combustion chamber type, and is so arranged that the two halves of the sphere are formed in the cylinder block and cylinder head, respectively.

The Perkins engines, used for motor cars and light commercial vehicles, employ the combustion head shown at G, Fig. 217. In this case the fuel is injected into the air charge passing through the throat; the charge burning as the piston commences to descend, for the air rushes out through the throat and meets the fuel spray.

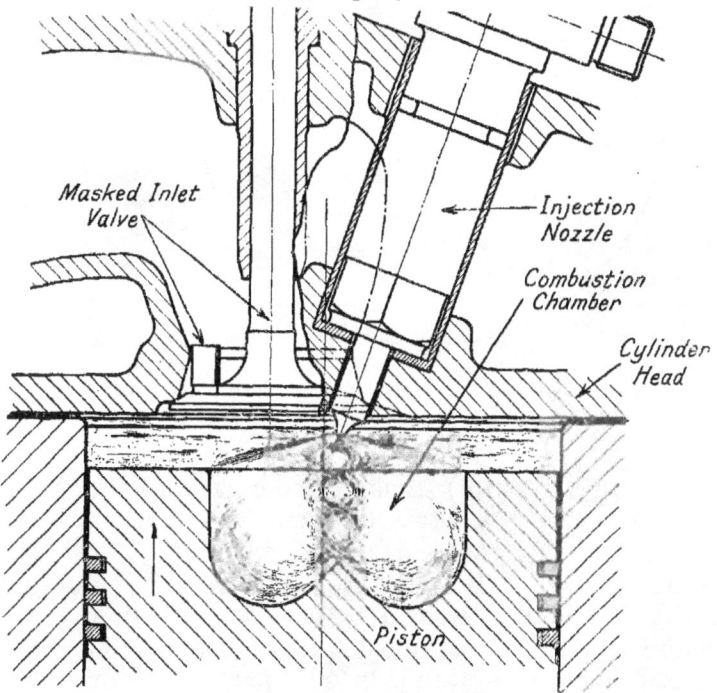

Fig. 218.—The A.E.C Direct Injection Cavity Piston Engine.

The A.E.C. Direct Injection Engine.—A recent design of A.E.C. engine as used on commercial and passenger vehicles employs the direct injection method and has a special design of cavity piston and flat cylinder head, of the type shown in Diagram D, Fig. 215. The air charge is given a certain amount of controlled turbulence prior to and during the fuel injection period. The injection nozzle is arranged centrally, but inclined to the cylinder axis in order to give maximum fuel penetration and admixture with the air. A sectional view of the head of the A.E.C. engine is given in Fig. 218. It has six cylinders each of 105 mm. (4·13 in.) bore

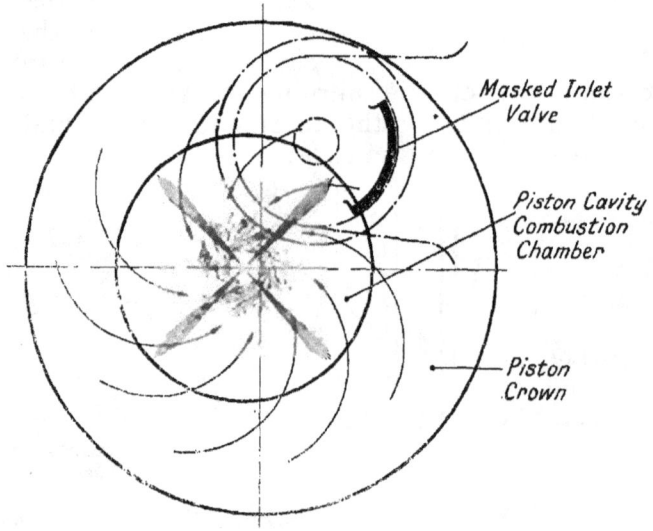

Fig. 219.—Plan view showing direction of air flow into piston cavity at end of the compression stroke.

and 146 mm. (5·75 in.) stroke, with corresponding cylinder capacity of 7·58 litres (462 cu. in.). The engine is rated at 41 H.P. and develops 90 B.H.P. at 1,700 R.P.M. and 113 B.H.P. at 2,000 R.P.M., giving a B.M.E.P. at 1,000 R.P.M. of 108 lb. per sq. in., and a maximum torque of 300 lb. ft. at 800 R.P.M. The lowest fuel consumption is 0·36 lb. per H.P. hour. The compression ratio used is 16 : 1.

Special features of this engine include the use of masked inlet valves; high camshaft for operating the overhead valves; two cast iron cylinder heads; chain-driven camshaft, fuel-pump, vacuum pump and generator; dry cylinder liners of heat-treated alloy cast iron; aluminium alloy pistons with three compression and two oil control rings; floating pattern hollow gudgeon pins; lead-bronze lined big-end bearings; main bearings having top halves of white-metal and bottom halves of lead-bronze; water circulating pump

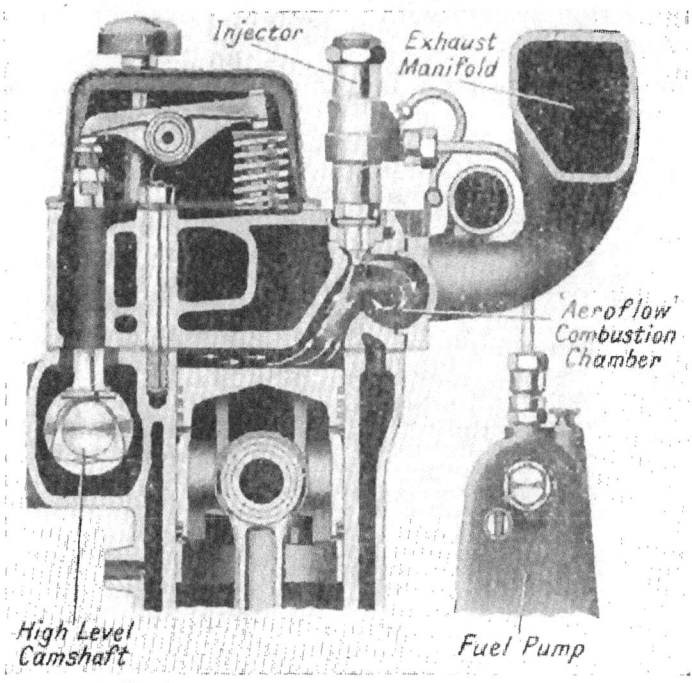

Fig. 220.—The Perkins C.I. Engine.

driven in tandem with 24-volt generator and vacuum pump for operating the servo brakes and also for gear changing. The engine with flywheel weighs 1,374 lb.

The Perkins P. 6 Engine.—This six-cylinder engine, of 88·9 mm. (3·5 in.) bore and 127 mm. (5·0 in.) stroke, with a cylinder capacity of 4·73 litres (77·6 cu. in.) has been fitted to commercial vehicles, whilst the smaller P.4 model of similar cylinder dimensions has been used experimentally on motor cars with promising results.

K*

These engines employ the Perkins modified "Aeroflow" system of combustion (Fig. 220). The P.6 engine develops 36 B.H.P. at 1,000 R.P.M. and 85 B.H.P. at 2,600 R.P.M. At 1,500 R.P.M. the value of the B.M.E.P. is 100 lb. per sq. in.; the R.A.C. rating is 29·4 H.P. The engine weighs 580 lb., without flywheel, so that the corresponding weight per H.P. is about 6-8 lb. The fuel consumption under the conditions of best economy is about 0·4 pint per H.P. hour. The C.A.V. fuel injection system with pneumatic governor control is employed.

The Victor Engine.—This is a two-cylinder opposed water-cooled C.I. engine of 80 mm. (3·15 in.) bore and

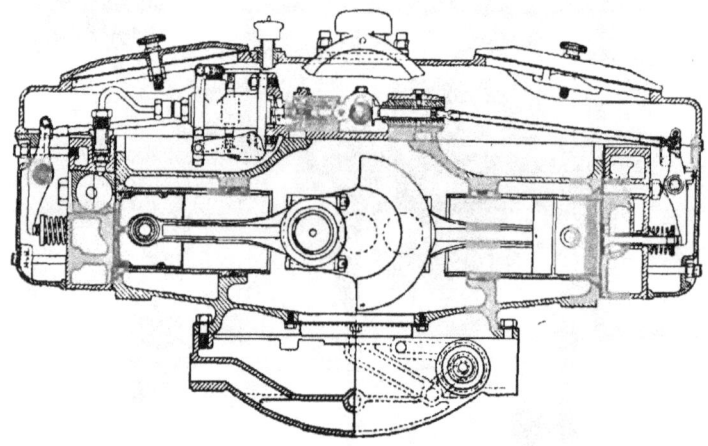

Fig. 221.—The Victor Two-Cylinder Opposed C.I. Engine.

100 mm. (3·94 in.) stroke, with a cylinder capacity of 1,000 c.c. (61 cu. in.). It develops 20 B.H.P. at 3,000 R.P.M. and consumes 0·4 lb. of fuel per H.P. hour. Owing to its excellent balance this engine will operate at speeds up to 3,000 R.P.M. without appreciable vibration. It employs separate cylinder liners of the wet type and has detachable cylinder heads. The crankshaft is of the built-up type the central portion being of nickel chrome steel of 70 tons per sq. in. tensile strength and the outer portions of nickel chrome molybdenum steel of about 75 tons per sq. in. tensile strength. The connecting rods of nickel chrome

steel are fitted with roller bearings each having 48 hardened steel rollers. The small end bearings are of phosphor bronze and the main bearings are heavy duty roller ones. The crankcase is an Alpax aluminium alloy casting. C.A.V. fuel injection equipment is employed. *For starting purposes* the engine has a spring-operated decompressor cam which holds the

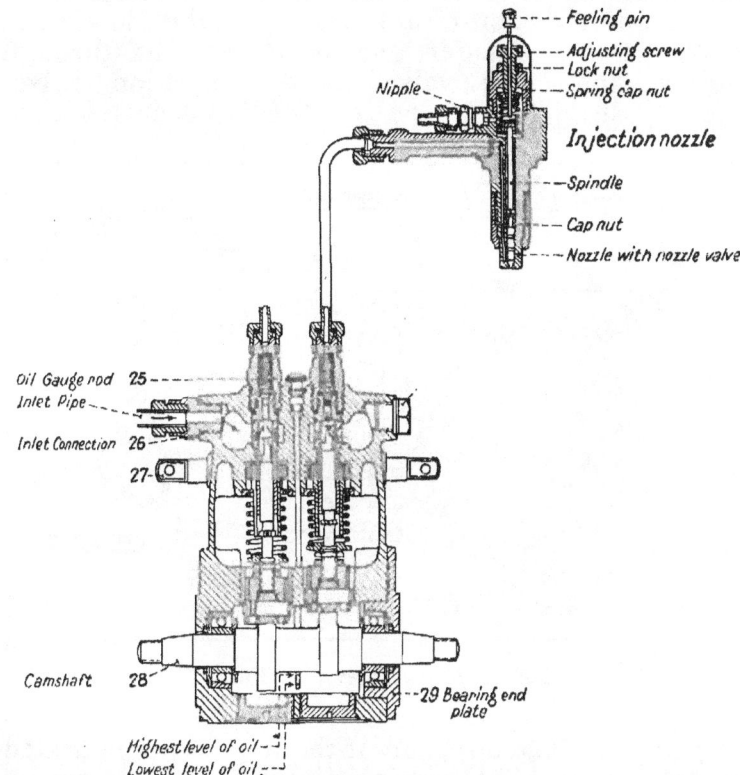

Fig. 222.—The C.A.V. Two-Cylinder Fuel Pump and Injector. showing the Principal Parts.

exhaust valves open until sufficient momentum has been developed by cranking, when the full compression is restored by releasing the decompressor gear and the momentum of the flywheel carries the engine over compression for starting. The Victor engine has been used on automobile tractor units such as the " mechanical horse " and on motor car chassis units.

The Fuel Injection System.—There are several alternative methods used for injecting the fuel, but the most commonly favoured system is the mechanical pump and hydraulically-operated injection nozzle shown diagrammatically in Fig. 213.

On the left is shown the plunger which is lifted by the engine-driven cam and closed by the external spring, indicated. Fuel is sucked into the pump chamber through a non-return valve on the downward stroke of the plunger and is forced out through another non-return valve to the injection valve, through the delivery pipe, on the upward stroke of

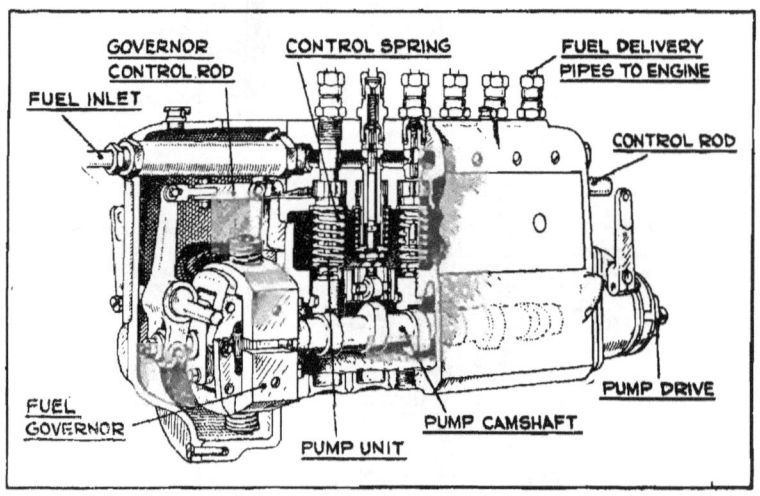

FIG. 223.—The C.A.V. Six-cylinder Fuel Pump with Governor.

the plunger. The direction of the fuel flow is indicated by the arrows. The fuel injection nozzle is shown on the right, fuel being forced down the small inclined passage on the left to the space near the bottom of the nozzle plunger. The latter is kept on its seating by means of a spring; the latter is indicated by the dotted lines above.

Usually there is a separate pump unit for each of the engine cylinders, but all of the units are enclosed in a common casing. A four-cylinder engine will have a four-pump unit, a six-cylinder one, six, and so on.

Apart from its purpose of pumping the fuel to its appropriate injector, the fuel pump must be designed to deliver varying small quantities of fuel at exactly the right moment of injection.

To do this, each pump unit is provided with an adjustment for varying the quantity of fuel pumped per stroke of the plunger. All of these adjustments are connected to a single control, so that by moving the latter all of the deliveries from the separate units are regulated together. Most fuel pumps have a control rod (as shown in Fig. 223), in the form of a toothed rack, the teeth engaging with corresponding teeth cut on the cylindrical part of the lower end of each plunger.

When the rack is pulled endwise it rotates each plunger, so that spirally-cut portions at the upper end of each are moved up or down in order to control the quantity of fuel.

The cams operating the pump plungers are cut from the solid on a single camshaft—just as in the case of petrol engine camshafts; the cams are, therefore, arranged at their correct relative angles, these being the same as those of the cranks on the engine crankshaft.

Fig. 222 shows the C.A.V. twin cylinder fuel pump and injection nozzle system, whilst a complete six-cylinder engine C.A.V. fuel pump, with automatic governing device is illustrated in Fig. 223.

A complete fuel injection feed system is shown in Fig. 224 for the four-cylinder Dennis commercial engines. The arrows show the direction of travel of the fuel. Thus, from the main fuel tank below the fuel is drawn up the right hand fuel pipe by the suction action of the mechanically-operated diaphragm fuel pump. Thence the fuel is pumped upwards to the large capacity fuel filter, shown on the upper right hand side. It then passes to the second fuel filter which has a pressure relief valve for by-passing the fuel back to the fuel tank should the pressure become excessive in the filter. The fuel from the second filter is drawn into the fuel injection pump on the left and thence is measured and distributed to each of the four injection nozzles, which are inside the cylinder head casing. Any leakage of fuel past the stems of the injection nozzles is taken away from leak-off unions, on the upper portions of the nozzle bodies, and thence through pipes to the common leak-off pipe shown in Fig. 224; this pipe leads the fuel back into the fuel tank.

The filters are provided with air venting screws which enable the air within the filters to escape when the fuel system is " primed ", i.e., filled with fuel. In this con-

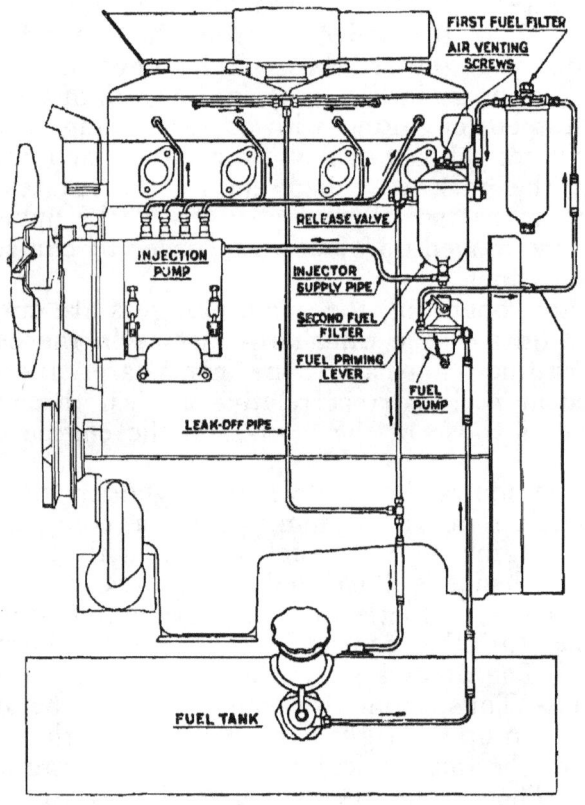

Fig. 224.—A complete Fuel Injection System.

nection it is very important to get rid of any air in the fuel system, otherwise the injection pump will not function correctly.

Regulating the Power Output.—In place of the usual throttle control of the petrol engine, the C.I. engine employs a device on the fuel pump for increasing or reducing the period of time during which fuel injection occurs.

It is usual to " time " the fuel injection so that it has an advance (analogous to petrol engine ignition advance) of 10° to 15° of crank angle. When the engine is idling the fuel is injected for a crank angle period of 5° to 10°; for maximum power the injection period would be extended up to 25° or 30° of crank angle.

The driver of the C.I. engine vehicle has an accelerator pedal control which in effect varies the amount of fuel injected. In addition, most fuel pumps have a centrifugal governor device for limiting both the high and also the low, or idling speeds of the engine.

CHAPTER VII

LUBRICATION OF THE ENGINE

THE lubricating system is one of the most vital elements of the high speed engine; if the oil supply to the working parts is stopped the engine will seize up after a short time. The modern engine contains a relatively large number of components rotating, sliding or reciprocating at very high speeds, and in attaining these speeds with engines which are light and compact, the bearing loads and rubbing velocities used are much higher than those employed in steam engine and machine practice. It is therefore essential that these parts be kept well lubricated, and that the lubricating system must be absolutely reliable. In passing, it may be mentioned that the majority of engine mechanical breakdowns may be traced to failure of lubrication, and that the greater wear in the working parts of some engines is usually due to their inefficient oiling systems.

The essentials of an efficient lubrication system are: (1) reliability; (2) proper lubrication of all working parts; (3) minimum oil consumption, otherwise much oil is lost by leakage past the pistons, forming carbon deposit in the combustion chamber and gumming up the valves; (4) accessible replenishing filler, strainer and oil drainer; (5) conspicuous indicator or pressure gauge showing whether the oil is circulating properly.

The Principles of Lubrication.—When two components of a bearing, such as the shaft and its bearing, are adequately lubricated with a suitable oil the two metallic surfaces are not in metallic contact, but are separated by an oil film, and the resistance to relative motion between the shaft and its bearing therefore depends largely upon the physical properties of the oil, namely, its viscosity; it is independent of the frictional

304

coefficients of the bearing metals. This type of lubrication is termed the *"fluid"* one. There is, however, another form of lubrication liable to occur in the case of bearings and sliding surfaces under inadequate lubrication conditions, or in instances where appreciable surface roughness of the components prevents the maintenance of "fluid" lubrication. In such cases the oil film is insufficient in thickness to prevent occasional contact of the bearing surfaces, with the result that the frictional resistance is increased appreciably; usually the frictional coefficients under these " *boundary* " lubrication conditions are from ten to twenty-five times greater than for fluid lubrication, but are less than the " dry " coefficients. It should here be explained that if the bearing surfaces were perfectly, or chemically, clean the coefficient of friction would be much higher, but a trace of oil on the surfaces reduces this value appreciably. Thus, under the conditions of boundary lubrication the film of oil is infinitely thin, but the actual value of the frictional coefficient will depend not only upon the nature of the oil, but upon the metals used for the shaft and bearing The friction effect in this case is proportional to the load on the bearing, so that the coefficient of friction is actually independent of the load.

There is a certain property associated with lubricating oils, which, for want of a better name, is termed " *oiliness.*" It is difficult to define, but it can be stated that it is a surface effect produced by the lubricant upon the metallic surface with which it is in contact. In this connection fatty oil, i.e., oils saponifiable or those containing " fatty" ingredients, such as castor, rape or olive oil, exhibit a greater degree of " oiliness " than purely mineral oils, so that the frictional coefficient at any given temperature is lower under severe conditions of loading and slow speeds. Under extreme conditions of loading and high rubbing speeds, seizure between the shaft and its bearing would be less likely to occur with the lubricant of greater " oiliness." It is mainly for this reason that *compound oils,* i.e., mineral oils blended with a proportion of fatty oil are preferred by many engine users. In this connection it has been found that when pure hydrocarbon (mineral) oils are blended with fatty oil, the former constituent appears to eliminate the normal tendency of the fatty oil to oxidise and thicken under engine operating temperatures.

In the case of pistons and cylinder walls it is very probable that the boundary conditions of lubrication, previously mentioned, exist, the most favourable circumstances for this condition being during the maximum cylinder pressure and initial expansion pressure period.

Lubricating Oils.—There are two principal kinds of lubricating oil employed, namely, the mineral hydrocarbon, and the vegetable class, of which castor oil is an example. The former oil, derived from distillation treatment of crude petroleum, is the more widely employed for ordinary engines, whilst the latter, on account of its high viscosity at the working temperatures in the cylinder, is sometimes used for racing automobile engines, and also for aircraft engines—chiefly for running-in purposes.

A good lubricating oil should, at all temperatures, form an oil-film between the two sliding parts. If the oil becomes very thin at high temperatures, i.e., of low viscosity, the pressure between the sliding parts may cause it to break down, or be squeezed out, when the metal parts will make contact and may then seize up. The oil must therefore have sufficient viscosity at the engine working temperatures, but on the other hand it must not be too viscous (or thick) at atmospheric temperatures, or it will be difficult to start the engine. The viscosity falls off fairly rapidly with temperature increase, so that those engines which tend to run " hot " should be lubricated with the more viscous oils. For these reasons, the oil used on air-cooled engines is always " thicker " than for water-cooled ones, and for any engine the oil used in the winter (or in cold climates) is less viscous than that used in the summer (or in hot climates). Many reputable lubrication oil firms supply " winter " and " summer " grade oils, and a special series of oils of different viscosities to suit various engines.

Most automobile and aircraft engine oils are of the mineral type derived from higher boiling constituents of petroleum by distillation processes. The actual properties of these mineral oils are to some extent governed by the origin or character of the crude petroleums from which they are derived. Thus, oils obtained from Russian crudes are usually naphthenic; from Texas and other sources, asphaltic, and from Pennsylvanian sources, paraffinic. The latter have the lowest densities and are less liable to oxidation; the asphaltic oils are the heaviest and more readily oxidised.

Compound oils consisting of mineral and vegetable oils, e.g., castor oil, are also employed, but with modern high piston temperatures and speeds the gumming effects on piston rings is a serious drawback, since it reduces the periods between engine overhaul.

The more modern mineral oils, produced by special processes of refining and blending have proved fully satisfactory for engines of high performance, there being no difference in wear or lubrication as compared with castor base oils.

More recently, special oils, including the doped, hydrogenated, volatilized, etc., varieties, produced by new processes have appeared and important claims in regard to non-oxidising, gumming and sludging tendencies have been made for such oils.

Changing the Engine Oil.—If an oil is used in an engine with badly fitting pistons, or for a long period, it blackens and becomes diluted with petrol (or other fuel used). This causes a serious loss of lubricating value, apart from the fact that particles of carbon and other solid matter are also suspended in it. It is therefore essential for the best results to change the oil periodically. In the case of new engines, where solid matter, chippings or fragments of metal are liable to occur, it is advisable to drain out the sump after 250 to 500 miles of road-running. Afterwards, the sump should be drained once every 1,500 to 2,500 miles of running, although the car will run for much greater distances, but actually not so efficiently, and with the risk of excessive wear.

Lubricating Systems.—Automobile engine lubrication systems that have been or are in present use may be classified under definite headings, as follows: (1) *The Fly-wheel Splash Type;* (2) *The Assisted Splash and Constant Level Type* (3) *The Low Pressure System;* and (4) *The High-Pressure System.* Before referring to these systems in detail let us consider for a moment the principal working parts which require lubrication. In order of importance these are as follows: (1) The Cylinder Walls and Pistons. The

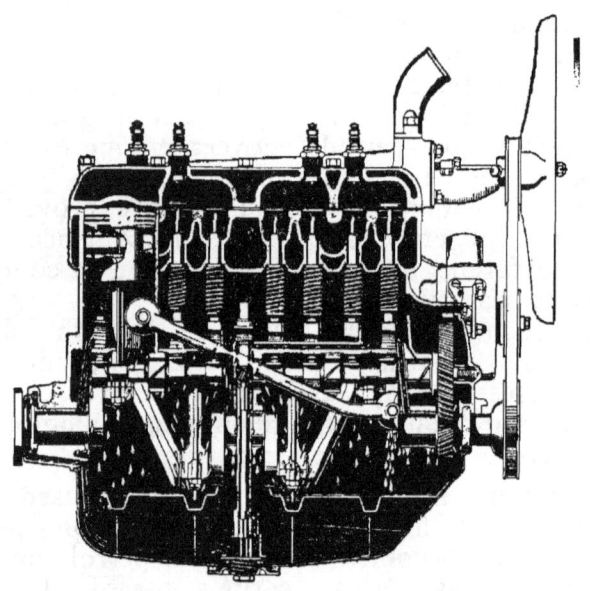

Fig. 225.—The earlier Ford Engine Lubrication System.

piston of a modern engine has an average bearing pressure due to piston thrust of 250 to 300 lbs. per sq. in. Its crown temperature is about 250° C. to 370° C. (2) The Connecting-Rod Big End Bearings and Crank-pin. White-metal is used for solid bearings, but roller bearings give rather less friction and require less lubricant. (3) The Small End Connecting-Rod Bearing. This rod end rocks a little only on its pin, but the temperature of the piston is relatively high, so that the lubrication of these parts is important.

(4) The Main (Crank-shaft) Bearings. (5) The Cam-shaft, Cams and Bearings. (6) The Valve Tappets and Guides. (7) The Timing and other Gears.

Most lubricating systems aim at oiling the main and crank-pin bearings first, and allow the surplus oil to be splashed about so as to lubricate the cylinder walls, pistons and other parts. Where the former are liable to be over-lubricated, a special oil-scraper ring, or groove, is arranged in the lower part of the piston.

The Fly-wheel Splash System.—In this method which was once used on Ford engines the fly-wheel has below it an oil-collecting sump, or trough, this being the lowest part of the enclosed crank-chamber, so that all the oil drains down into it. The fly-wheel rim runs in this oil, and the oil which is flung off the rim tangentially by centrifugal action is caught by specially shaped troughs or channels cast in the upper part of the crank-chamber, whence it gravitates to the main bearings and gearing. Below each connecting-rod, in its lowest position, an oil trough is arranged, into which an oil scoop, or projection, on the lowest part of the rod just dips; the oil thus caught is led through suitable holes to the big-end bearings. These troughs fill up automatically.

This system is not now used in automobile engines, having been superseded by the splash and pressure methods.

The Assisted Splash and Constant Level System.—In this case an oil pump replaces the fly-wheel oil-flinger of the previously described system, and oil is pumped from an oil reservoir or sump, situated at the lowest part of the crank-chamber, to the connecting-rod troughs (Fig. 226), whence it not only feeds the big-end bearings, but is also splashed, or flung about in the form of a spray to all the other parts, including the cylinder walls, to oil troughs above the main bearings, timing gears and cam-shaft bearings (whence the oil flows by gravity to these parts).

Fig. 225 shows the lubrication system employed in the earlier Ford engines, subsequent to the fly-wheel lubrication system previously fitted. In this case a submerged type of oil pump in the sump, driven by a vertical shaft, forces oil to the main bearings, whence the oil drains back into the sump. The connecting-rod big-ends dip into the oil troughs at the bottoms of their strokes, and thus lubricate the big-end bearings. The oil from the sump is first pumped up to a trough above the cam-shaft, whence it gravitates downwards along specially arranged passages to the main crank-shaft bearings.

The splash system, although now superceded by the high pressure system was very efficient and reliable; it also possessed the merits of simplicity, durability and cheapness. The oil pump required for this system was

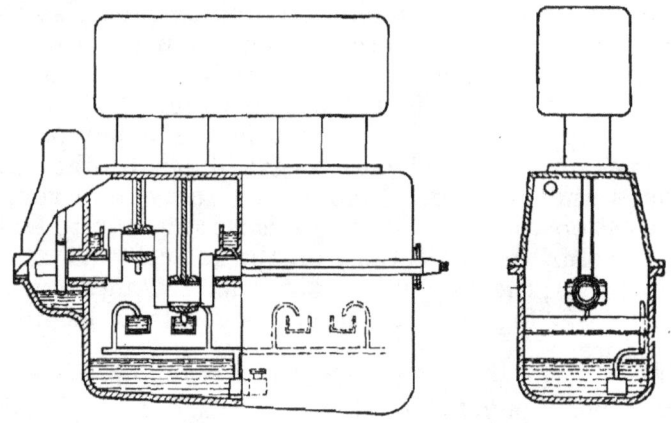

Fig. 226.—The Splash System of Librication, showing Oil Sump, Connecting-rod, Oil Troughs, and Oil Supply Pipes.

a low-lift, non-pressure type, since it had only to raise the oil to the troughs. Any surplus oil draining down the sides of the crank-chamber flowed through wire gauze gratings or sieves usually on a level with the bottoms of the troughs, whence it drained into the sump below; this enabled all solid matter to be trapped by the gauzes.

In this system, an oil level indicator, either of the two-tap type described, the float and wire, or the ordinary gauge glass type, was employed. It was also usual to fit some form of indicator in the system to show that the oil-dump was working. The simplest indicator was a small tap situated in the delivery side of

the pump.   On opening this oil flowed out if the
pump is working.   Another indicator consisted of a
brass fitting on the lower part of the dashboard, in
which a small plunger and wire pulsated in synchronism
with the strokes of the pump; a wire moved upwards
and remained there when a rotary pump was used.
Another readily observable indicator was an ordinary
oil gauge-glass on the dash, which showed the oil
streaming down from a central jet inside.   Sometimes

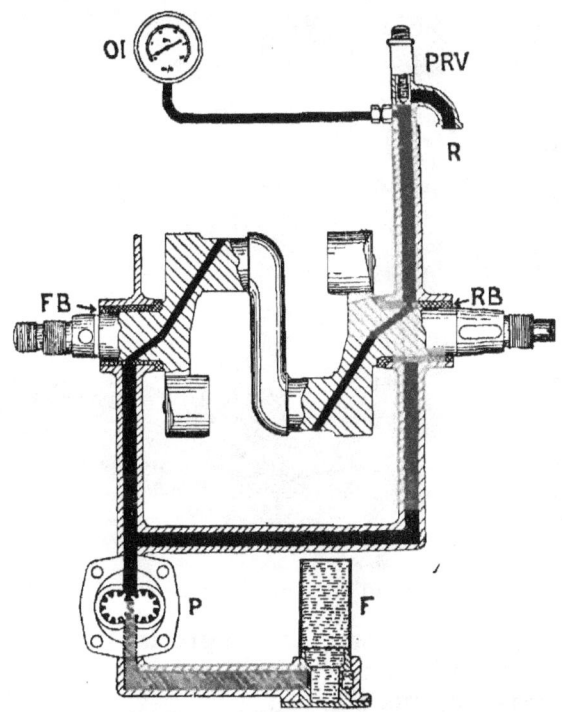

FIG. 227.—The Jowett Pressure Oiling System.

the delivery oil was made to impinge upon a vane, or
small impeller, situated in a small brass box provided
with a glass window; the movement of the vane or
impeller at once showed that the pump was working.

The Semi-Pressure System.—This system consists in
pumping oil under a pressure (of 5 to 10 lb. sq. in.) to
the main bearings, whence it passes through holes
drilled in the crank-shaft to the big-end bearings,

impelled partly by the pump pressure and partly by centrifugal action. The oil escapes from the big ends and is flung out, whence it lubricates by splash the cylinders, gudgeon pins and other parts. Sometimes the oil was supplied under pressure (from 5 to 8 lbs. per sq. in.) to the main bearings only, and oil troughs were used for the connecting-rod big end lubrication. It is essential in the former case that the oil ways be kept quite clear, otherwise any choking will cause the

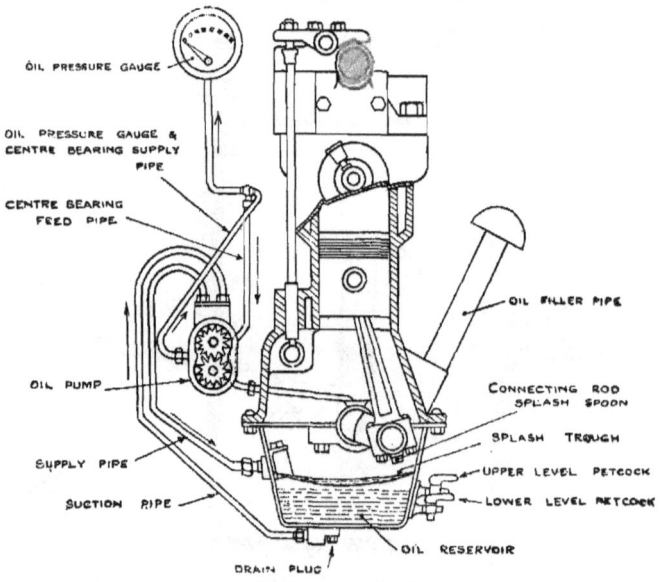

OIL PRESSURE GAUGE

OIL PRESSURE GAUGE & CENTRE BEARING SUPPLY PIPE

CENTRE BEARING FEED PIPE

OIL FILLER PIPE

OIL PUMP

CONNECTING ROD SPLASH SPOON

SPLASH TROUGH

SUPPLY PIPE

UPPER LEVEL PETCOCK

SUCTION PIPE

LOWER LEVEL PETCOCK

OIL RESERVOIR

DRAIN PLUG

Fig. 228.—A Semi-Pressure Lubrication System.

system to fail. Similarly any appreciable wear in the main bearings will cause the oil to leak away, and the big ends may become starved.

Fig. 227 illustrates the 7 H.P. Jowett two-cylinder opposed engine lubricating system. This diagram will serve also to show the various items referred to in the preceding section. Oil is drawn by the gear-wheel type pump P from the sump, through a fine brass gauze filter F. From the pump it is delivered through holes or passages cast integral with the crank-case to the front main bearing FB, whence it flows through the drilled crank-shaft to the left-hand crank-pin, and there lubricates the big-end bearing of the corresponding connecting-rod. The black lines show the paths of the oil under pressure. At the same time oil is delivered from the pump to the right-hand main bearing *RB*, and thence to the right crank-pin and its big-end bearing.

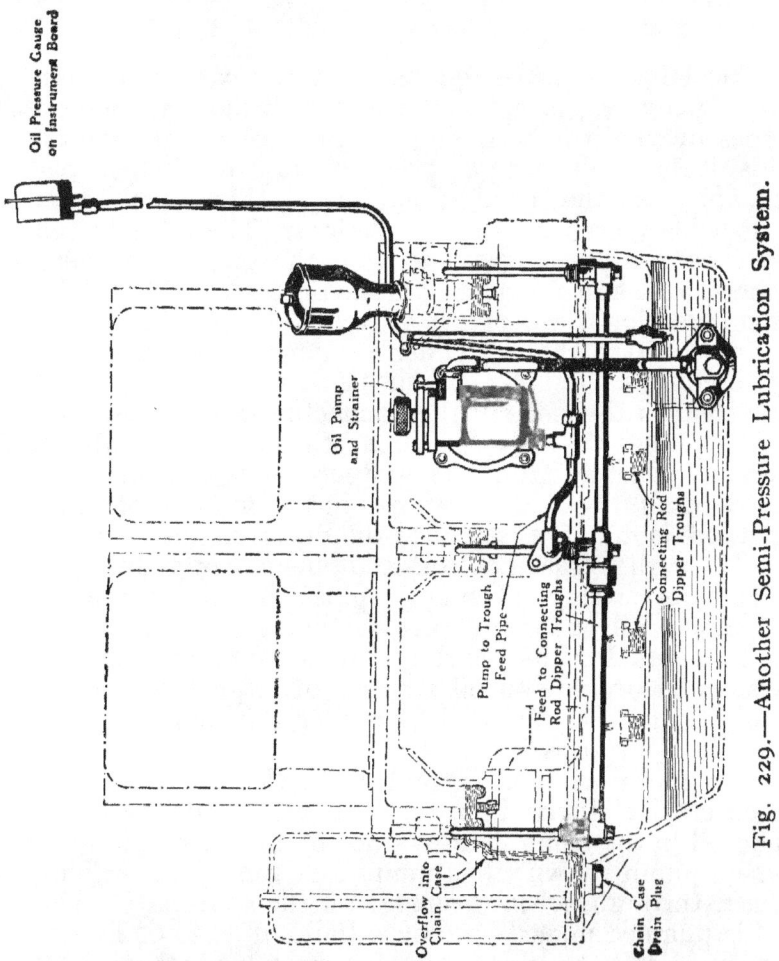

Fig. 229.—Another Semi-Pressure Lubrication System.

From *RB* the oil is also led to the oil-pressure gauge *OI*, and to the pressure release valve *PRV*, situated at the rear of the engine in front of the fly-wheel. This valve is adjusted to prevent excessive oil pressure, any excess of oil returning by the pipe *R* *via* the crank-shaft rear bearing plate, which is drilled for the purpose. The oil pressure gauge reads about 4 to 6 lb. sq. in. at idling speeds, and 14 to 16 lb. sq. in. at maximum engine speed.

In common with most oil pressure systems the gauge reads high (22 to 25 lb. sq. in.) when the engine is first started up with the oil cold.

The High-Pressure System.—In the case of certain high-speed engines, and engines fitted in modern types of car, it is becoming the practice to feed the oil positively under a high pressure (25 to 60 lbs. per sq. in.) to the friction surfaces. This method is also widely used on aircraft engines. The oil is forced by means of a pump, usually of the plunger or gear wheel type, and submerged in the oil sump, direct to the main bearings, whence it travels through the drilled crank-shaft to the big ends. Thence it is led to the gudgeon pins, usually by means of small copper tubes attached to the sides of the connecting-rods or by rifle-drilled holes through the connecting-rod webbs themselves. The oil from the gudgeon pins is sometimes used to lubricate the pistons, but a more preferable method is to have separate oil ducts to the cylinder walls, so that fresh cool oil is supplied. In addition to the main bearing oil supply, pipes are also arranged to supply oil under pressure to the camshaft bearings and oil-jets for the gearing. It is essential with this system to provide an oil strainer of large area around the oil pump supply, and to render this accessible for cleaning purposes.

A pictorial view of a modern lubrication system as used on the Morris Four engines is given on Fig. 231. The oil in the sump is drawn up, by means of the gear wheel pump shown; this pump is driven by an inclined shaft, through helical gearing, off the cam-shaft. The oil is pumped under a pressure (hot) of 30 to 60 lb. per sq. in. direct to the main bearings and thence by drilled passages in the crank-shaft to the crank-pins and big-end bearings.

The cam-shaft bearings are also pressure-lubricated from the same supply, and there is a separate feed to the timing chain. The cylinder walls are lubricated by splash from oil escaping at the big-end bearings.

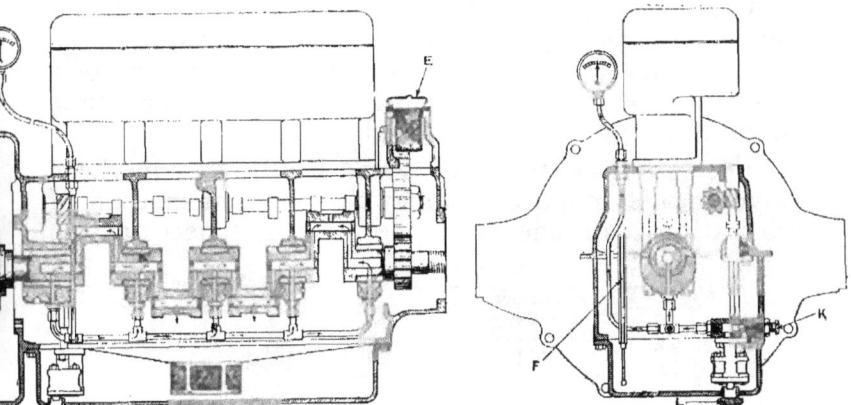

Fig. 230.—A Typical Pressure System, showing Drilled Crank-shaft, Oil Filler (E), Dipper Rod for Oil level (F), Pump (K), and Drain Plug (L).

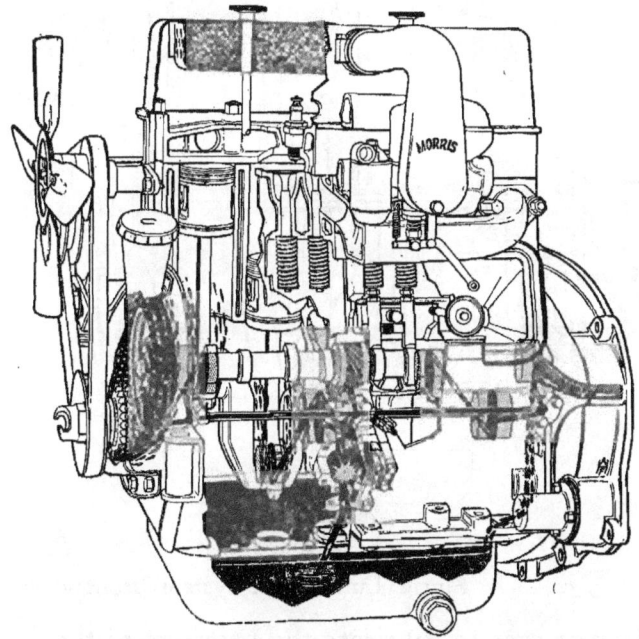

Fig. 231.—Morris Lubrication System

The surplus oil from the cylinder walls, big-ends, etc., drains back into the sump, having to pass on its way a large tray-type filter; this removes any solid matter.

When starting from the cold it is usual to obtain a high gauge reading, but this falls as the oil warms up.

The oil level in the sump is indicated by means of a dip-stick, on the left side of cylinder. The sump contains about one gallon of oil.

Lubrication of Overhead Valve Engines.—Overhead valve engines require a somewhat different method of lubrication than side-valve ones, for it is necessary in

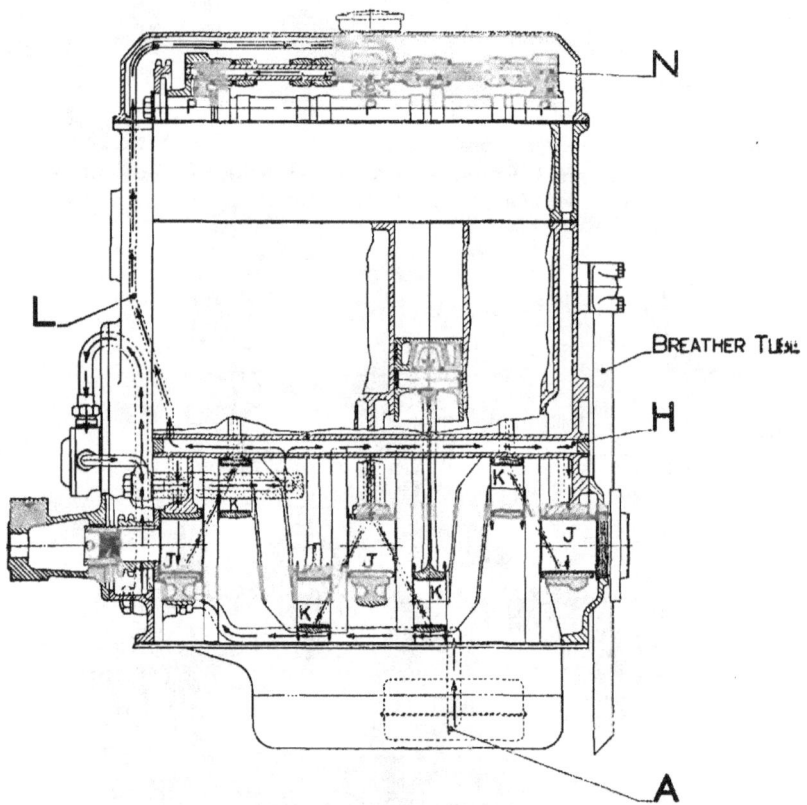

Fig. 232.—The Singer Lubrication System (front view).

the former case to lubricate the bearings of the rocker-arm shaft; in most instances the valve stems and cam contacts with the valve stem ends are also lubricated.

A typical lubrication system, viz., that of the **Singer** engine, is shown in Figs. 232 and 233 in front and side views, respectively.

The latter illustration also indicates the positions of and connections to the oil-pressure gauge and to the Tecalemit felt-filter pattern oil cleaner.

The fresh oil is poured through a filler cap on the engine top cover, and it makes its way down passages through the cylinder head and block castings into the sump. Oil is drawn from the sump through a large

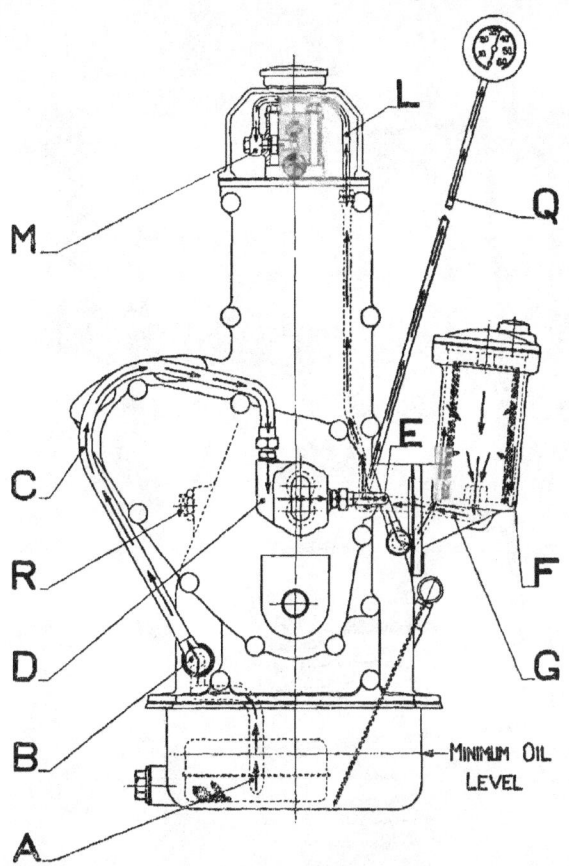

Fig. 233.—The Singer Lubrication System (side view).

area wire gauze filter up pipe (A) into union (B), and through pipe (C) to the pump (D).

It is then discharged through pipe (E) into the filter housing (F), where it is forced through a fabric filter element into the main oil gallery (H). From here the oil passes through leads to the three main bearings

(*J*), and through the crank-shaft to the big-end bearings (*K*), surplus oil from the big-end bearings being thrown on to the cylinder walls.

An upward lead (*L*) is taken from gallery (*H*) to the rocker shaft (*N*) for the purpose of lubricating the valve

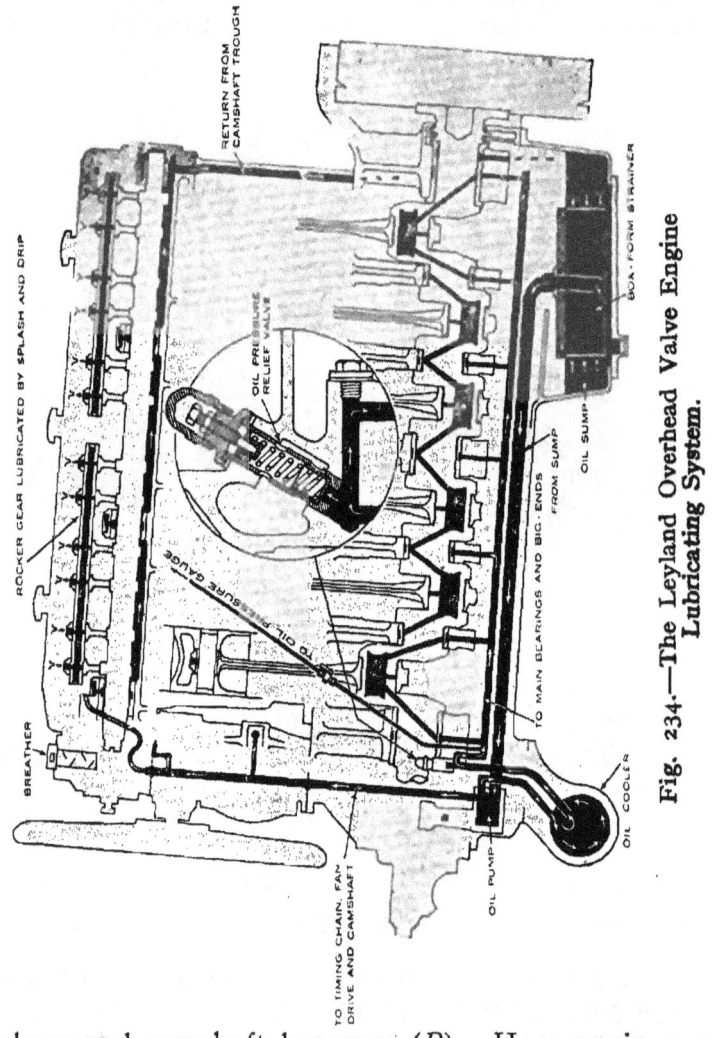

Fig. 234.—The Leyland Overhead Valve Engine Lubricating System.

rockers and camshaft bearings (*P*). Here again, surplus oil from the valve rockers and cam-shaft bearings drains back into the sump and is, of course, filtered before again passing into circulation.

A lead (Q) is taken from gallery (H) to the pressure gauge and an oil release valve (R) is incorporated in the system for oil pressure adjustment.

The oil pressure should be 20 to 25 lbs. per sq. in. when the oil is warm.

Another example of a well-designed high pressure lubrication system is that of the Leyland six-cylinder overhead valve engine illustrated in Fig 234. From the oil pump on the lower left-hand side filtered oil drawn through the box-form strainer in the oil sump is forced through the ribbed oil cooler and filter unit below the crankcase and thence delivered under pressure to each of the seven main crankshaft bearings.

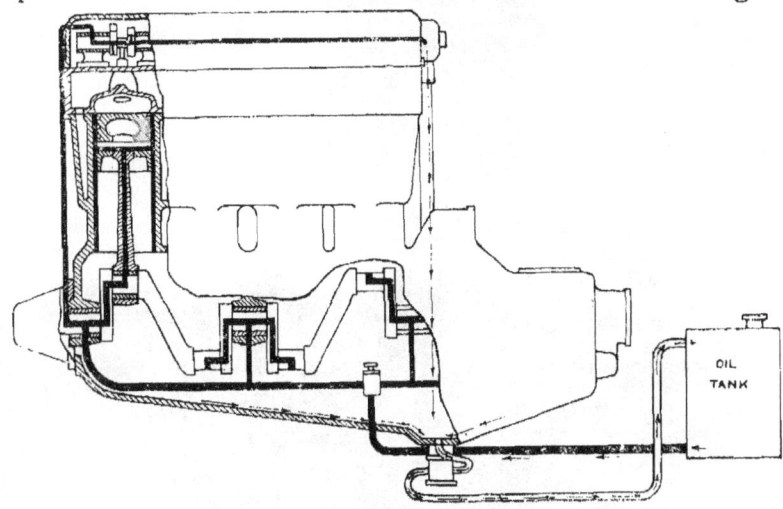

Fig. 235.—Illustrating the Dry Sump Lubrication Method.

This oil then proceeds through holes in the main journals into the hollow crankshaft and finally escapes through holes drilled in the crank pins so as to lubricate the big-end bearings. Another supply is taken vertically to the top of the engine, lubricating on its way the timing chain and fan drive. It lubricates the overhead gear by splash and drip and the surplus oil then returns from the camshaft trough down a vertical passage on the right and into the oil sump below.

Between the oil cooler and the main crankshaft bearing oil passages an oil pressure relief valve is arranged—as indicated by the central diagram and direction arrow. The pressure regulation screw and its locknut are enclosed within a metal cap seen above the valve itself.

The oil pressure gauge connection is taken from a point on the outlet side of the oil pressure relief valve,

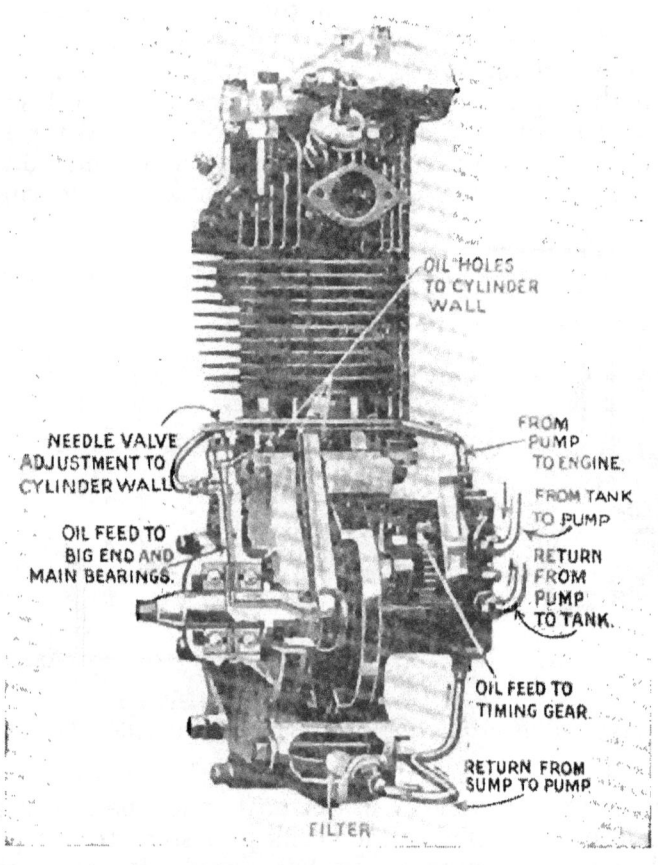

OIL HOLES
TO CYLINDER
WALL

NEEDLE VALVE
ADJUSTMENT TO
CYLINDER WALL

OIL FEED TO
BIG END AND
MAIN BEARINGS.

FROM
PUMP
TO ENGINE.

FROM TANK
TO PUMP

RETURN
FROM
PUMP
TO TANK.

OIL FEED TO
TIMING GEAR.

RETURN FROM
SUMP TO PUMP

FILTER

Fig. 236.—The Dry Sump Method of Lubrication used on a Motor Cycle Engine.

between the latter and the main crankshaft bearing oil passage.

The Dry-Sump Method.—In the previous system the main oil supply is kept in the sump below the engine crank-case. There is another system which is some times used on the more expensive car, and also on most aircraft, engines in which there is no oil sump in the crank-chamber, but the supply is contained either in a separate tank on the dash-board, or in a separate reservoir, provided with air-cooling fins on the side or below the engine. Two pumps are employed in this case, one to force the oil under (high) pressure as in the preceding case, to the main, big-end and other bearings, directly, whilst the other, or "Scavenger," pump, which is the larger in capacity, sucks up the oil which drains down to the bottom of the crank-chamber, and returns it to the oil reservoir. The principal advantage of this arrangement is that the oil is cooled in its circulation, and therefore has a better lubricating value. The oil is also controlled in its circulation.

Motor Cycle Engine Lubrication.—Motor cycle engines now employ either the *Wet* or *Dry Sump* lubrication systems, somewhat on car engine lines.

In the former method the oil is contained in the crank-case sump, whence it is drawn by a mechanically driven pump and forced to the crank-shaft bearings and cylinder walls. It then drains back into the sump again.

In the latter method, which is the most favoured, two mechanical pumps are used and the oil for lubrication is contained in a separate tank. The *pressure pump* draws the oil from this tank and supplies it to the various engine parts under pressure; the used oil drains back into the sump. The *suction, or scavenging, pump* then returns this oil back into the oil reservoir, or tank, for use again.

The dry sump method is the more satisfactory one from the point of view of oil economy and efficient lubrication; moreover, it is independent of the inclination or position at any time of the engine.

Fig. 236 illustrates a typical dry-sump lubrication system of a motor-cycle engine.

Cylinder Wall Lubrication.—As mentioned earlier in this Chapter the usual method of lubricating the cylinder wall is by means of oil splashed from the rotating parts, e.g., the cranks and connecting rod ends.

L

The more recent tendency is to provide positive lubrication of the cylinder walls by means of an oil jet or jets from small holes drilled through the upper half of the big end bearing and a corresponding hole in the crank pin.   These holes come into line once every revolution and owing to their positions allow the high pressure oil supplied through the drilled crankshaft to spurt upwards on to the cylinder walls. Usually in modern engines, a surplus of oil, above that actually required for lubricating the walls, is supplied to the walls and the oil control rings then scrape off most of this oil leaving the correct quantity for lubrication; this ensures adequate oiling of the walls, reduces cylinder wear and also tends to cool them more efficiently than would otherwise be the case.   The oil consumption, however, is not increased owing to the efficient scraping action of the oil control rings.

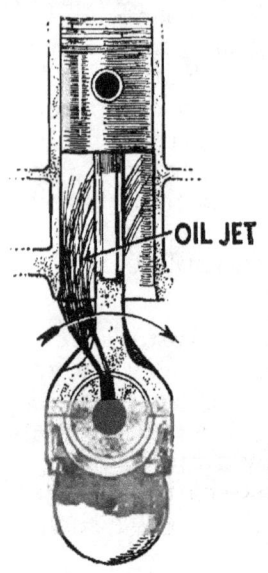

Fig. 237.—Method of Lubricating Cylinder Walls (Vauxhall).

In one or two instances a positive supply of oil has been led by means of a special pipe or oil passage to ensure efficient lubrication.   Sometimes this additional oil is employed only whilst the engine is starting from the cold and warming up.

Fig. 238 illustrates the Tecalemit " Lubrostat " device for admitting oil to the upper part of the cylinders during the initial warming up period.   It will feed small predetermined amounts of oil to the inlet manifold or cylinder walls direct until the oil in the engine sump attains its proper working temperature.   The time of this oil admission varies from about 30 minutes in winter to five minutes in summer, and the period is controlled thermostatically by the oil temperature. The device comprises a valve body A with valve B controlled by a bimetallic U-shaped element C housed in a cap D.   The latter is inserted in a hole in the engine

sump and held by the nut E which makes a joint with
the washer F. Thus the sump oil temperature will
control the degree of flexing of the bi-metallic element
which in turn operates the valve B. The oil enters
through the union G from any convenient source, such
as a T-piece on the oil pressure gauge line, and passes
through the filter H into the body A and cap D and
flows past the valve B and a needle valve I, which is
adjustable and held in position by a spring J. The oil
is then led by a small bore copper pipe (about $\frac{3}{16}$ in.

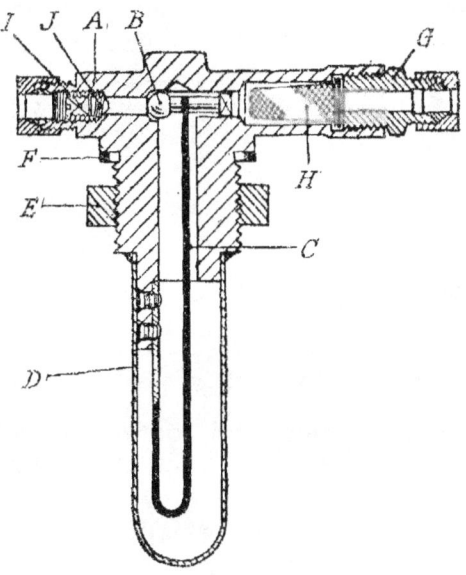

Fig. 238.—" Lubrostat " device for
Cylinder Lubrication.

outside diameter) to the inlet manifold or base of each
cylinder; it is found that good results are obtained if
the oil pipes are led to holes drilled through the cylinder walls opposite the gudgeon pin centre when at its
lowest position. When the engine sump oil has
warmed sufficiently the thermostatic element C shuts off
the valve B so that no more oil can pass to the engine
through this device.

Another efficient upper cylinder lubricating scheme
is the S.U. " Thermoil " lubricator shown in outline in

Fig. 239. It works in conjunction with the S.U. thermostatic carburettor control. When cold starts are made, namely, at temperatures below 35° C., the carburettor thermostat which provides automatically for the rich starting mixture also arranges to supply a small amount of oil (such as Adcaynes upper cylinder lubricant), to the inlet manifold. At about 35° C. when the thermostat switches the engine over to the normal fuel mixture the supply of oil is cut off. The Adcayne fluid container is mounted on the back of the dashboard

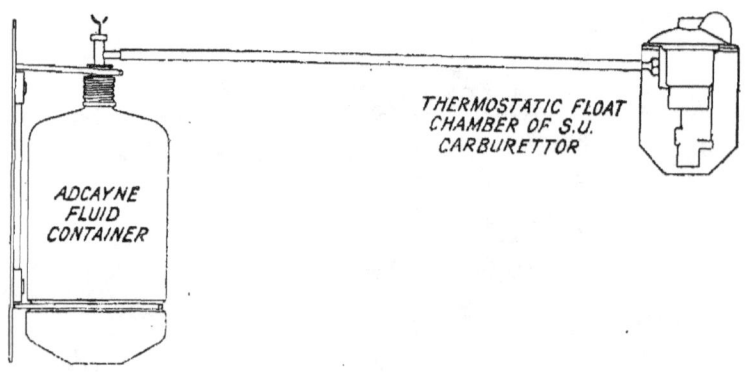

FIG. 239.—The S.U. Thermoil Cylinder Lubrication System.

under the engine bonnet and contains sufficient oil for 2,000 cold starts, i.e., for about 10,000 miles of road running .

Overhead Valve Gear Lubrication.—The method of lubricating overhead valve gear, whether of the camshaft or push rod and rocker arm pattern, now follows the general plan of the example illustrated in Figs. 232, 233 and 234, and all modern engines include provision for such positive lubrication. In certain instances the oil for lubricating the overhead valve mechanism is supplied from a relief valve in the main oil system which provides oil at a pressure of a few pounds per sq. in. instead of at the full pressure used for the main and big end bearings.

A typical example of a well-designed lubrication system for a push-rod and rocker arm mechanism is shown in Fig. 240. The oil from the low pressure system previously referred to enters near the lower left side and

passes upwards through the hollow stem of the push
rod, through a hole drilled in the ball end of the latter
and along another passage in the rocker arm to the main
bearing of the latter. It thus lubricates both the ball
and cup joint and also the rocker arm bearing. The
oil escaping from these bearings flows downwards
through the space formed between the push rod and
another larger diameter tube—which is fixed— and
thence back to the engine sump.

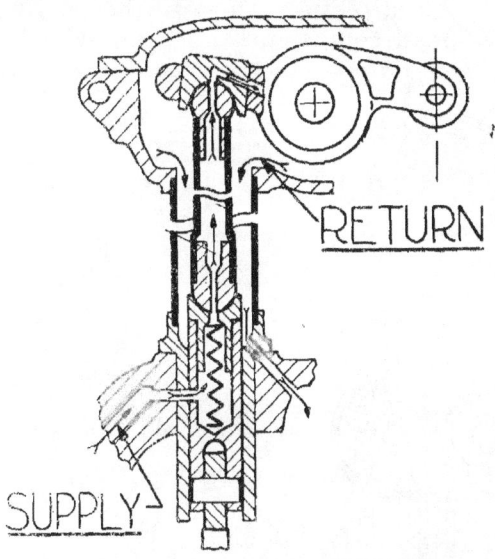

Fig. 240.—Overhead Valve Lubrication Method (Wright).

Oil Pumps.—Except with dry sump systems, the oil
pump can be made of quite simple design. For the high
pressure and assisted splash systems the gear-wheel
type of pump illustrated in Figs. 227, 228 and 241 is
quite satisfactory. It will run almost indefinitely, and
will supply an adequate amount of oil at pressures up
to 8 or 12 lb. per sq. in., if necessary. The suction is
on the side where the gear-wheel teeth move away, and
the delivery on the side where the teeth approach.
Each tooth-space carries round an equivalent volume
of oil and discharges it on the delivery side. One
gear-wheel only is driven, the other meshing with it is
thereby driven. It is usual to run this type of pump at
one-half engine speed, and to submerge it in the oil

sump so that the spaces are always full of oil, and no "priming" is therefore necessary. For a bigger delivery, three gear-wheels may be employed.

The rotary eccentric type of pump occasionally used for low pressure lubrication, consists of a driving drum set eccentric relatively to the cylindrical pump casing, and provided with a pair of sliding vanes pressed outwards by means of a spring, so that the other ends of the vanes are always touching the casing. The oil is forced in the direction shown, the sliding vanes acting more positively than the gear-wheels. It is

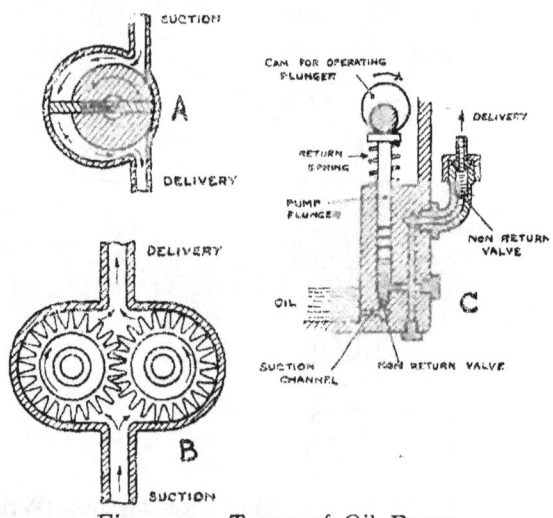

Fig. 241.—Types of Oil Pump.

A—Rotary.   B—Gear Type.   C—Plunger.

possible to attain pressures of 30 to 50 lbs. per sq. in. with suitably designed pumps of this type. The wear of the blades is the one disadvantageous feature of this pump.

The plunger type pump is occasionally employed with medium and high-pressure lubricating systems. This pump is quite positive, and has a much higher suction lift than the preceding types, so that it can, if necessary, be placed above the sump. As a rule, however, the pump is driven either by one of the valve cams, or by a special eccentric cam, or eccentric sheath (Fig. 241) on the cam-shaft. On the upward stroke of the plunger oil is sucked from the sump past a non-return ball-type valve opening inwards; during the down stroke the oil cannot pass this valve and is therefore forced past another ball valve, opening outwards, into the delivery pipe to the bearings.

Fig. 242 shows the earlier Morris plunger pattern oil pump which is placed vertically within the crankcase

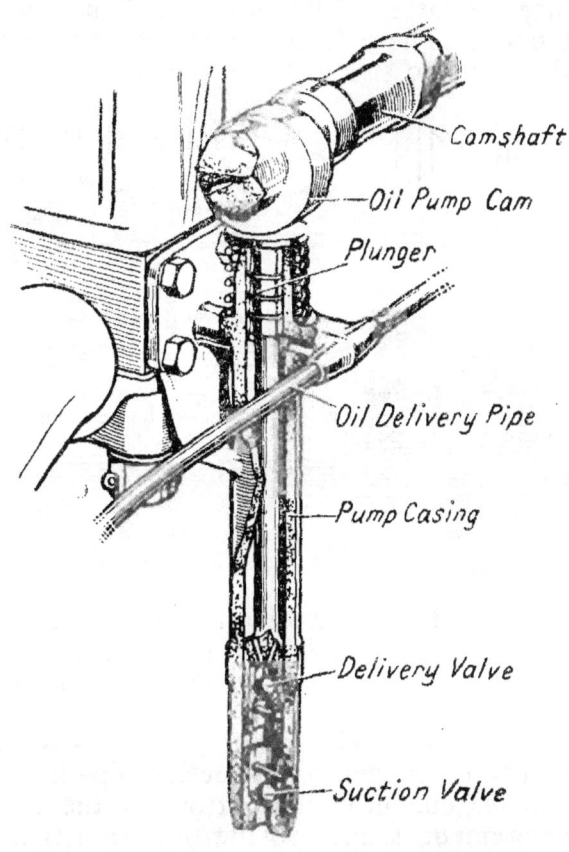

Fig. 242.—Morris Plunger-type Oil Pump.

and operated by means of an additional circular eccentric cam formed on the main camshaft. The upward stroke of the plunger draws oil from the sump past the lower spring-loaded ball valve whilst the next downward stroke closes the bottom valve and forces the oil upwards past the upper delivery valve, through holes in the plunger stem and thence to the oil pressure delivery pipe seen above, whence the oil is led to the main bearings.   In later model Morris engines (see Fig. 231) the gear-wheel pump is employed.

It is usual to surround the whole suction side of the pump with a cylinder of wire gauze, of ample dimensions, to trap any solid matter; this is an important point since the latter might otherwise cause the valves to stick open.   There is also a gauze strainer on the top of the sump to strain the surplus oil which returns

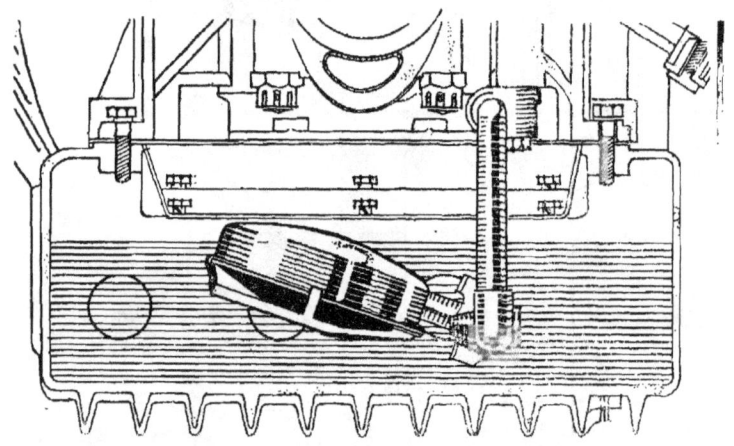

Fig. 243.—Constant Level Oil Intake Device.

to the sump, and a gauze also in the oil filler.   Sometimes the oil pump is made as a separate unit attached to the outside of the crank-chamber, where it is readily accessible.

Constant Level Oil Intake Device.—With the ordinary type of oil pump the suction pipe is arranged with its lower end near the bottom of the oil sump. This arrangement is open to the objection that as the level of oil in the sump changes so does the flow of oil to the bearings, since the pump is working under a varying suction head.   Another disadvantage is that if there happen to be any very fine impurities in the oil these will be found near the bottom of the sump and tend, therefore, to be drawn into the oil pump.

A method of overcoming these difficulties that is employed in certain modern car engines, of which the

Wolseley is an example, is to provide a float mounted on a hollow arm that can hinge about a horizontal axis, so that as the oil level in the sump changes the float moves up or down and the arm moves with it. The lower face of the float carries a gauze filter inlet which allows clean oil to flow along the hollow hinged arm and up the vertical suction pipe to the oil pump. With this method the oil flow is always constant and is unaffected by surging effects due to road inequalities, cornering or acceleration effects. The Wolseley oil intake is shown in Fig. 243, and it may be mentioned that a finned aluminium oil sump is employed for assisting in cooling the oil returned in the sump; the latter has a capacity of 2½ gallons for the 14 H.P. rating engine.

Pressure Indicators and Relief Valves.—It is necessary with pressure systems of lubrication, whether semi- or full-types, to provide a pressure gauge (of the ordinary Bourdon type) on the dash-board to indicate the oil pressure on the delivery side, as shown in Figs. 227 and 228. In cold weather and at starting the oil pressure will often be much lower than when the engine is thoroughly warmed up. In other cases the pressure is often higher for a short period after starting up, due to the inertia and resistance of the oil in the system, and as the oil becomes warmer, the pressure will fall, as the oil experiences less resistance; this is often the case with plunger-pump systems. The pressure gauge should be connected to the pump delivery side before any other pipes are led off. It is the usual practice to mark, conspicuously, the normal working pressure graduation on the gauge so that abnormal reading can be quickly detected. It is also necessary in high-pressure systems to fit a relief valve to obviate any excessive oil pressure due to an obstruction in the system. All that is required is a spring-loaded valve of the conical or ball type, opening outwards, and placed in the delivery side near to the pump. The spring is made adjustable, by means of a screw cap, so that the valve can be arranged to open at any given pressure. The oil thus discharged is led directly to the suction side or to the sump again.

L*

Adjusting Pressure Relief Valve.—If it should become necessary to regulate the oil pressure the adjustment should be made to the screw cap controlling the spring pressure of the relief valve. Before making this adjustment, however, it is important to have the oil at its normal working temperature and also to run the engine at a speed corresponding to about 25-30 M.P.H. in top gear. Usually, after loosening the lock nut of the screw adjuster the latter is screwed inwards to increase the oil pressure and outwards to reduce it; the lock nut should be tightened securely after making the adjustment.

Upper Cylinder Lubrication.—In order to ensure adequate lubrication of the cylinder walls above the piston a special mineral lubricating oil may be mixed with the petrol; this procedure is recommended by some car manufacturers. In the case of new engines it is generally advisable to provide for this upper cylinder lubrication. Suitable oils for the purpose include Mixtroil and Miracle Oil; colloidal graphite containing mineral oils is also recommended.

The quantity of oil used for this purpose is usually in the proportion of about $\frac{1}{8}$-$\frac{1}{4}$ pint per gallon of petrol and it can either be mixed with the petrol or introduced into the induction manifold by means of a special device marketed for this purpose. If colloidal graphite is used the recommended amount is $\frac{1}{2}$ oz. per two gallons of petrol.

Oil Failure Indicators.—Certain makes of car and commercial vehicle engine are now provided with a device in the form of an oil-pressure operated diaphragm, which, when the lubrication system is working properly, breaks a pair of electrical contacts in a circuit containing an electric bulb placed behind a red window on the dashboard. When the ignition is switched on and the engine is stopped or idling the circuit is " closed " by a spring so that the red window lights up. When the engine is speeded up, however, the diaphragm is deflected by the oil pressure so that the light goes out. Should the red light appear under ordinary working conditions, however, this is a sign of oil pressure failure and the engine should at once be stopped for investigation.

Oil Cleaners.—After the lubricating oil has been in use for some time it becomes dark in colour, due principally to suspended carbon particles. Although gauze strainers are fitted to both the oil filler and oil pump suction parts, these are too coarse in mesh to deal with finely suspended matter, so that special oil cleaners are now fitted to get rid of the latter.

There are two principal types of oil cleaners, namely, (1) The Centrifugal, and (2) The Absorbent Type.

In the centrifugal types the hot oil, after its passage through the engine, is forced, under pressure, into a circular chamber whence, by means of spiral vanes, it is swirled around at a high speed, when all solid matter, being heavier than the liquid oil, is flung outwards by centrifugal action. This solid matter is in the form of a fine sludge and is caught in special channels, whence it is removed by hand at intervals.

Sometimes the crank-pins are made hollow and dirt-traps provided for solid matter thrown outwards by centrifugal action; in other cases the sludge has been caught in the flanged rim of the fly-wheel, whence it was removed every 15,000 miles or so.

One notable advantage of oil cleaners is that they reduce the amount of wear on the cylinders, pistons and bearings; they do not, however, get rid of crank-case dilution products.

Felt Type Oil Filters.—This type of lubricating oil-cleaning device is placed in the oil circuit so that all of the oil circulating around the engine is constantly filtered, thereby, extracting all solid deposits in the oil.

The felt element type of oil filter has a vertical cylinder of star section made of special felt material and the dirty oil in passing from the outside of the element to the inside

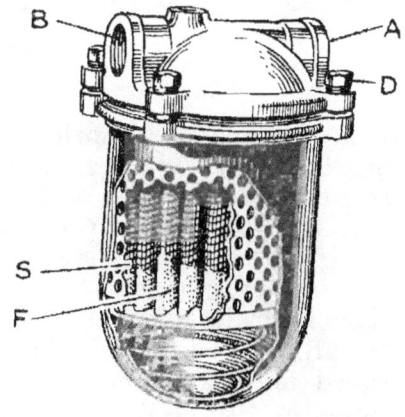

Fig. 244.—Typical Felt Element Type of Oil Cleaner.

has the solid matter removed by the felt. It is necessary to clean this type of filter every 2,000 to 3,000 miles and to discard the felt element after 10,000 to 15,000 miles.

The later types of oil filter, of which the Tecalemit (Fig. 244) and Purolator (Fig. 245) are good examples, employ metal supported fabric-covered elements of relatively large area to trap the fine solid matter, so that clean oil only passes into the central space, whence it is supplied to the engine.

Referring to Fig. 244, the dirty oil enters at A and passes into the space between the outside of the filter element and the inside of the outer metal container. The filter element consists of fabric F mounted on a gauze S and it is held against a cork joint in the head by the spiral spring shown. As long as the filter is

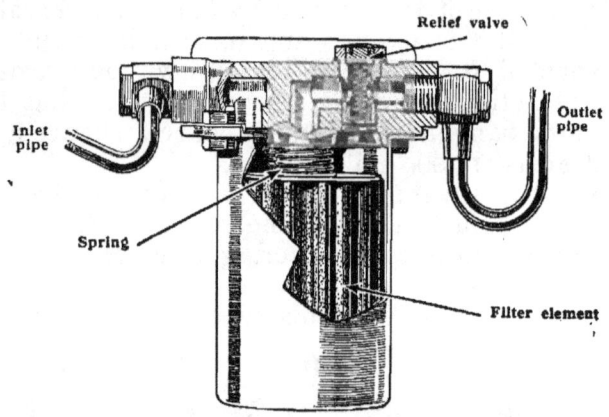

Fig. 245.—The Purolator Oil Filter.

working properly the spring retains the element in position, but should it become choked with solid matter the pressure developed on the inside of the element will cause the spring to depress when the oil is by-passed through the holes provided for this purpose, the engine will therefore not tend to become starved of oil.

The Purolator filter (Fig. 245) also has a fabric filter element; the dirty oil enters at the top and is cleaned in passing through the filter element from the outside to the inside. The clean oil leaves the filter through a central aperture below. This type is made with a spring-loaded ball release valve to prevent

failure of engine lubrication should the filter element
become choked. In the case of the two types of filter
described the element should be cleaned every 2,000-
3,000 miles in petrol or paraffin after which, if the fabric
is not perforated, it can be used almost indefinitely.
New elements are, however, advisable after about
15,000 miles.

Another class of oil filter is that known as the Autoklean self-cleaning one, illustrated in Fig. 246.   In this
pattern of filter the
dirty oil is led to the
annular   space   be-
tween the inside sur-
face of the outer
vertical   cylindrical
container   and   the
Monel   metal   fine
mesh   gauze   cloth
c y l i n d e r within,
which is carried in
a   perforated   metal
cage.   This   gauze
element removes the
larger dirt particles.
The oil then passes
between a series of
thin   metal   discs
arranged axially one
above   the   other.
The clearances be-
tween these discs are
very small, but the
total surface of entry
for the oil is suffi-
ciently great to pre-

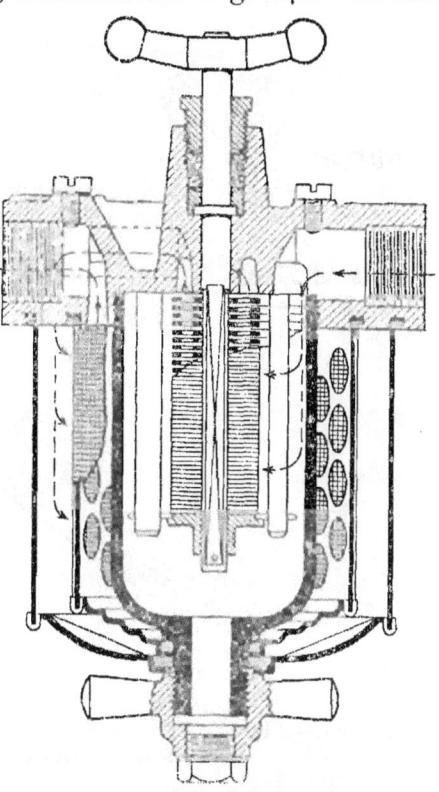

Fig. 246.—The Autoklean Oil Filter,
with Handle Cleaning Device.

vent appreciable resistance to the oil flow.   The
finer particles are trapped between the plates and the
cleaned oil passes upwards through holes in the plates
to the oil outlet above.   The plates are cleaned by
rotating the handle seen above the filter in Fig. 246,
and the solid matter then falls to the bottom of the

central chamber, whence it is removed periodically through an orifice having a handle operated plug. The filter will also deal with any water in the oil. The Monel metal gauze filter will operate for at least 2,000 hours before it is necessary to remove it for cleaning.

In a recent development of this type of filter the rotation of the filter element plates is performed auto-

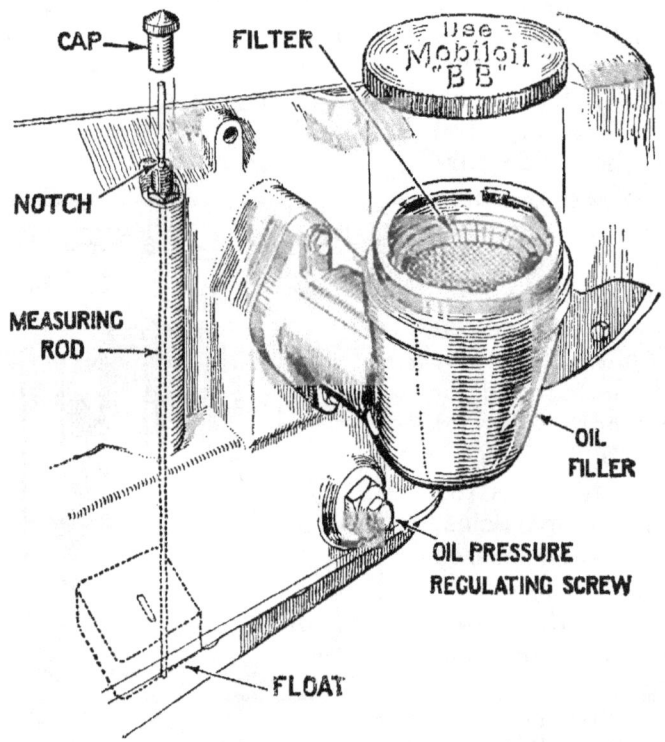

Fig. 247.—The Austin Oil Filler, Pressure Regulator and Oil Level Indicator.

matically by means of a small oil-operated motor which utilizes the difference of pressure between the suction and delivery sides of the oil pump of the lubrication system.

Oil Coolers.—It is well known that the higher the temperature of the engine oil, the less its viscosity and lubricating properties. Since in the majority of engines the oil is heated considerably, and is used to lubricate the working parts in this condition, it is advisable to arrange, in some convenient manner, to keep the average working temperature fairly low. For this reason some manufacturers make the crank-case sumps of aluminium alloy, and provide cooling ribs on the outsides (Fig. 243). In other cases oil from the engine is led to coolers of the radiator element type; these in one or two instances are neatly embodied in the water-cooling radiator block so as not to detract from the appearance of the car. With suitable means for oil cooling the engine will work more efficiently and wear better.

Oil Fillers, Indicators and Filters.—Automobile engines are now provided with oil filling pipes or openings situated in convenient positions and provided with dust-tight covers. All fresh oil added to an engine should be poured through a gauze filter; in most cases there is a cylindrical type of filter embodied in the oil filler orifice or pipe.

Further, the oil used for lubricating the cylinder walls and the crank-shaft bearings drains down the sides of the crank-case and passes into the sump below through gauze-covered openings, which extract any solid matter. These gauzes should be cleaned once every 8,000/10,000 miles.

In addition, the oil pump suction pipe is always provided with a gauze filter of generous proportions; this filter should be removed and cleaned periodically.

The level of the oil in the crank-case sump must be maintained at a certain minimum level to insure correct lubrication and operation of the oil pump.

Various types of oil-level indicator are now fitted, the most widely used type being the *dip-stick* one. In this case a graduated metal rod is used to ascertain the oil level.

Float-type gauges, glass gauges, and, more recently, electric oil level gauges, reading on a dial on the instrument board, are also included in automobile oil-level indicators.

Oil Consumptions. — The ordinary motor-cycle engines use appreciably more oil per horse-power than car engines. It is usual to obtain a mileage per gallon of from 500 to 1,000 in most cases. Two-stroke engines, using the "Petroil" system, average about 1,000 miles per gallon.

Modern light cars of 10 to 14 H.P. (R.A.C.) will give mileages of 1,200 to 2,000 when new, and 700 to 1,000 when worn appreciably.

A good engine should show an oil consumption of about ·05 lb. per B.H.P. per hour, for air cooling, and about ·03 lb. for water cooling.

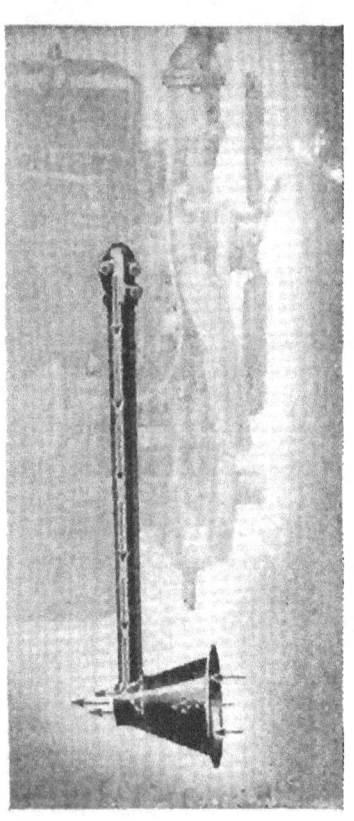

Fig. 248.—Crankcase Ventilating Device.

Ventilating the Crankcase.—In order to prevent contamination of the oil within the crankcase by any petrol vapour or the burnt gases that may escape past the pistons it is usual in some modern engines to provide a special ventilator which arranges for air to be drawn into the crankcase and then ejected; the air stream thus carries off the lighter vapours and gases. In many instances the air inlet for this purpose consists of a cone facing the front of the car so that the additional air pressure due to the forward motion of due to the forward the car promotes the necessary circulation of air. In other cases the ejector principle (Fig. 248) is employed, utilizing the current of air due to the forward motion of the car, to draw out the vapours from a pipe leading to the crankcase. A typical example of such a ventilator is that employed on the Vauxhall engines; the fresh air enters at the top of the valve mechanism casing and is ejected from a pipe, thus creating a suction within the crankcase which draws the vapours into the vent pipe; the lower end of the latter experiences a suction effect due to the motion of the car.

Small End Bearing Lubrication.—A method used in the case of the more expensive types of petrol engine to lubricate the gudgeon pin bearing consists in rifle-drilling a hole from the big end to the small end bearing so as to convey lubricating oil from the hollow or drilled

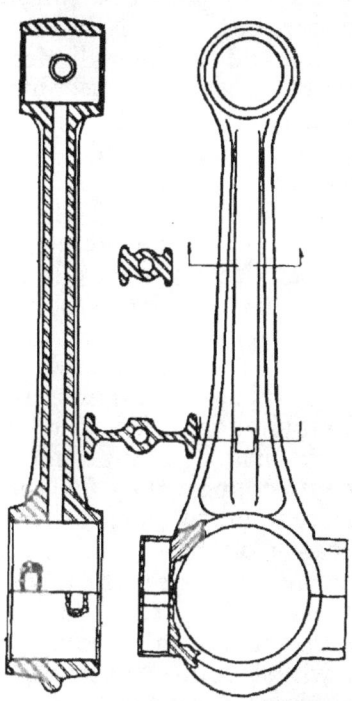

Fig. 249.—Method of Lubricating the
Gudgeon Pin Bearing.

crank pin through the big end bearing and up to the small end bush.

An example of a connecting rod provided with this method of lubrication is given in Fig. 249. The sectional views of the connecting rod indicate how the web has been thickened at the centre to allow for drilling the oil hole in question.

CHAPTER VIII

THE COOLING OF AUTOMOBILE ENGINES

It has been shown that about one-quarter of the heat of combustion of the petrol passes through the cylinder walls in an ordinary engine. Now this quantity of heat is considerable, and if the cylinder were a plain metal barrel, it would very soon become red hot, and the engine would cease to operate, since the incoming charge would ignite whilst the inlet valve was open, and moreover, as the lubricating oil on the cylinder walls would all be burnt away, the piston would certainly seize up. When it is remembered that the temperature of the exploding gases is about 1500° C. to 2000° C.,* and that of the exhaust gases from 600° C. to 800° C., it will be evident that the average temperature in the cylinder is very high. It is therefore imperative to provide some means for keeping the temperature of the cylinder down to reasonable limits.

In this respect the use of aluminium alloy (such as the 8 to 12 per cent. copper one) for the cylinder and piston enables the temperatures to be kept down more readily than with cast-iron.

Aluminium is a much better heat conductor than cast-iron or steel. For the same temperature differences it will conduct away about three times the quantity of heat, whilst copper is better still, and will conduct about six times the heat.

Applied to petrol engine construction, aluminium alloys enable the use of higher engine compressions, facilitate the engine cooling and lighten the engine parts, since aluminium has only about one-third of the weight (density) of cast-iron.

Concerning strength properties, the L 8 aluminium alloy, which contains from 9 to 12 per cent. of copper,

* Platinum, one of the highest melting point metals, melts at 1750° C., iron at 1530° C., and aluminium at 657° C.

has a tensile breaking strength in the cast state of
9 to 11 tons per sq. in., and is thus as good as cast-
iron; this alloy is now much used for cylinders and
pistons.   Other stronger casting and forging alloys
used are R.R. and " Y " alloys.

Methods of Cooling the Engine.—In order to ensure
that the cylinder temperature is not excessive, the
heat conducted through its walls must be disposed of
continuously.   This can be done in either of the
following ways: (1) By providing cooling or
radiating fins of sufficient area, and arranging for
an air current to blow
past these fins, in order
to carry off the surplus
heat—this process is
termed *Air Cooling*.
(2) By providing a
water jacket around
the hotter parts of
the cylinder barrel and
the combustion head,
and circulating water
around same. (3) By
arranging for the oil
supplied for lubrication
purposes to cool the
cylinder barrels, as in

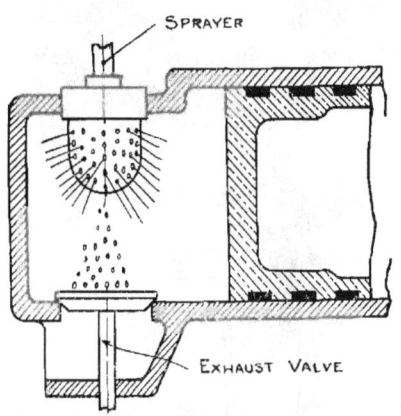

Fig. 250.—The Hopkinson System
of Internal Cooling.

the Bradshaw engine; and (4) By injecting a water-
spray into the cylinder. Professor Hopkinson employed
this method for cooling gas engines. The water-spray
was arranged to occur during the combustion and
expansion processes, the water being converted into
steam; the heat required for this conversion was
absorbed from the hot gases.   Part of the spray
reached the exhaust valve and the combustion chamber
walls, also, and tended to cool these parts directly.

In addition to the cooling methods mentioned, it is
usual to combine these, and to employ, say, water
cooling for the combustion head and valve passages
and seatings, and air or oil cooling for the cylinder
barrels. Alternatively, as in the Bradshaw engine, the
latter were oil-cooled, and the combustion chamber

and valve seatings air-cooled. Again, in some two-stroke engines the barrels and valve seatings are water-cooled and the detachable heads air-cooled. The lubricating oil also assists in the cooling of most engines. Steam cooling of cylinders is another method.

Air Cooling.—The amount of heat carried off by the cooling air depends upon the following items: (1) The total area of the fin surfaces; (2) the velocity (and amount) of the cooling air; (3) the temperatures of the fins and of the cooling air. The greatest

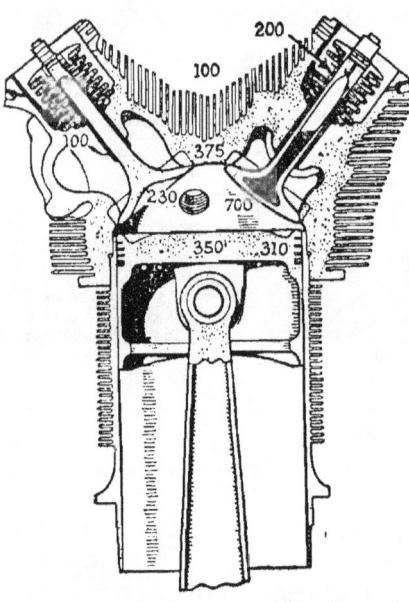

Fig. 251.—Air-cooled Cylinder Temperatures.

amount of heat will be carried off in the case of engines having a large area of cooling fin surface well disposed around the combustion head and valve surfaces, for which the speed, and cross-sectional area, of the cooling air stream is a maximum. The temperature of the air should be low for the best cooling effect.

It is, of course, quite possible to overdo the cooling, and to keep the cylinders *too cool*, so that the engine will lose in petrol economy and output. Most automobile engines have to run at light loads for the major part of their existence, and are apt to be over-cooled during these periods. It is a well-known fact that the petrol-consumption of an engine is appreciably greater per B.H.P. hour (say) when running light than when on three-quarter or full load.

Cooling Fins.—These are usually made of about the cylinder wall thickness at their roots (or junctions with the cylinder wall) tapering down to about one-half the root thickness. The length of the fins varies from one-quarter to one-third of the cylinder diameter; their pitch (i.e., the distance between the fin centres) is about one-quarter to one-third of their length; the total length of finned cylinder barrel is from 1·0 to 1½ times the cylinder bore, as a rule. Another rule based upon experimental considerations is to allow from 1½ to 2½ sq. ft. of cooling fin area per horse-power. This gives about the correct cylinder temperature at 30 to 40 m.p.h. air speed.

In the case of aircraft air-cooled engines, where much higher cooling air speeds are utilized the cooling areas for the high output per litre engines work out at about 17 to 20 sq. in. per B.H.P. or 75 to 85 sq. in. per litre; the cooling air speeds past the cylinder fins are 150 to 200 M.P.H.

The tendency in regard to cooling fins is to reduce the pitch and to increase the length of these. It is not, however, advantageous to reduce the pitch below about ¼ in. The usual fin lengths for high output radial engines are 1½ to 2½ in. for the aluminium alloy heads and ½ to ¾ in. for the steel cylinder barrels.

A typical aircraft engine cylinder unit of the air-cooled pattern is shown in Fig. 251, with the average working temperatures of the various parts. The inlet valve, on the left, has a relatively low temperature of 230° C. due to the cooling influence of the incoming mixture, whilst the exhaust valve on the right which is in the stream of outflowing hot gases has the highest temperature, namely, 700° C. The valve shown is of the hollow sodium-filled type. It is well known that the limiting factor governing the maximum compression ratio that can be used without detonation occurring is the temperature of this valve.

Applications of Air Cooling.—The simplest air-cooling system is that of the ordinary motor-cycle in which the air circulation, or draught, due to the forward motion of the machine is sufficient to carry off the surplus heat. The fins are now made much larger in area than hitherto, to ensure adequate cooling when running up long inclines on low gear, in following winds, and at low road speeds.

In the case of two-stroke engines, since these develop about 50 to 70 per cent. more heat, at the same engine speeds, it is necessary to provide a relatively larger cooling area, usually by deepening the fins and increasing their number.

In earlier cycle-car engines of the two cylinder opposed air-cooled types, similar to those of the Rover Eight and A.B.C., it was arranged for the combustion heads of the cylinders to project through the bonnet of car, and to provide cup-shaped air-scoops facing the direction of motion of the car; in this way adequate cooling of the heads was ensured.

Fan Cooling.—This is employed on the larger air-cooled engines, more particularly on light and other cars. A two or four-bladed fan is driven either at engine, or twice engine-speed, and the stream air

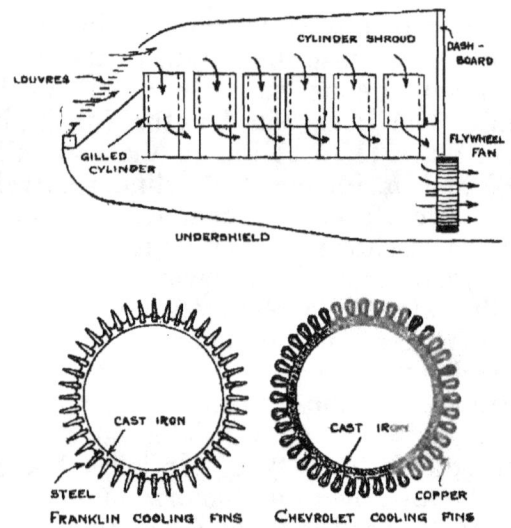

Fig. 252.—Showing Principle of Air-cooling of Franklin Engine, and Method of arranging the cooling Fins.

is directed on to the cylinder heads. In fan-cooled engines the cooling depends chiefly upon the engine-speed, and not upon the forward speed of the car. These fans absorb only a small proportion of the power output, namely, about 1 H.P. for every 15 to 20 H.P. output, but it must be remembered that with their aid much better running conditions and a higher net output are obtained. In the case of one well-known light car engine of 12 H.P. (rated) the total output at 1,000 r.p.m. was 9·0 H.P., of which the fan absorbed about one-half H.P. The brake m.e.p. obtained, however, was 114 to 118 lbs. per sq. in. The cooling air speed was 20 m.p.h. and the maximum temperature of the cylinder head was 216° C. At 2,000 r.p.m., about twice the H.P. was developed.

In the case of small single-cylinder engines, running in enclosed situations, an excellent arrangement is that of a fan of about the same dimensions as the fly-wheel diameter (i.e., about twice the stroke) mounted on the main shaft, and enclosed in a metal casing so arranged that the air is drawn in at the centre and

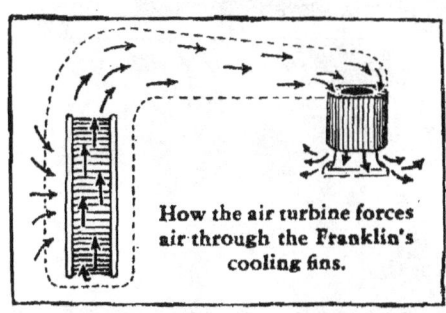

How the air turbine forces air through the Franklin's cooling fins.

Fig. 253.—Showing principle of later model Franklin Air-cooled Engine.

expelled peripherally through a bell-mouthed duct directing it on to the exhaust side of the cylinder.

For small air-cooled engines, the blower type fan which "blows" its air on to the cooling fins, works quite satisfactorily if suitable guides and ducts are provided for the air stream. The system usually adopted for larger engines is to shroud, or surround each cylinder with another barrel, and to *suck* the air, by means of the fan, through these barrels, commencing at the crank-shaft side. By placing the cooling system on the suction side of the fan, a more positive circulating effect is obtained. Sometimes the fly-wheel itself is designed to function as a cooling fan, and air is discharged backwards through it, after having been drawn past the cylinder barrels. The early Franklin engine had 6 cylinders of 3⅝ in. bore and 4 in. stroke, each cylinder being provided with a series of

56 steel cooling fins running lengthwise (i.e., parallel to the cylinder axis). Each cylinder, with its fins, was enclosed in an outer casing, just fitting the fin tips, and air was drawn upwards through this casing by a 20 in. diameter fan, situated at the rear of the engine. One of the standard four-seater Franklin cars on a test fitted with this engine performed a journey of 98 miles on its lowest gear, at an average speed of 11 m.p.h.; this was followed by a climb up a hill 6,300 ft. high, with a final grade of 27 per cent.

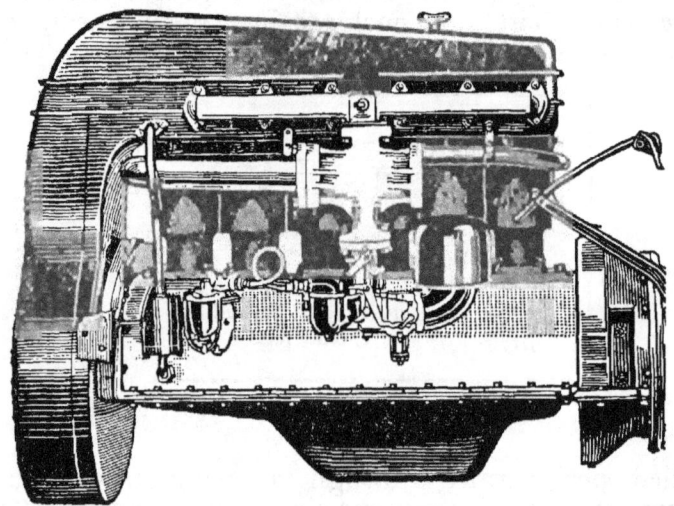

Fig. 254.—The Franklin Air-Cooled Six-Cylinder Car Engine.

(about 1 in 4), during which no signs of overheating were observed.

The later model of the Franklin car engine has a different cooling system (Figs. 253 and 254). In this case the cooling fan is situated at the front of the engine, and the air is actually forced through a suitable duct and flows past the cylinder heads, down the barrels and out past the crank-case. This system is more efficient, and enables the unit construction of gear-box to be employed. In the later design the air passage above the jackets is formed of rigid aluminium castings so shaped that the air speed is uniform throughout; all air leaks have also been eliminated. The new system gives a circulation about $2\frac{1}{2}$ times as great as before. There is an open grill in front of the bonnet which allows a free passage for considerably more air than is used by the blower, the excess air being used to cool the outside of the cylinder jackets and the aluminium crank-case. The engine in question has 6 cylinders of $3\frac{1}{4}$ in. bore and 4 in. stroke (giving about 3,260 c.c.

Fig. 255.—The Krupp Air-Cooled Vee-eight Engine.

capacity). Each cylinder head is a single casting, with the steel cooling fins cast in position, so that there is a good connection between the two metals to enable the heat to be conducted through.

Other special features of this engine are its die-cast aluminium pistons, duralumin drop-forged connecting-rods weighing only $1\frac{1}{4}$ lb. as against $2\frac{1}{4}$ lb. for the steel rods previously used, overhead valves operated by push-rods and case-hardened crank-shaft.

German Air-Cooled Engines.—The air-cooled engine appears to have received much attention in Germany in recent years, for there are several makes of air-cooled automobile engine, including 2, 4, and 8 cylinder models.

A good example of the modern air-cooled type is the Krupp four-cylinder opposed compression-ignition engine. This has a cooling fan fitted at the front end, and it is engine driven. It forces the cooling air through a casing around the front end of the crankcase and thence to the horizontal cylinder barrels which are ribbed and enclosed in rectangular casings.

Another more recent example is the Krupp eight-cylinder Vee-type petrol engine, which has very similar cooling arrangements. (Fig. 255).

Overhead valves in cast-iron cylinders, with aluminium heads, are employed. All of the cooling fins are parallel with the cylinder axis, and are encased in metal covers. The blast of air from the centrifugal-type blower on the front end of the crank-shaft is forced upwards through the channels thus formed.

The two banks of four-cylinders are inclined to each other at 90°. The bore and stroke are 85 mm. and 110 mm., respectively; the engine develops 95 B.H.P.

General Considerations on Air Cooling.—Air-cooled engines in the past have been noted for their higher oil-consumption, due to some of the lubricating oil being burnt in the combustion chamber. They have not given the same petrol economy as water-cooled engines, and are still undoubtedly noisier in action.* Their working temperatures are, on the average, higher than those of water-cooled engines (the cylinder heads being at 250° C. to 300° C.) and the maximum allowable compression ratios some what less.

The additional noise of the air-cooled engine, apart from the valves and silencer, may be due to the fact that there is no outer jacket and water content to damp down the noise as with the water-cooled types, so that the surface vibrations are of larger amplitude, and the emitted sounds therefore greater.

* The single sleeve valve air-cooled engine is an exception

Water Cooling.—In the case of water-cooled engines, it is necessary to dissipate the heat conducted through the cylinder walls by circulating the water through a cooling device, known as the *Radiator*. Fig. 256 illustrates a typical car engine water-cooling system. The water is made to circulate through a large number of very small section tubes in the radiator, having large cooling surfaces. Air is forced past the outer surfaces of these tubes, and carries off the surplus heat. This system is therefore an air-cooling process, with the intermediary of water to carry the cylinder heat to the cooling device, or radiator. There are three principal methods of circulating the

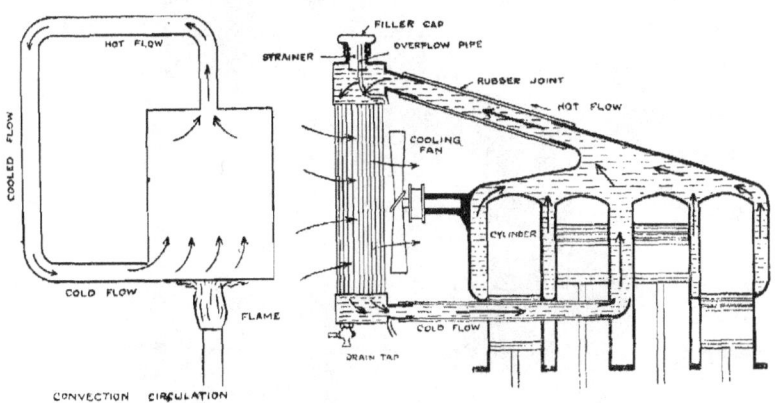

Fig. 256.—The Thermo-syphon System of Cooling.

water in present use, namely: (1) The Natural Convection, or Thermo-Syphon system, (2) Impeller-Thermo-Syphon, and (3) Pump Circulation.

The principle of thermo-syphon circulation depends upon the fact that if a vessel of cool water is heated, the hot water, in virtue of its lighter density, will tend to rise, and its place will be taken by cooler water, thus causing a definite circulation. In the case of the automobile engine, the hottest part is the space around the cylinder head, and the hot water rises from this part to the top of the radiator, thence down the numerous passages in the latter to the bottom, and along to the lowest part of the water-jacket.

In order to obtain the maximum benefit from thermo-syphon cooling the engine cylinder jackets should be placed as low as possible, relatively to the radiator, so that the coolest water will always be available from the radiator. If the lower pipe from the radiator to the jacket be placed horizontal, as shown in

Fig. 257.—The 8 h.p. Ford Cooling System.

Fig. 256, an excellent cooling system results. The advantage of thermo-syphon cooling lies in its simplicity and automaticity. It requires, however, a larger capacity radiator than when the water is circulated by pump, and the radiator must always be kept filled well above the level of the pipe from the cylinder to the top of the radiator.

It is usual to arrange for a capacity of the complete thermo-syphon system, of 1 gallon of water to every 5 to 7 B.H.P. of the engine. The width of the water spaces around the cylinders is also often made from 20 to 40 per cent. greater than for pump circulation.

In order to assist cooling of the water in the radiator, a two or four-bladed suction-type fan is mounted

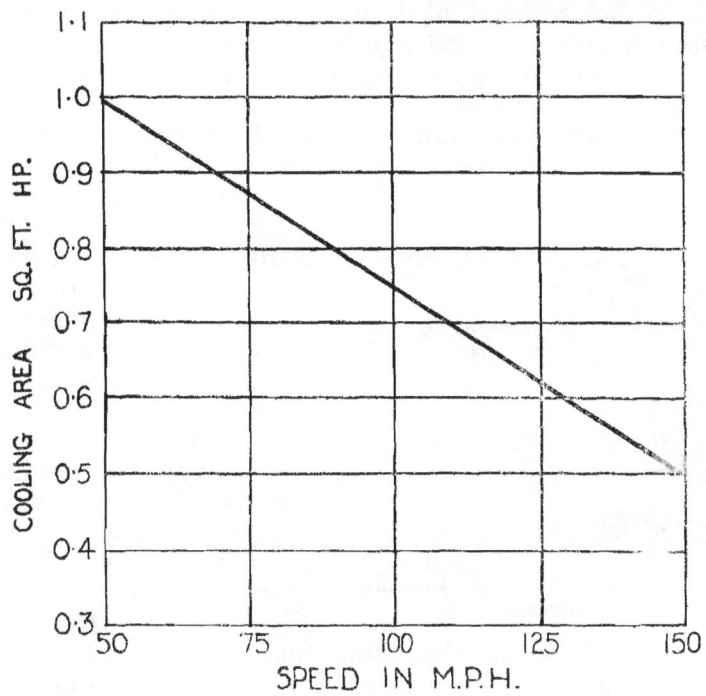

Fig. 258.—Radiator Cooling Areas and Speeds.

behind, and is driven at 1 or 1½ times engine speed, so that the amount of air drawn through is roughly proportional to the engine speed.

Heat Dissipation of Radiators.—The amount of heat that can be dissipated or got rid of by a radiator depends upon a number of factors of which the principal ones are as follows:—(1) The relative wind velocity. (2) The cooling surface. (3) The ratio of tube depth to diameter. (4) The air density. (5) The water and air temperatures. (6) Design of radiator and disposition. (7) Conductivity of metals used.

The area of the cooling surface is inversely proportional to the air velocity; this is illustrated graphically in Fig. 258, from which it will be seen that at 150 M.P.H. the cooling area is 0·5 sq. ft. per B.H.P.; at 100 M.P.H., 0·75 sq. ft.; and at 50 M.P.H., 1·0 sq. ft.

A honeycomb radiator of 1 sq. ft. frontal area and 1 in. depth of tube gives about 8 sq. ft. of cooling surface, whilst one of 6 in. depth gives 48 sq. ft., although unless the ratio of length to diameter of tube is the same it will not necessarily give the same cooling efficiency.

The rate of heat transfer from the cooling water to the air through the radiator tubes is shown to be dependent upon the ratio of the tube length to its diameter. In this connection the results of some tests* upon film or cellular matrix radiators of the construction shown in Fig. 259 are of interest. Tests were made upon radiator blocks of different depths up to 5 in. and at different velocities from 20 to 80 M.P.H.

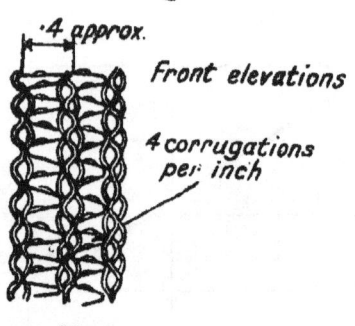

Fig. 259.—Radiator Cooling Elements.

The results given in Fig. 260 indicate that the *rate of heat transfer* does not increase linearly with the depth as would be the case for free air flow through the air spaces, but falls away continuously as the depth increases. There is, of course, a progressive increase in the *total heat dissipated* on account of the increase in cooling area with core depth. It should be mentioned that the usual depths of radiators with the cell dimensions of Fig. 259 range from 2 to 4 in.; the smaller depth gives a better rate of heat transfer but requires a larger frontal area in order to obtain the same cooling surface.

In the case of supercharged engines a larger cooling area is necessary since the heat loss to the cooling water

* Automobile Cooling.—C. S. Steadman, Proc. Inst. Autom. Engrs. Aug. 1939.

increases in proportion to the supercharging pressure; the radiator area should therefore increase at about the same rate. Thus, increasing the H.P. of an engine of 3·5:1 compression ratio, 50 per cent. by supercharging, results in a 20 per cent. increase in losses to

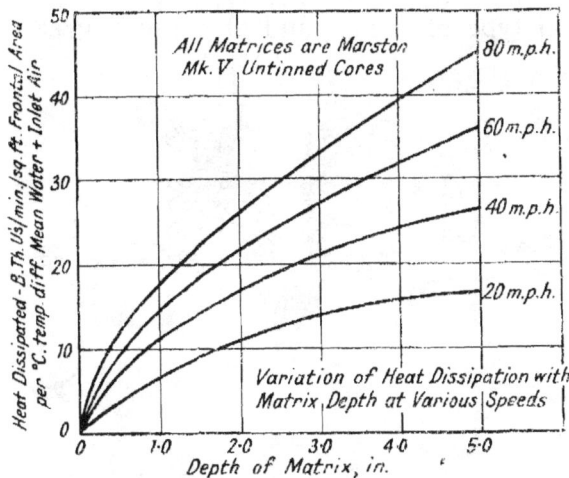

Fig. 260.—Radiator Cooling Data.

the cooling water; increasing the H.P. of an engine of 7·5:1 compression ratio by 50 per cent. results in an increase of 34 per cent. in losses to the cooling water.

Improved Liquid Cooling.—In order to reduce the weight of cooling liquid carried and also that of the radiator, it has been proposed to employ some liquid other than water, which has a higher boiling point. Examples of such liquids are glycerol (glycerine), Prestone, which boils at 290° C., and ethylene glycol (195° C.). The latter liquid is actually used on certain modern aircraft engines, such as the Rolls Royce Merlin. It is possible to use a radiator of less than one-half the size of a water-cooled type. The engine operates at about 130° C., but loses about 2½ per cent. of its maximum power through working at this temperature; with water-cooling the temperature is 80° to 85° C.

Pump Circulation.—In the case of light cars, the additional size of the radiator and the extra weight of cooling water required with the previous system does not matter much, but for medium and large car engines, thermo-syphon cooling would usually necessitate large radiators and an appreciably greater weight of cooling water. It is therefore the rule to circulate the water positively, by means of a centrifugal type of pump, and at a much greater rate

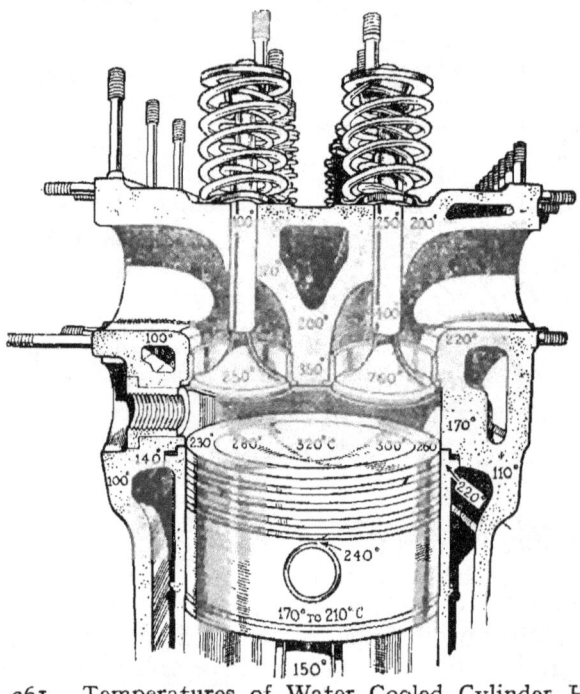

Fig. 261.—Temperatures of Water Cooled Cylinder Heads, Valves and Piston.

than is possible with the previous method. The water capacity of the system and the size of the radiator can then be reduced appreciably.

The quantity of water allowed for pump circulation is about one gallon for each 7 to 10 B.H.P. Thus a nominal 14 H.P. engine (which will develop from 25 to 40 B H.P.) requires from $3\frac{1}{2}$ to 4 gallons of water, and a nominal 20 H.P. engine, 6 to $6\frac{1}{2}$ gallons. The centrifugal type pump employed is usually of from 2 to 3 inches (impeller) diameter, and draws the water from the lowest part of the radiator in at its centre, and ejects it by centrifugal action at its periphery into the lowest part of the cylinder jacket, whence it is forced through into the top of the radiator. It is driven at from one-half times the engine speed, usually by means of the same shaft driving the electric generator and magneto. A watertight gland is required where the driving shaft enters the pump. This is usually packed with asbestos or hemp string soaked in tallow, a union nut being provided for tightening purposes.

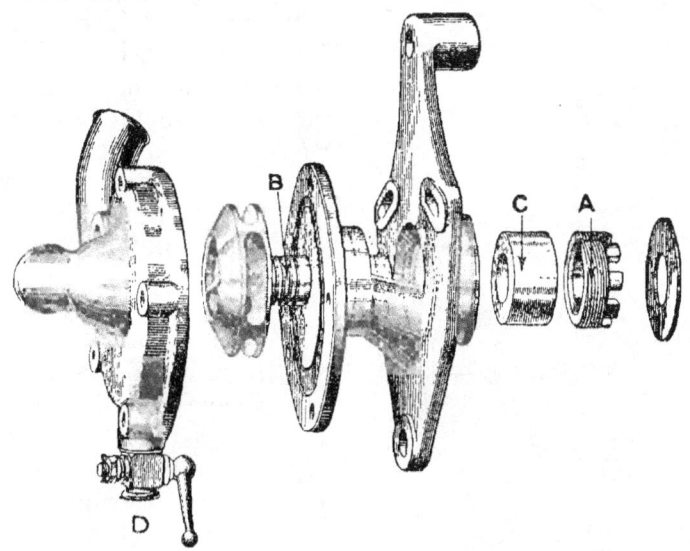

Fig. 262.—Typical Water Circulating Pump.

Greasers or oilers are supplied to both the pump spindle and to the fan bearing; these parts should frequently be oiled or greased.

Fig. 262 illustrates a typical centrifugal water circulating pump as used on Austin engines. The rotor B is driven by means of a shaft from the engine and leakage of water along the shaft is prevented by a special packing ring C, which is held in place, and maintained so as to provide a water-tight gland by means of the nut A. In later designs of pump the packing ring C is *a graphite disc* held in position by means of a compression spring

M

The pump casing is in two parts bolted together by means of studs and nuts. A drain cock D is provided for letting out the water in the pump casing when the radiator is drained in frosty weather.

In order to obviate the necessity, on the part of the car driver, of greasing the pump impeller spindle, many recent pumps are fitted with a special graphite or carbon disc thrust element having a light spring on one side to maintain it in contact with a machined face on the pump casing. This method has proved effective on both car and commercial vehicle engines and the carbon disc has been shown to last for very long periods without attention. The carbon disc is usually held in a steel sleeve and bears against a slightly convex seating.

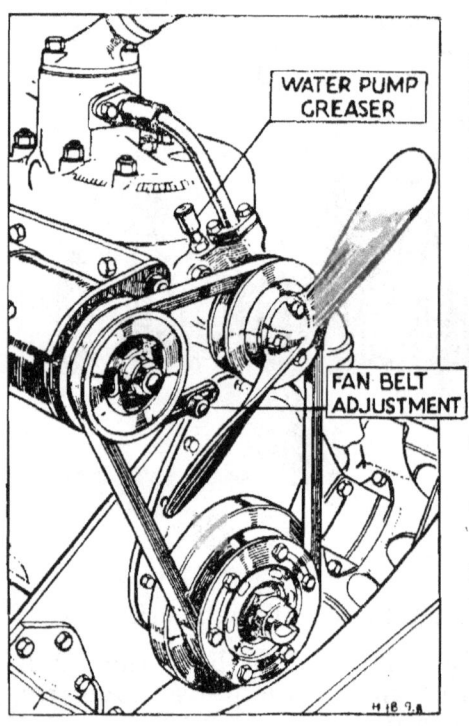

Fig. 263 illustrates the method of driving the circulating water pump and also the dynamo by means of a Vee-belt and pulleys from a pulley on the engine crankshaft in the case

Fig. 263.—Method of Driving Water Pump and Dynamo (Austin).

of the Austin 18 H.P. engine. It shows also the water pump spindle greaser and the method of adjusting the tension of the belt by swivelling the dynamo and its bracket about two bolts, one of which is seen just above the dynamo pulley on the right. The slotted plate enables the dynamo unit to be rotated about these bolts and locked in position, when the three bolts in question (the other one of which is seen on the slotted plate) are tightened.

The rate of circulation of the pump should not be less than 1 pint per B.H.P. per minute; for aircraft engines it usually exceeds twice this rate.

Water Temperature.—Experience has shown that the hotter an engine can work without detonation, pre-ignition, or loss of power, the better. Not only is the thermal efficiency higher, but the engine's frictional losses, as a general rule, are lower with the result that the petrol consumption is lower, and the output greater.

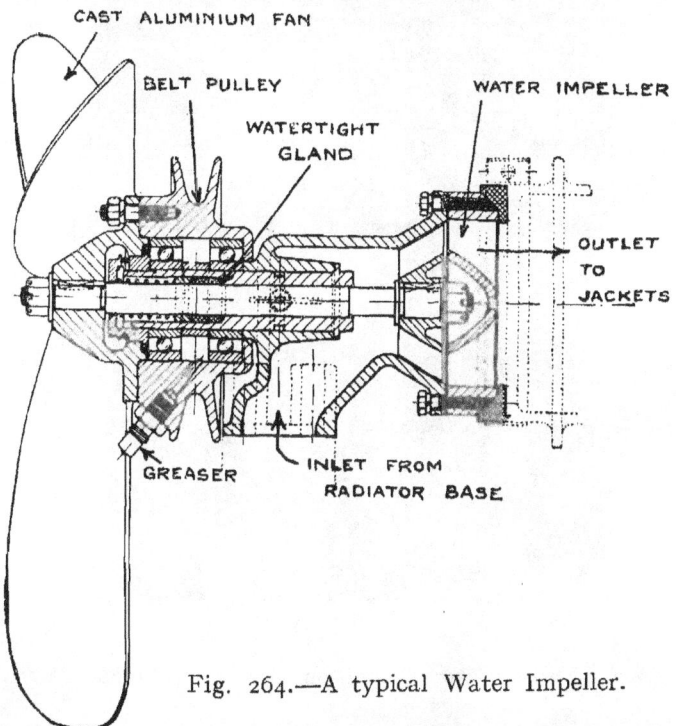

CAST ALUMINIUM FAN

BELT PULLEY

WATERTIGHT GLAND

WATER IMPELLER

OUTLET TO JACKETS

GREASER

INLET FROM RADIATOR BASE

Fig. 264.—A typical Water Impeller.

For normal running, the temperature of the cooling water at the top of the radiator should lie between 75° C. (167° F.) and 85° C. (185° F.). This allows a sufficient margin for temperature increase when climbing long hills on lower gears.

It is useful to fit a radiator thermometer of the radiator or dash type in order to keep a check on the water temperature. These instruments are useful in detecting "over cooling" in the winter, and overheating in the summer. It will sometimes be found, more particularly with American and Colonial-type cars used in this country, that they are over-cooled in winter, the running temperature being from 55° C. to 65° C. Where other means, such as radiator-shutters, or thermostats, are not provided for regulating the temperature, it is possible to blank off a portion of the radiator with a strip of wood covered

Fig. 265.—Deep Water Jacket on Modern Cylinder Barrel.

with felt placed behind the radiator inside the bonnet. Alternatively the fan can be disconnected during winter.

Impeller Thermo-Syphon Circulation.—A method which has found much favour is a combination of the two previous ones. A small straight-bladed impeller fitted usually on the fan-shaft spindle in a circular casing is connected at its centre, or suction, side with the water inlet pipe from the lower part of the radiator, and discharges the water into the cylinder jackets. It will be evident that as there is a free passage for the water past the straight blades of the impeller, the natural, or convectional circulation of the water can occur, but the impeller assists this circulation, although not so rapidly as in the case of the centrifugal pump. (Fig. 264).

In one typical instance, the impeller casing is bolted direct on to the water-jacket, and the Vee-pulley used is provided with an adjustable screwed flange, to tighten the Vee-rubber and canvas belt.

A good deal of progress has been made in recent

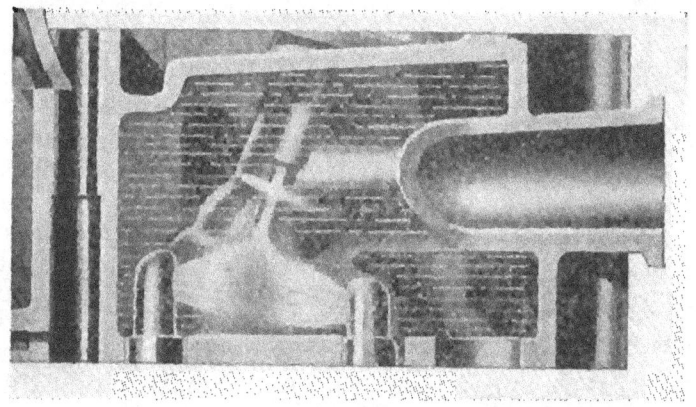

Fig. 266.—Water Jet Method of Cooling Exhaust
Valve Seating

years in connection with the cooling of water-cooled cylinders and exhaust valves by improvements in the design of the water jackets and other means. One notable result of these developments is the provision of water jackets for the full lengths of the cylinder barrels instead of the previous method of partial jacketing. In this connection the design of the Vauxhall cylinder jacket shown in Fig. 265 illustrates the method adopted for ensuring more uniform cooling of the cylinder barrel.

Another development is that of the exhaust valve seating which is now liberally jacketed for cooling purposes. A recent improvement in exhaust valve seat cooling, shown in Fig. 266 consists in the provision of nozzles within the water jacket for directing streams of cool water from the radiator on to the valve seatings and also the sparking plug seating.

Thermostats.—It has been stated that there is a definite mean temperature for the cooling water, at which an engine works most efficiently   When an engine is started from the cold, it frequently takes an appreciable time before it attains its correct working temperature, as all of the water in the system has to be heated up.   In cold weather, in the ordinary way the engine will run too cool, whilst in hot summer it may run too hot, if adjusted for normal conditions. From tests which have been made, it has been shown that there is a difference of 40° F. to 60° F. in winter and summer petrol engine temperatures. When driving with and against a strong wind, also, there will be an appreciable difference in the radiator temperature. It follows, then, that with a fixed area radiator, it is impossible to keep the mean circulating water temperature constant under all circumstances.

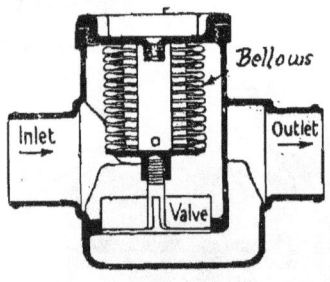

Fig. 267.—Showing a typical Thermostat Cooling Water Temperature Regulator.

For these reasons, it is necessary to employ a device, known as a *Thermostat*, the object of which is always to keep the temperature of the jacket water constant. This is usually effected by cutting out the radiator when the engine is cold, and by varying an opening in the outlet water pipe to the radiator according to the temperature of the engine. The jacket water temperature, in this case, operates a thermostatic element, which opens or closes a valve.

Fig. 267 illustrates a typical thermostat used on automobile engines. It consists of a series of thin copper discs, made in the form of bellows, somewhat like the pressure element of the ordinary aneroid barometer. The interior of the bellows is sealed and contains a volatile liquid, such as ether, and its vapour. When such a bellows is heated up, the vapour pressure increases and causes the bellows to expand in the direction of its axis. One end of the bellows is anchored rigidly, and the other is attached to a flat or cone-seated valve. As the water temperature on the jacket side on the bellows side increases the bellows expand and lower the valve from its seating, thus allowing the water to flow from the jackets to the radiator. Thus when the engine is heated up, the radiator is fully connected up, but whilst the engine is cool, it is disconnected, and can therefore warm up quickly. It is usual to fit the thermostat between the cylinder jackets and the radiator, in the outlet pipe as shown at B, Fig. 272.

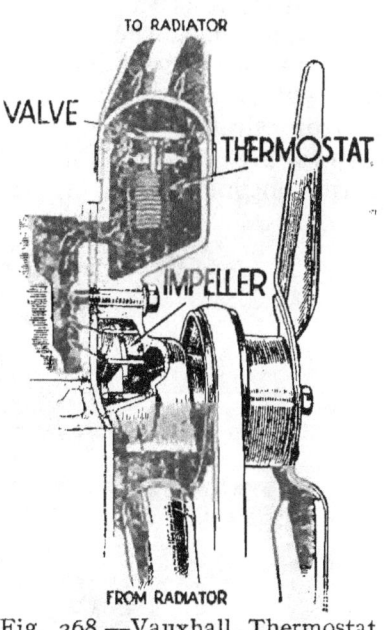

Fig. 268.—Vauxhall Thermostat and Water Impeller.

Fig. 268 shows the Vauxhall thermostat and circulating water impeller arrangement and illustrates the location of the former element between the cylinder head and upper part of the radiator.

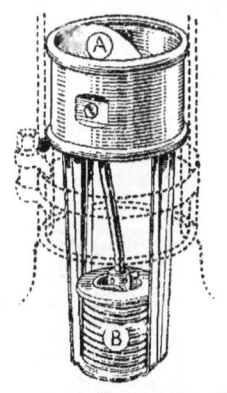

Fig. 269.—Butterfly Valve-operated Thermostatic Control.

In certain Austin engines the water outlet pipe from the cylinder head contains a thermostat of the bellows pattern, as shown at B in Fig. 269. The centre of this unit is connected by a rod to an arm on a butterfly valve shaft so that as the bellows expands whilst the water warms up, the butterfly valve A is opened gradually until at normal engine water temperature it is fully opened.

In some cases the thermostat is placed in the top of the radiator, and its movement operates an elbow lever, which in turn works a series of Venetian type radiator shutters. When the engine is cold the shutters are closed, thus blanking off the radiator. As the engine warms up the hot water flowing into

Fig. 270.—Thermostat-operated Radiator Shutters.

the radiator causes the thermostat to expand, and to operate the lever connected to the hinged members of the shutters, causing these to rotate so as to admit more and more air to the radiator.

It is an advantage to employ one of the radiator thermometers now available, in conjunction with the thermostat, as a check on its working, and for adjustment purposes.

Fig. 270 illustrates a typical thermostatically-operated radiator shutter arrangement. In this case there is a series of thermostatic disc elements, clamped securely at one end and attached at the other end to a hinged lever arrangement connected by means of a rod to a lever operating the vertically pivoted radiator shutters. The thermostatic discs are arranged in the radiator and expand as the water heats up, thus opening the vertical shutters through the agency of the mechanism described.

Improved Shutter Thermostat.—The principle of an improved type of radiator shutter control is shown in Fig. 271. In this case provision is made not only for

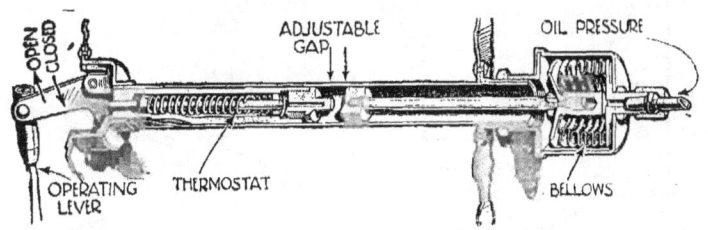

Fig. 271.—Thermostat for Controlling Radiator Shutters.

operating the shutters as the water temperature changes, but also when the engine stops. The advantage of this scheme is that the heat of the engine is conserved by closing the shutters when the engine stops, whereas with the ordinary thermostat the shutters would remain open until the temperature of the water fell to the air temperature.

In addition to the thermostat there is a bellows which is connected to the main lubrication system.

When the engine is running the oil pressure expands the bellows and the gap between the plunger and thermostat is immediately taken up. When the engine stops the bellows contract through the loss of oil pressure and the gap is re-established with the result that although the thermostat may be expanded owing to the heat of the engine the expansion is compensated for by the gap, and the radiator shutters are at once closed by their springs.

M*

There is a thermometer on the instrument board, the normal engine temperature being 160° to 180° F.

Radiator Construction.—The purpose of the radiator is to dissipate, or get rid of the heat of the cooling water as rapidly as possible, and with the lightest and least bulky form of construction. Modern radiators are now very efficient and light, yet they will stand up to their work for long periods.

The principle of most radiator designs lies in the provision of a large number of small sectioned water spaces extending from top to bottom, and a maximum external cooling surface area exposed to the cooling

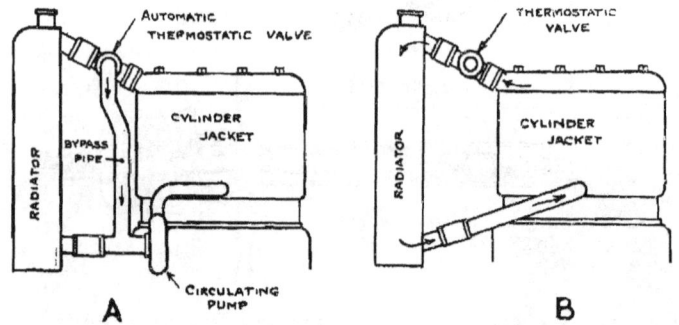

Fig. 272.—Showing Alternative Thermostat Arrangements. In A the cooling water is by-passed, and in B the flow is stopped when engine is cold.

air. There is a water reservoir at both the top and bottom of the radiator, into which the outlet pipe from, and the inlet pipe to, the cylinder jacket are led, respectively. The radiator tube system connects these two water reservoirs, or headers, so that when the engine is working there is a continuous series of water streams flowing from top to bottom. In addition the radiator is provided with a filling cap at the top, an emptying tap or plug at the bottom, and an overflow, or steam pipe inside. The latter is a small copper or brass tube (of about ¼ to ½ in. diameter) extending from near the top of the radiator, or filler, to the bottom of the radiator, either outside or inside, and to the air. If the radiator is filled to too high a level, the surplus flows out of this tube. Similarly when the engine is started up, the water as it warms up expands, and any surplus flows out through this overflow pipe. If the water boils under any running circumstances, the steam escapes through this pipe.

There are three more common methods of constructing radiators (and also a number of special

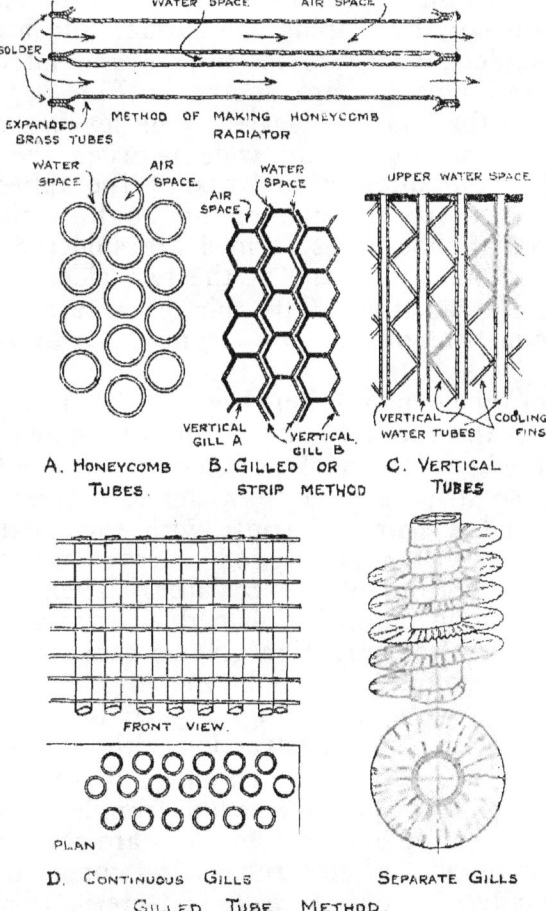

Fig. 273.—Illustrating Different Systems of Radiator Construction.

methods which space prevents mention of here) as follows: (1) The Honeycomb Tube, (2) The Film or Corrugated Strip, and (3) The Gilled Tube Systems.

The honeycomb tube is perhaps the most efficient

type. It is used in the better classes of cars, and externally resembles the ordinary honeycomb of the bee. It is made up of a large number of thin brass tubes, having their ends expanded outwards, as shown in Fig. 273 (A). The tubes are assembled with their expanded ends all in the same planes and the ends are soldered (usually by dipping the surface into a molten bath of solder), so as to fill up the end interstices; the soldered portions are shown in Fig. 273 (top). In this way the ends are sealed, thus leaving narrow water-spaces, about 1 mm. wide between the tubes, but leaving the interior surfaces of the tubes quite open and free for the cooling air to flow through. The blocks of tubes thus formed are soldered to the top and bottom tanks, and at the two sides, to form the complete radiator. The thin brass used for (1) and (2) constructions is only from $\frac{5}{1000}$ to $\frac{10}{1000}$ in. thick.

The film type construction (Fig 273 (B)) consists in building up the tubes by means of corrugated strips, with their edges (or ends) expanded as in the former case. These strips extend from top to bottom of the radiator, and a pair of strips, with their expanded ends, form a long ziz-zagged narrow water-space from top to bottom of the radiator. Their external surfaces form hexagonally shaped air-spaces. This cheaper mode of construction gives a radiator element resembling the true honeycomb type.

The gilled tube system (Fig. 273 D) is perhaps the oldest type of radiator construction, although it is still much used, principally on commercial vehicles, on account of its strength and freedom from leakages. In this method the top and bottom tanks are connected by means of a number of straight vertical copper or brass tubes, usually of about $\frac{5}{16}$ to $\frac{7}{16}$ in. external diameter, thus providing a corresponding number of vertical paths for the water; in this case the water flows *inside* the tubes. Each tube has a large number of annular rings or fins pressed firmly over its outside surface, and spaced at uniform intervals, somewhat on the lines of the cooling fins of an air-cooled engine. These fins conduct and radiate the heat from the water inside the tube, the air current flowing through the spaces between the gilled tubes.

In the case of a car of, say, 12 H.P. rating, about 80 to 100 gilled tubes of $\frac{5}{16}$ in. inside diameter, with external fins of $\frac{3}{4}$ in. diameter would be used, there being 20 to 25 tubes per row of four rows. The length of the tubes would be from 16 to 20 in. The $2\frac{1}{2}$ ton Dennis chassis has 118 tubes, in six rows, bolted joints being employed at each end.

The Galley radiator construction shown in Fig. 274 represents the modern automobile and aircraft radiator

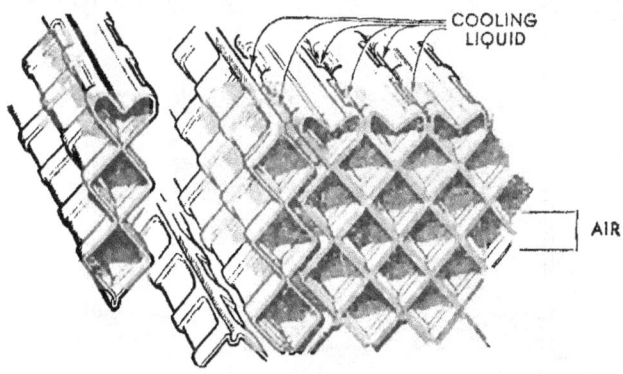

COOLING LIQUID

AIR

Fig. 274.—Galley Radiator Construction.

method of providing the maximum cooling effect from a given air velocity. The air spaces are roughly of square form and the water film spaces between the air orifice portions run in serrated fashion from top to bottom of the radiator. The materials used for the radiator cooling fins include brass, copper, steel and nickel-copper alloys.

The ratio of the fore-and-aft length of the air orifices or tubes (when the honeycomb radiator construction method is used) to diameter ranges from 30 : 1 to 80 : 1, the modern tendency being to use higher ratios of length to diameter for minimum frontal areas. The thickness of the metal used for honeycomb radiator tubes varies from 0·005 in. to 0·010 in.

In the case of certain commercial vehicle radiators, including those used for compression-ignition engines, vertical tubular elements consisting of brass tubes with spiral fins or gills—similar to that shown in the lower right hand illustration of Fig. 273—are often employed The upper and lower connections to the radiator tanks are made with rubber bushes pressed into place in the tank plates, as shown in Fig. 275 (A). This construction not only protects the tubes against severe vibration effects but also enables them to be replaced without difficulty.   An alternative method of mounting the

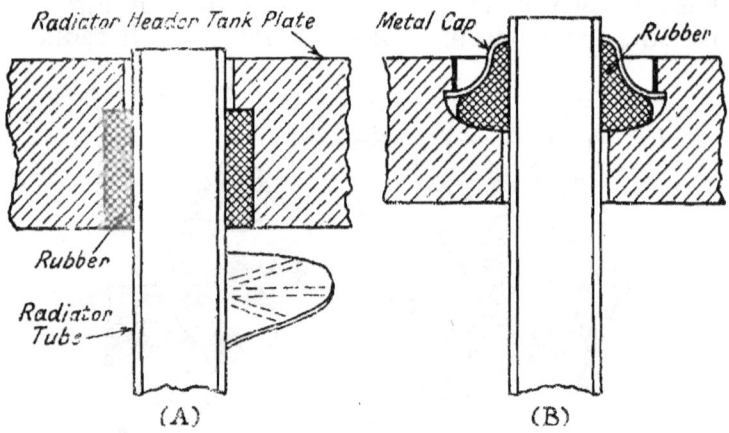

Fig. 275.—Methods of securing Radiator Tube Ends.

radiator tube in order to provide increased flexibility is the Heenan and Froude one illustrated in Fig. 275 (B). These two methods have been shown to give satisfactory results over a period of several years of road service.

The radiator tubes, instead of having spiral corrugated gills may, alternatively, have spirally-coiled wire cooling surfaces arranged in a similar manner to the spiral gills.

The Steam Cooling Method.—In the ordinary pump and thermo-syphon methods of cooling an engine the water passing through the cylinder jackets has its temperature increased by some 20° to 30° Fah.   Each 1 lb. of water then deals with 20 to 30 heat units.*   If however, the same weight of water were converted

*British Thermal Units

into steam it would be able to deal with no less than
966 heat units, that is, about 33 to 48 times the
quantity.

It will be evident, therefore, that if the cooling system
is so arranged that instead of merely raising the
temperature of the water in the jackets, it is converted
into steam, a very much smaller quantity of water will
be required in the system. It will, of course, be neces-

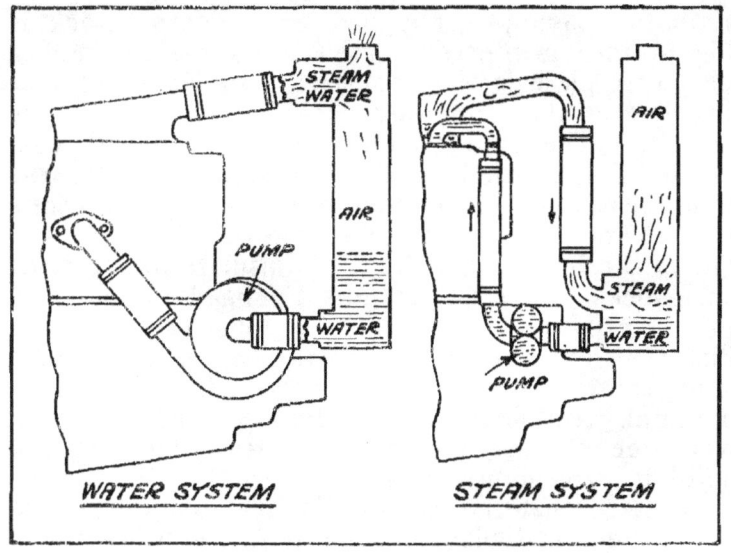

Fig. 276.—Illustrating Steam Cooling Methods.

sary to employ a " radiator " to act as a condenser
of this steam, the condensed water being returned to
the cylinder jackets.

This method has been used with satisfactory results
on motor-car and aircraft engines and has resulted in
an appreciable saving of weight.

As compared with the cooling water method it gives
a rather higher cylinder temperature, and in most cases
a gain in efficiency.

Fig. 276 illustrates the ordinary and the Rushmore
steam cooling systems used on some American motor
vehicles. In this case the top water outlet on the
engine, instead of being connected to the top of the
radiator, is carried down and enters the lower tank a

few inches from the point where water enters the line leading to the pump. A header-pipe lies in the bottom of the lower tank and has holes drilled through it to distribute the steam through the width of the radiator.

A gearwheel type of water pump draws the condensed water from the base of the radiator and delivers it to the bottom of the cylinder jackets. In this system of cooling it is necessary to fit a safety-valve in the top of the radiator to prevent steam pressure occurring.

Radiator Casings.—The ordinary radiator block is soldered into a pressed plated brass casing, and a back cover plate is then soldered into place. a space being left at the front and back for the air to flow through. The casing is usually provided with cast brass holding down brackets, for attaching it to the chassis members. Sometimes ordinary angle brackets are used, and a block of fibre, canvas or rubber inserted between the bracket and frame to insulate the radiator from shocks. In some designs a hinged fork or ball joint is provided at each side, thus relieving the radiator of any stresses due to the two side members of the frame distorting differently. In other cases horizontal trunnions are provided, one at each lower side of the radiator. These trunnions are housed in capped bearings bolted to the chassis side members, so that the radiator is free to tilt fore and aft; sufficient end play is allowed for endwise movement. Each bearing is provided with a greaser.

A type of radiator construction much favoured in America (used on the Ford, Chevrolet, Buick and other cars), and which has now been adopted in this country, consists in making the radiator block, complete with its headers and water pipes, and inserting the whole in another separate chromium-plated casing, designed for appearance purpose only. This method enables the radiator block to be removed without its casing, and cheapens the cost of replacement.

In more recent American-type cars the radiator casings are dispensed with and louvres made in the inside portions of the mudguards or the lower front parts of the engine cowling or bonnet allow the cooling air to reach the radiator block within the cowling.

Nickel Alloy Casings.—There is a tendency to replace the usual metal casings employed in radiator construction with radiator shells made of nickel-silver, or cupro-nickel. The grade of nickel-silver used contains from 15 to 20 per cent. nickel, the rest being copper and zinc. Cupro-nickel contains 20 per cent. nickel and 80 per cent. copper. This alloy retains its bright polished appearance for long periods, whereas the nickel-plated brass shells often wear right through to the brass, and the nickel itself becomes dull. White metal and aluminium have also been used for casings

Fig. 277.—A Typical Sectional Radiator.

Chromium-Plated Casings.—The more recent developments in radiator construction include the external plating of the casing with pure chromium—usually over nickel-plate. This metal is extremely hard and will not oxidise or dull under atmospheric conditions. It is rather whiter than nickel and requires only an occasional rub with a dry cloth to remove sedimentary deposits from raindrops; no metal-polishing preparations are needed.

Sectional Radiators.—There is another design of radiator which is much used on commercial vehicles, in which the complete radiator is made up of a number of quite separate tube-elements, usually from four to six, bolted to the top and bottom headers; rubber packing is used to make the water-joint. Fig. 277 illustrates a typical commercial sectional type radiator, which has six separate honeycomb elements, held top and bottom to the aluminium casing with rubber joints. This construction not only reduces the leakage tendency, but in cases of accident damaged sections can be removed bodily, the openings sealed effectively, and the car run home on the undamaged sections.

Care of the Radiator System.—As the water passages in the honeycomb and film type radiators are of very fine sections, it is necessary to prevent any solid matter in the water from reaching them. All water should therefore be strained before filling the radiator, and " hard " water, which is likely to deposit magnesia or chalky matter on the surfaces, should not be used. It is best to use rain-water, or to " soften " the water with one of the boiler water-softening compounds sold for the purpose. Should any deposit have formed in the tubes, it can be removed by filling the radiator with a strong, hot solution of washing soda unless there is any aluminium alloy in the cooling system components when such a solution should not be used, since it attacks aluminium.

Leaky radiators should be repaired by re-soldering or with one of the proprietory makes of " cold solder " sold in lead tubes. The use of bran and the compounds sold for stopping leaky radiators is not recommended, as the cooling efficiency is invariably reduced. In frosty weather the water should be drained off completely to prevent freezing of the radiator and cylinder jacket waters, otherwise a leaky radiator or cracked cylinder will result. Where pump circulation is used, the pump casing must be drained of water; a tap is usually provided for this purpose on the casing. (Fig. 262.)

To test the temperature of the cooling water, the hand should be placed at the *top* of the radiator—not at the bottom, where the water is coolest.

Anti-Freezing Solutions.—To obviate emptying the water system in cold weather, an anti-freezing solution can be used in place of pure water. Glycerine mixed with water will lower its freezing-point several degrees. Thus 10 per cent. of glycerine will freeze at 2° F. below freezing-point, 20 per cent. at 7° F. below, and 40 per cent. at 32° F. below (i.e., at 0° F.). It is not advisable, however, to use more than a 20 per cent. solution, or the circulation will tend to become choked up, and the rubber connections attacked. Alcohol, in a 15 to 20 per cent. solution (using methylated spirits), has a freezing-point several degrees below that of pure water. Another mixture useful for preventing freezing is that of 5 lb. of calcium chloride ($CaCl_2$) and a handful of washing soda dissolved in a gallon of hot water. It should be strained after dissolving. In the case of this solution, and with alcohol, any loss by evaporation should be made up for by the addition of mixture, *not* water alone. With glycerine, water only need be added.

A more recent anti-freeze solution is that containing *ethylene glycol* (Grade AF). 20 per cent. of this in water has a freezing point of −3° Fah., i.e., 35 Fah. below the freezing point of water.

Special radiator and garage heaters are obtainable from motor accessory firms for preventing freezing in cold weather.

Radiators for Tropical Use.—Cars used in tropical climates are often fitted with condensing tanks or reservoirs, where any water vapour, or steam formed in the radiator itself is led, and condensed for use again. The Citroen-Kegresse caterpillar tractors which cross the Sahara desert use this system. It is also employed on several American cars.

The radiator cooling area is usually about 25 to 35 per cent. greater for tropical, and also for mountainous, country use.

The steam cooling system also has practical possibilities for tropical purposes on account of its water economy.

CHAPTER IX

TESTING OF AUTOMOBILE ENGINES*

Practical Tests.—Without the use of special testing apparatus it is only possible to make rough qualitative tests, but nevertheless an experienced engineer can usually tune up his engines satisfactorily. Assuming that an engine has been overhauled completely, the crank-shaft re-ground and main bearings re-fitted, or adjusted, cylinders re-bored and new pistons fitted, valves skimmed and ground, new valve springs, big-ends adjusted, new small-end bearings and other minor repairs, the first item is to check the clearances of the working parts, such as the valve tappets and contact-breaker. The engine will usually feel quite " stiff," and must be run in for a period at light loads.

The modern garage will have a special running-in electric motor or belt drive from the machine-shoɔ shafting. A temporary wooden pulley is fixed to the crank-shaft, or the fly-wheel itself, can be used in some cases, and the engine is motored around with the petrol supply and ignition switch off. The engine should be given an excess of lubricating oil, preferably with a fatty ingredient such as castor oil, and the speed of running-in should correspond to a road speed of 20 to 25 m.p.h. (engine speed 1,000 to 1,500 r.p.m.); the running-in period is from 3 to 6 hours.

The engine should then be run light under its own power for about an hour, and then, if satisfactory, accelerated gradually for a short period, to a speed corresponding to a top gear speed on the road of about 30 to 35 m.p.h. The engine is then throttled down and note is made of any points requiring attention, such

* A much fuller account will be found in the author's " Testing of High Speed Combustion Engines and Automobiles " (Chapman and Hall).

as oil leakage, exhaust leaks, excessive noise in the tappets or gearing, and similar items. It will usually be found necessary to re-adjust the engine tappets after these tests; this should be done with the engine warm.

Running in the Engine on the Road.—In the case of new engines the machined surfaces of the cylinder walls, pistons and the various bearings are usually left with very fine tool feed marks on them; even with ground bearing surfaces there are fine irregularities. It is, therefore, necessary to remove the crests of these machining marks before one can say that the surfaces are properly bedded down for heavy duty purposes.

Whilst it is possible by special methods, such as lapping or superfinishing, to obtain the equivalent surface finishes to those of the bedded down ones, these methods are usually too expensive for the ordinary mass-production engine manufacturer to use. The engine, after a relatively short running in period, therefore has to complete its bedding down process by a period of road running. For this purpose, the new owner is instructed to drive his car on top gear at speeds not exceeding about 30 m.p.h. for the first 500 miles or so, with correspondingly reduced speeds on the lower gears; most mass-produced engines are fairly " stiff," when new, for this reason. To ensure that the driver does not over-run his engine many manufacturers sometimes fit restriction washers in the carburettor-inlet pipe joint. These are removed after 500 miles.

Road Tests.—When the engine is sufficiently run in the car should be tested for: (1) Petrol consumption on the level at normal running speed, say 20 to 25 m.p.h. (2) Maximum speed on the level. (3) Hill-climbing capabilities. (4) Acceleration.

The petrol consumption may be checked by filling the main tank up to a given fixed mark, with the car on level ground, and then running the car for a known distance (by speedometer or milestones) and noting how much petrol must be poured into the tank in order to fill it to the same fixed mark on level ground.

Sometimes it is more convenient to rig up a small auxiliary tank of a few pints capacity on the dash-board, and to run the engine off this. If a glass or float gauge be fitted, the quantity of petrol consumed for a given distance of running is readily deducible.

Petrol Consumption.—Thus if x pints of petrol be used whilst the car covers y miles the consumption $= \dfrac{8y}{x}$ miles per gallon. For example, if $1\frac{3}{4}$ pints are consumed for a road distance of 7·4 miles, the consumption will be

$$\frac{8 \times 7\cdot4}{1\frac{3}{4}} = \frac{8 \times 7\cdot4 \times 4}{7} = 33\cdot8 \text{ m.p.g.}$$

Instruments are now on the market, known as mileage-per-gallon or consumption meters, which read the consumption direct. Of these the Milegal, Gallometer, and Zenith flowmeter are typical examples. Fig. 278 illustrates the Gallometer device. It consists

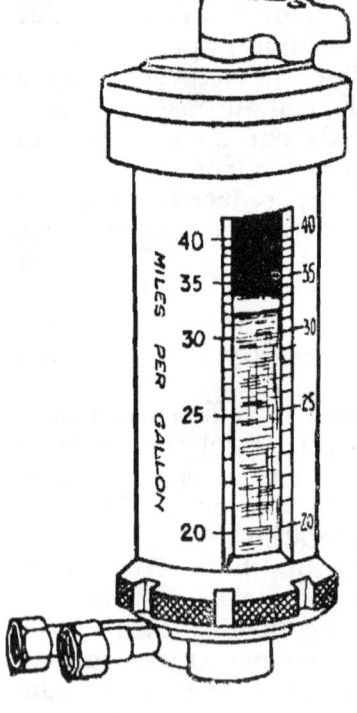

of a vertical glass gauge graduated with a scale reading direct from 20 to 40 m.p.g. It has a three-way tap below, and is used as follows: The gauge is first filled by turning the tap to the "filling" position, a slight overflow indicating when it is full. The tap is then turned to the running position when the car passes a milestone or the speedometer mileage index shows an even number of miles. The car is run at a constant speed for exactly one mile. when the tap is turned off, the petrol supply going direct to the engine. The petrol will have sunk in the glass gauge during the mile test, and the reading opposite its level will give the miles per gallon.

It is essential to keep the car speed uniform during a

fuel consumption test, since the consumption is different at different speeds; in this way the most economical speed of travel can be found.

The petrol consumptions of average cars should be approximately as follows:

TABLE No. 6

Engine Capacity. cu. cm.		Miles per Gallon on Level.
700 to 1,000	......	40 to 45
1,000 to 1,500	......	35 to 40
1,500 to 2,000	......	26 to 35
2,000 to 3,000	......	22 to 26
3,000 to 4,000	......	17 to 22

If it is possible to ascertain the B.H.P. of the engine, the fuel consumption can be checked from the fact that a good design of water-cooled petrol engine uses only 0.50 to 0.55 lb. of petrol per B.H.P. per hour.

Thus a 20 H.P. engine, running at full throttle, will use, say $0.50 \times 20 = 10$ lb. of petrol per hour, and since 1 gallon of medium grade petrol weighs 7.6 lb., it will therefore use $\frac{10}{7.6} = 1.31$ gallons per hour.

Ton Mileages.—For accurate comparisons of car performance, the weight of the car, as well as the fuel consumption, should be taken into account. Most of the fuel consumption tests results on cars made by the R.A.C. are expressed in this way.

If the weight of a car be W tons, and C = its fuel consumption in m.p.g., then $W \times C$ = ton mileage per gallon.

The highest values recorded to date are those for heavy commercial vehicles, namely, 120 to 130 ton miles per gallon. Most cars range from 40 to 60.

"Road" Tests in the Laboratory.—It is also possible to test the engine, gears, and transmission of a car in the test house by arranging a large diameter metal drum under each of the rear wheels, and providing a spring balance from the rear axle to a fixed post at the back. The engine is started up, clutch engaged with the gears in the gear-box in mesh, when of course the rear wheels will revolve, turning the drums. The car cannot move forward owing to the spring balance tension cable. The power delivered to the drums can readily be measured, electrically or otherwise, and the

efficiencies of the gears and transmission computed.
Messrs. Heenan and Froude provide a special car (and
also a motor-cycle) testing plant on these lines; by
using drums having irregular surfaces, road inequalities can be reproduced in the tests. It is possible
also to attach the two rear axles each to a separate

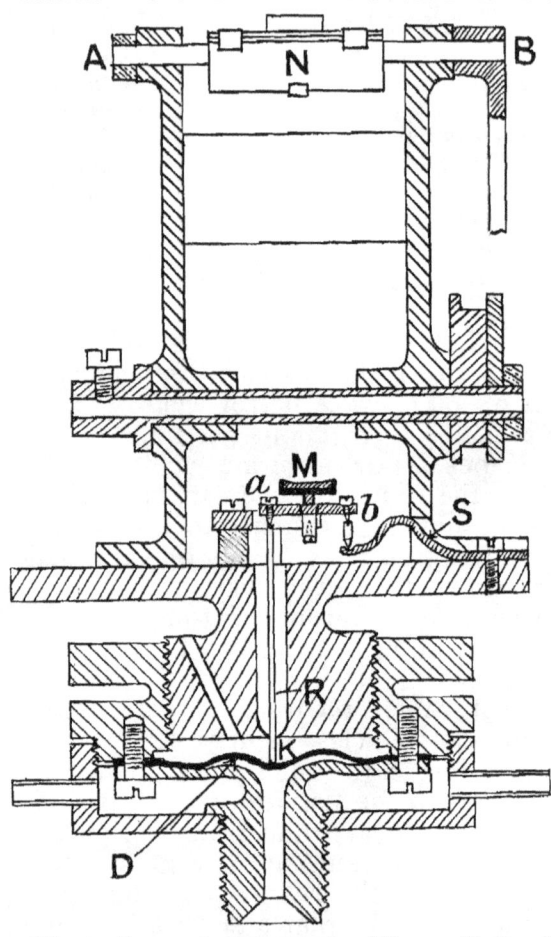

Fig. 279.—Illustrating principle of the Watson-Dalby Indicator.

dynamometer, such as the Froude hydraulic type, and
to measure the power delivered at the rear axles, on
all the gears. Much useful information can be
obtained without actual road tests. The Sprague
electric branch of the G.E.C. in America have supplied

a large number of chassis testing plants of this type, with swinging field dynamometers for absorbing and measuring the power at the wheels.

Measuring the Indicated Horse Power.—It is necessary, in the case of new type and special engines, to know what power is developed in the cylinders; this

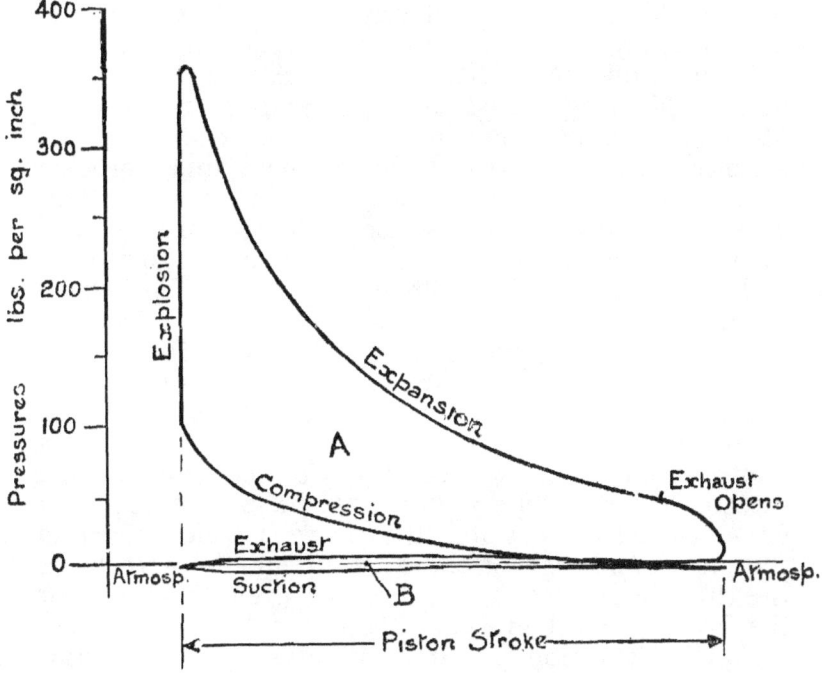

Fig. 280.—A Typical Petrol Engine Indicator Diagram.

involves the measurement of the pressures at various parts of the working cycle. It is difficult to measure these pressures, owing to the high speeds and temperatures. Instruments, known as *Indicators*, have now been evolved, however, which draw miniature diagrams of pressure on a piston stroke or crank-angle base, whilst the engine is working. Some employ optical mirrors to reflect a ray of light, in order to minimize the weight of the rapidly moving parts. Fig. 279 illustrates the Watson Dalby indicator of this type, which has been used for accurate pressure records. The lower threaded portion or plug screws into the valve cap of the cylinder, and the pressure is received on a spring steel, circular diaphragm K, corrugated in section (so as to give a uniform scale of deflections), and gilded on its lower side to prevent corrosion by the burnt cylinder gases. As the cylinder pressures increase or decrease, so the diaphragm deflects, and in doing so pushes the light steel rod R up; this causes the concave mirror M to rock sideways about its pivot *b*. The mirror M thus rocks proportionately to the pressure on the diaphragm. A second plane mirror N is given a rocking motion in a plane at right angles to the first, by means of an eccentric and rod, driven by a chain and sprocket system, off the engine and at engine speed. This gives a motion to N, in phase with, and proportional to, the motion of the piston. The two motions of M and N are combined by allowing a beam of light from an electric bulb filament to fall first on M, whence it is reflected to N, and then brought to a focus on a ground glass screen. When properly adjusted the beam traces out a diagram of pressures (or Indicator Diagram) similar to that shown in Fig. 280. If a photographic plate be substituted for the ground glass screen, a permanent record can be obtained. It is very instructive to watch the changes which occur in the shape and size of the diagram on the screen, as the throttle is opened, the mixture strength varied, or the ignition level moved. More can be learnt about the cylinder action in a few minutes than in months of theorizing. It is only necessary to measure the area of the indicator diagram and divide this area by the width parallel to the base line (or stroke) in order to find the mean height, that is, the mean effective pressure during a working cycle. If the diagram be divided into, say, ten parts of equal width by lines parallel to the pressure axis, and the average heights of these sections be measured, added together and then divided by 10, the mean height can readily be obtained. Alternatively there is an instrument, known as a Planimeter on the market, which will measure the area automatically.

The Farnboro' Indicator.—The optical type of high-speed indicator is open to the objection of possible errors on account of the inertia, or weight, effects of the rapidly moving parts of the mirror actuating mechanism; it also required highly-skilled operators. Although attempts have been made to overcome this drawback by using extremely small or "Micro" indicators and tiny celluloid recording plates, a more robust indicator was found necessary for works' and routine test purposes.

The Farnboro' indicator made by Dobbie McInnes Ltd., of Glasgow, fulfils this need, and it has been fairly widely used for high-speed petrol and C.I. engines, with satisfactory results.

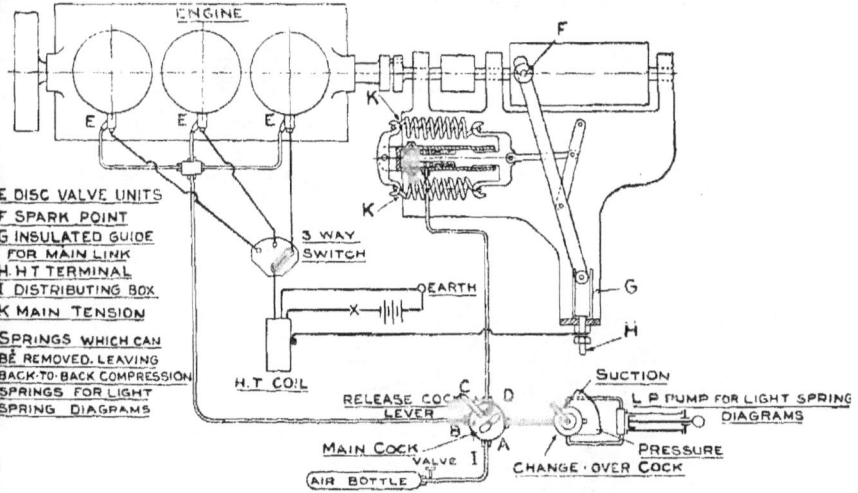

Fig. 281.—Showing Layout of the Farnboro' Indicator.

The principle (Fig. 281) employed is to use a small stainless steel disc with a limited amount of movement, confined in a small chamber, one side of which is open to the combustion chamber of the engine under test. The other side of the disc chamber is connected to a source of compressed air. The disc is free to move about ·01 inch, and can float between two faces of its chamber; it is guided by a spindle which is electrically insulated from the rest of the body. The spindle is connected to the primary circuit of a high tension coil and the two faces of the body between which it floats acts as the "earth" of this circuit; in effect the disc acts as a contact-breaker for this electrical circuit. When the engine is working, air at a gradually increasing pressure is applied to one face of the disc. Each time, during a cycle of operations, that the air pressure equals the cylinder pressure, the disc will vibrate backwards and forwards, giving rise to a series of sparks in a gap formed in the secondary circuit.

It only remains to cause these sparks to record themselves on a piece of black paper wound on a

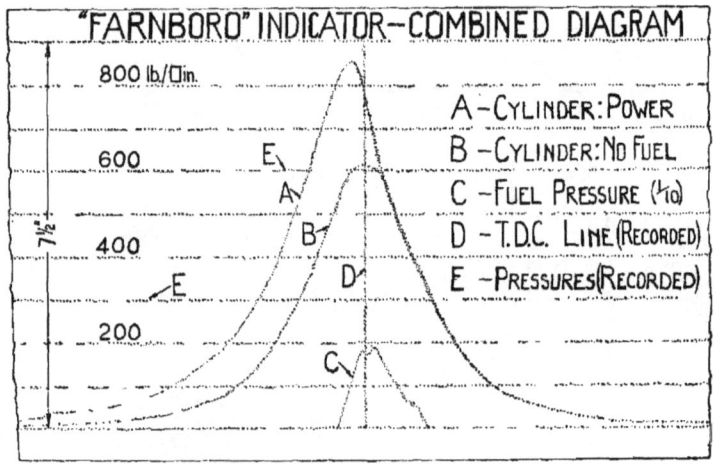

Fig. 282.—Much reduced Pressure-Crank-Angle Records, obtained from High Speed C.I. Engine.

synchronized cylinder drum rotated at engine speed. This is done by operating a pointer by means of a pressure piston from the air pressure supply, so that the position of a sparking point at the end of the pointer is proportional to the air-pressure value on the disc.

This also gives the pressure values in the cylinder, whilst the rotation of the drum gives the piston position—or more accurately the crank-pin position or time base.

In this way a pressure-time diagram is built up by means of the spark punctures or dots, and the diagram can be drawn in through these dots (Fig. 282).

Modern Cathode Ray Tube Indicators.—An entirely new type of high-speed indicator which has been developed for petrol and C.I. engine purposes, utilises electrical methods of recording the pressure and time, or crank-angle base diagrams. Examples of these indicators are the Metrovick-Dodds, Sunbury and Cossor ones.

The former has a device, which is screwed into the cylinder head, for the purpose of converting the combustion chamber pressure changes into electrical variations. The latter are applied to a pair of parallel plates in the cathode ray tube (Fig. 284) where they cause the cathode ray—a luminous beam—to be

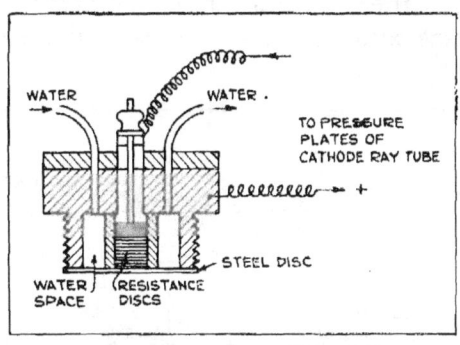

Fig. 283.—The Cathode Ray Tube Pressure Element.

deflected so as to describe a straight line on the fluorescent base of the tube. This gives the pressure line.

The time-base, or crank-angle line is obtained by applying voltage variations to another pair of plates having their planes at right-angles to the pressure plates. The cathode ray is thus given two movements, mutually at right-angles, proportional to the cylinder pressures and crank positions, so that it describes a pressure-time diagram on the fluorescent tube end.

It is possible to obtain pressure-volume diagrams also, by using a special contact-breaker mechanism.

The method of converting cylinder pressures into currents of varying voltage is as follows:—A special water-cooled plug (Fig. 283) screwed into the combustion head has a stainless steel disc, in contact with which is a stack of thin discs of a high (electrical) resistance material. The latter has the property of changing its resistance in proportion to the pressure to which it is subjected. It is only necessary, therefore, to connect this resistance element in an electrical circuit in order to produce changes of voltage, which are then applied to the " pressure " plates of the cathode ray tube.

The timing base is a combination of a condenser, a contact-breaker worked off the crank-shaft and a variable resistance; the electrical details are, however, too technical to discuss here.

There are other cathode-ray indicators working with steel discs and magnetic fields and piezo-electric

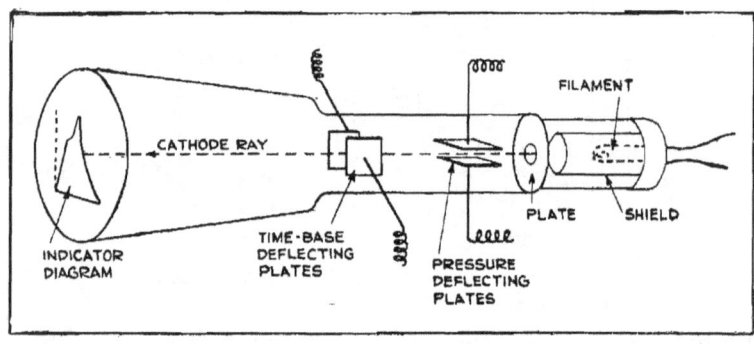

Fig. 284.—Showing Principle of Cathode Ray Tube Engine Indicator.

elements, for converting the cylinder pressures into voltage effects.

These types of indicator enable the pressure curves to be studied visually on the fluorescent base of the tube; photographic records can also be made. The spot of light is, however, often too faint for ordinary photographic purposes, and the base of the tube is not flat; moreover, the calibration of the pressure scale is not always constant.

These indicators are useful also for visual studies of the behaviour of C.I. engine combustion phenomena and fuel pump characteristics.

Finding the Horse Power.—Having found the M.E.P. from the indicator diagram, the *Indicated Horse Power* (I.H.P.) can at once be computed from the relation

$$\text{I.H.P.} = \frac{p_m \, l \, AN \, n}{33,000} \quad \text{where } A = \frac{\pi \cdot d^2}{4}$$

Where p_m = Indicated M.E.P. in lbs. per sq. in., d = diameter of cylinder in inches, l = stroke of piston in feet, n = number of cylinders and N = number of explosion strokes per minute. This formula is equally applicable to four-cycle and two-cycle engines.

A more convenient expression of the same relationship is as follows:—

$$\text{I.H.P.} = \frac{\text{I.M.E.P.} \times \text{Displacement} \times N}{504,230}$$

Where Displacement = total cubical capacity of engine in cubic inches. If the displacement, or cubical capacity of the engine is given in cubic centimetres it is useful to remember that 1 cu. in. = 16·387 cu. cm.

Engine Pressure Indicators. — The indicators described are usually expensive and complicated; some types require the services of a highly-trained assistant. In many cases all that is necessary, for test purposes, is a knowledge of the compression and the explosion pressures, for checking the performance of an engine.

It is not practicable to use an ordinary pressure gauge of the Bourdon type, for the pressures in the cylinder vary so rapidly that the mechanism of the gauge would quickly be rendered unreliable and inaccurate. The compression pressure should, of course, be taken when the engine is hot, and the cylinder in question not firing. It should be remembered also, that the compression value will depend upon the throttle opening, the maximum value being obtained at low-running speeds and at full throttle opening. At higher speeds, owing to wire-drawing of the charge, the compression pressure, for the same throttle opening, falls off somewhat.

Fig. 285 illustrates a useful pressure indicator which has been widely employed for automobile and other engines to measure compression and explosion pressures. This indicator screws into the valve cap, or sparking plug hole (for compression measurements). It contains an accurately fitting piston which can move in its cylinder against the pressure of a specially calibrated and adjustable helical spring, and a vibrating pointer to show when balance is obtained. To use the indicator for explosion pressures, the milled screw cap is rotated until the elbow lever on the left side, which magnifies the movements of the little piston, just ceases to vibrate; this indicates that the spring pressure just balances the explosion pressure. The value of the spring load and the area of the piston being known, the instrument is graduated directly in lb. per sq. in., a cyclometer type of indicator giving the balancing pressure. The latter can also be read off an axial scale on the barrel, against the screw cap piston.

Fig. 285.—The Okill Pressure Indicator.

Brake Horse Power.—The horse power obtained at the crank-shaft is less than that given in the cylinder (i.e., the I.H.P.) owing to the power lost, or absorbed, by friction at the working surfaces, namely, the piston and its rings, cylinder walls, connecting-rod, cam-shaft and main bearings, and in other ways. The horse power thus lost is termed the *Friction H.P.* The available horse power at the flywheel or clutch is termed the *Brake Horse Power* (B.H.P.), and we have the following simple relations:

Brake Horse Power = I.H.P. — Friction H.P.

$$\text{and Mechanical Efficiency} = \frac{\text{B.H.P.}}{\text{I.H.P.}} = \frac{\text{I.H.P.}-\text{Friction H.P.}}{\text{I.H.P.}}$$

When the B.H.P. has been measured, it is possible to calculate the Brake Mean Effective Pressure from the following relation:

$$\text{B.M.E.P.} = \frac{\text{B.H.P.} \times 504,230}{\text{Displacement (cu. ins.)} \times N}$$

where N = number of explosions per minute.

$$= \frac{\text{R.P.M.}}{2} \quad \text{for four-cycle engine.}$$

$$= \text{R.P.M. for two-cycle engine.}$$

The displacement is the total volume of the engine's cylinders, i.e., the piston area multiplied by the piston's stroke, multiplied by the number of cylinders.

Values of the B.M.E.P. for modern car engines vary from about 130 to 180 lb. per sq. in., at full throttle.

Mechanical Efficiency.—The mechanical efficiency of an engine should be as high as possible; the maximum efficiency is ensured by minimum friction losses. The friction losses can be reduced to a minimum by the use of well-lubricated surfaces, of ample, but not too large, a bearing area, the lightening of reciprocating parts by the use of high tensile steels and aluminium alloys, by the use of ball and roller bearings wherever possible in place of plain gun-metal or white-metal bearings, and by the balancing of the crank webs and the reciprocating parts, etc.

The mechanical efficiency of a well-designed engine ranges from about 90 per cent. at low speeds to 80 per cent. at high speeds. The falling off in efficiency is due partly to increased friction losses on account of increased friction with speed of rubbing, and partly to increased losses due to drawing in and expelling the charge.

Measuring the B.H.P.—It is better to measure the B.H.P. directly, however, with one of the dynamometers or power brakes now on the market for this purpose.

A Prony brake, for giving approximate results, is not difficult to rig up. In its simplest form it consists of two blocks of wood held together with spring tensioning bolts, around the fly-wheel rim, or a pulley

N

fixed temporarily to the crank-shaft. The frictional drag of these blocks opposes the fly-wheel's motion and absorbs the power output. The torque arm (Fig. 286) is loaded with weights W until there is no tendency for it to rotate with the fly-wheel. Stops are usually provided to limit its motion, and a pointer on the torque arm moving over a fixed scale enables the balance to be maintained.

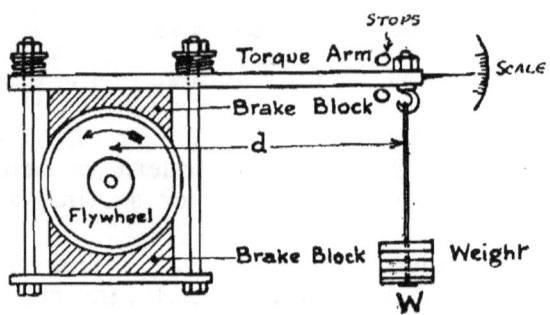

Fig. 286.—The Prony Brake for Horse Power Measurements.

Using the notation shown on Fig. 286, the Brake Horse Power is given by the relation

$$\text{B.H.P.} = \frac{2\pi N\ W\ d}{33,000}$$

where N = r.p.m. d is measured in feet, and W in lbs. $\pi = 3\cdot14159$.

With a weight of 25 lb. at the end of a 3 ft. torque arm it is possible to absorb 14·3 H.P. at 1,000 r.p.m.

This type of brake is simple and inexpensive to make; the fly wheel, however, becomes fairly hot, and there is a tendency on the part of the friction surfaces to "snatch." A water-cooled rim can be used.

The air, or fan, brake shown in Fig. 287 has been extensively used for tests on small engines. It consists of a pair of arms which bolt rigidly on to the crank-shaft. Each arm has a rectangular metal plate bolted at the same radius, the plates being normal to the direction of motion. The air resistance of these blades opposes the engine's torque, and absorbs the horse power. The plates can be bolted at different radii on the arms, and plates of different area can be fitted readily.

The size of plate and radius are chosen by trial so that the engine on open throttle attains the selected speed. Tables or curves are supplied with the brake which give the corresponding horse power for this speed, blade area and radius.

The advantage of this type is simplicity and low cost. The disadvantages lie in having to alter the blades or radii of fixing to vary the load at any speed. The brake's accuracy is also liable to be affected by draughts and air-currents due to extraneous causes. By enclosing it in a square or circular casing, provided

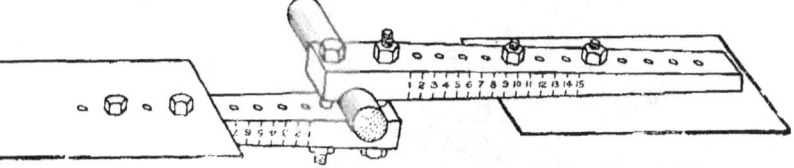

Fig. 287.—The Walker Fan Brake.

with a variable central inlet area, or outlet, a much wider range of control is obtained.

Absorption Brakes.—The majority of production-type automobile and also aircraft engines in this country are now tested for B.H.P. at various speeds by means of a water or hydraulic brake which absorbs the power given out by the engine by churning the water within the brake casing and thus heating it. This and the two types of brake previously described are termed *Absorption Brakes*. The alternative kind of brake, known as the *Transmission* type, transmits the engine power but enables the latter to be measured; the electric generator brake mentioned later is an example of this type. The principle of the hydraulic brake or dynamometer is illustrated in Fig. 288. The crankshaft extension or coupling of the engine is connected to the main shaft of the brake by means of a flexible coupling—which allows for small errors of alignment. The brake shaft contains a rotor in the form of a kind of paddle which rotates within a closed casing containing projecting vanes, just clear of the paddle blade tips, and filled with water. Glands are provided between the rotor shaft and casing to prevent water leakage. The whole of the casing is mounted on low friction bearings, so that if free it could rotate around the rotor shaft. When power is transmitted to the rotor the paddle churns the water and in doing so tends to drag the casing round in the same direction as its own rotation. Above a certain minimum speed of a few hundred r.p.m., below which slipping occurs, the torque or drag effect on the casing is equal to the torque applied to the rotor. The casing is prevented from rotating by means of a weighted arm, and if the casing is initially balanced, when at rest, the torque applied to the casing when the rotor is working is equal to the

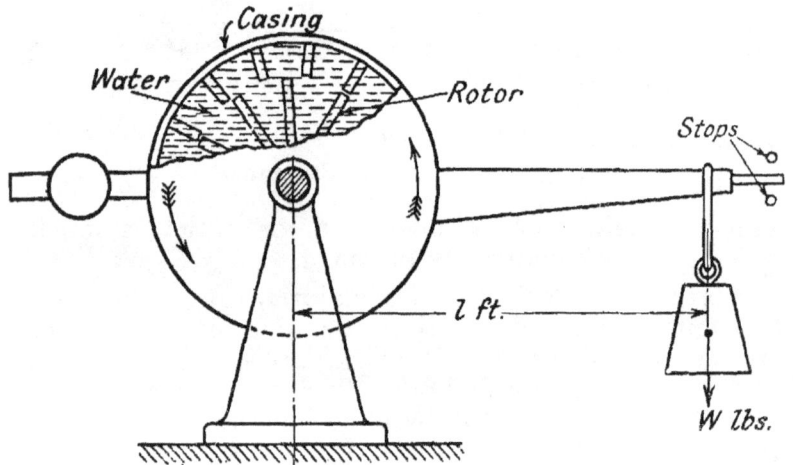

Fig. 288.—Principle of Hydraulic Brake.

product of W and 1 in lbs. ft., assuming that the arm of the casing is just floating between the stops shown.

It is then a simple matter to work out the horse power absorbed, for this is proportional to the product of the torque and speed of rotation.

$$\text{Thus B.H.P} = \frac{2\pi \, W. \, 1. \, N}{33,000}$$

where W = weight on torque arm in lbs. 1 = distance between C.G. of weight and shaft in feet, N = r.p.m.

For any particular machine the length 1 is constant, so that the expression $\frac{2\pi \, 1}{33,000}$ is also constant and can be denoted by the constant quantity k, so that the

formula may then be simplified as follows:—
$$B.H.P. = kW.N.$$
It is, therefore, only necessary to measure the weight values and speeds in order to know the horse power of the engine. In order to obviate the use of a number of weights a spring balance is usually employed, so that the value of W in the previous formula can be read off directly. In the case of a hydraulic brake having an arm of length 5 ft. $3\frac{1}{40}$ ins. long the value of $k = \frac{1}{1000}$ whilst for a metric machine having an arm of 1432·4 mm. length $k = \frac{1}{500}$.

The crude paddle arrangement shown in Fig. 288 is replaced in modern dynamometers by cup-like depressions on the rotor and inside the casing, to enable much higher power absorptions to be obtained for relatively small overall diameters of casings.

The Froude hydraulic brake shown in Fig. 289 is now largely employed for automobile engine tests. In this case the power is absorbed by hydraulic resistance. The engine shaft is connected to the shaft of this brake by means of a flexible coupling. To this latter shaft is keyed a rotor having a number of semi-elliptical pockets divided from one another by oblique vanes. The internal faces of the water-tight casing have similar pockets. The resistance offered by the water to the motion of the rotor reacts on the casing, which latter is free to rotate in its anti-friction bearings, but is constrained by weights hung at the end of a long torque arm attached to the casing. The product of the weight by the length of the torque arm, to the centre of the rotor, is equal to the mean torque of the engine. The friction of the bearings is included with the hydraulic resistance, so does not affect the results.

Since the water would get very heated if it were left in the casing, it is necessary to arrange for a continuous circulation from an outside source. It is necessary to provide an independent electric motor or belt pulley drive to start up the engine with this type of brake.

Another type of dynamometer of the transmission type much used for accurate engine tests is the electric swinging field type.   Here the armature of a motor is connected to the engine under test by means of a flexible coupling, and the field magnets and their casing are so arranged that they are capable of rotating about the same axis as the armature. The torque required to prevent this rotation is measured exactly as in the previous case, and is equal to the engine torque.   This form of dynamometer has the advantage that it generates

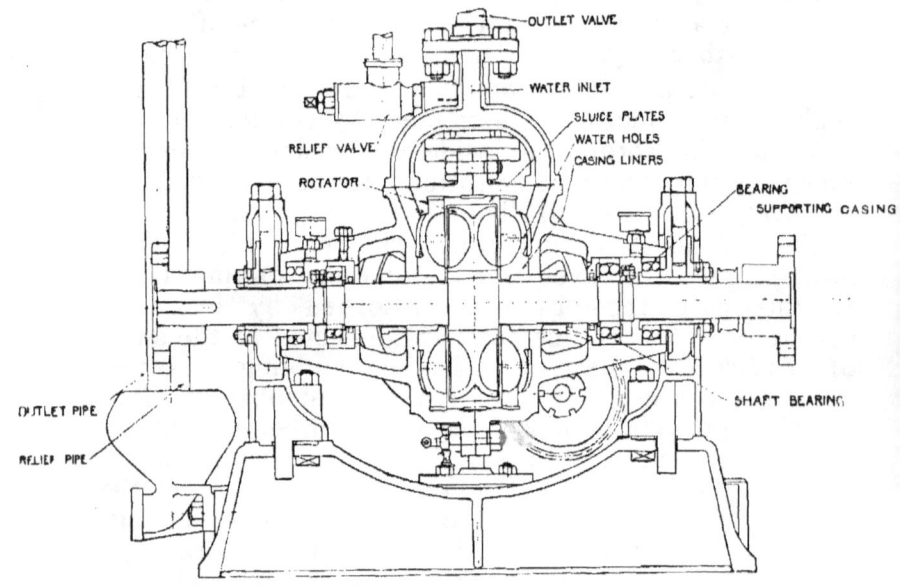

Fig. 289.—The Froude Hydraulic Dynamometer.

electric power which can be utilized for other purposes, and the dynamo can be converted to a motor for starting the engine, and for measuring its frictional losses, by making suitable switch connections.

Test Results.—It is usual to express the results of engine tests graphically by plotting them on squared paper, using the horizontal lines, or abscissæ, to represent speed values, and the verticals, or ordinates, the other quantities. Fig. 290 illustrates a typical test chart for an automobile engine.   In this case the I.H.P. increases from 5·3 at 500 r.p.m. to 13·1 at 1,200 r.p.m., whilst the B.H.P. ranges from 4·8 at the former speed to 11·0 at the latter. The mechanical efficiency drops from 91 per cent. at the former, to 84 per cent. at the latter speed. The horse power curves in this example refer to a low speed commercial engine, and are expressed for one cylinder only.

For most engine test house requirements in the case of production engines it is generally sufficient to

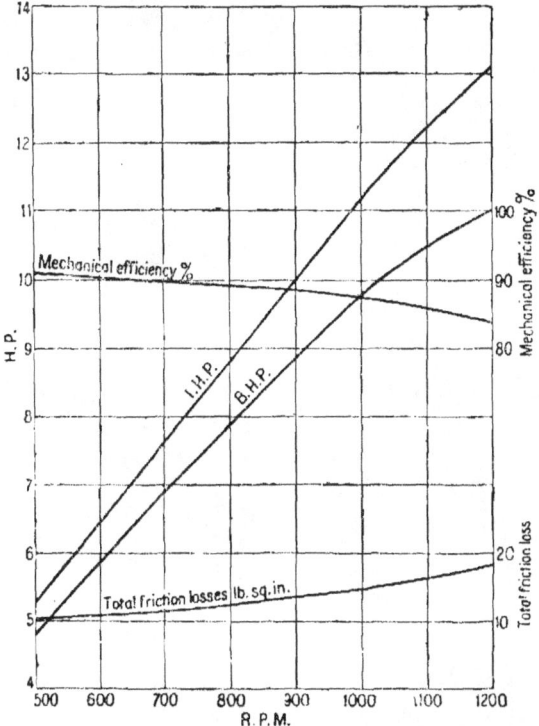

Fig. 290.—Showing Relation between Horse-Power, Efficiency and Speed.

measure the B.H.P.'s and fuel consumptions at various engine speeds and to plot the corrected readings in the form of graphs. The fuel consumption of an engine during brake tests is measured by means of an apparatus known as a *Flowmeter;* this has a scale against which readings of the height of the fuel in a vertical glass tube are read.   The corresponding quantity of fuel flowing to the carburettor can then be computed, readily.   This instantaneous method of reading or recording (by photographic means) the rate of fuel flow is more convenient and accurate than that of measuring the time necessary for a known quantity of fuel in a container to be used up by the engine.

CHAPTER X

MAINTENANCE AND TUNING

I⊤ is proposed to consider, briefly, the care and maintenance of the engine, dealing with this subject under a number of self-explanatory headings. No attempt will be made in the limited space available to go into the questions of repairs, overhauls and replacements; this subject is more fully dealt with in Volume 4 of this Series.

Starting the Engine.—A new engine should always be inspected superficially, and all nuts, bolts and terminals tested for tightness before attempting to start up. The oil level in the sump or reservoir should be examined, and, if necessary, oil added, through a strainer, to the required level. The radiator, in the case of a water-cooled engine, should be filled with filtered rain or soft water, and all joints and connections inspected for leakages; usually a turn or two on the tightening clip screws will stop such leakages.

Lubricate the engine parts not provided for in the automatic lubrication system. The overhead valve rocker bearings (if not autcmatically lubricated), fan bearings, magneto shaft (or distributor shaft with coil ignition), circulating pump shaft bearings, starting motor bearings (but *not* the worm and nut of the Bendix pinion), and other parts provided with lubricators or greasers must be attended to. It is a good practice to keep in a prominent place a list of parts to be lubricated (including the chassis) on a sheet of paper pasted to a piece of cardboard, or 3-ply, and afterwards varnished for preservation purposes.

The petrol may now be turned on, and the carburettor flooded, if necessary. The carburettor choke should be moved to enrich the mixture and the throttle opened slightly. The ignition switch being off. the engine should be cranked or motored around

for six to twelve turns so as to fill the cylinders with
mixture, and incidentally to ascertain whether every-
thing is satisfactory; any unusual knocks or noises
during the preliminary cranking should at once be
investigated. If *battery ignition* is provided, the
ignition timing lever should be *fully retarded*, the
switch turned " On," and a sharp pull up given by hand
from the bottom position of the starting handle. Alternatively the electric starter may be used, but when
starting from the cold, especially if the engine is new
or stiff, it is not always advisable to use the starter. A
good plain, in the case of modern engines is to crank
the engine with the choke set to give a rich mixture,
and the ignition switch " Off." Allow a few seconds to
elapse then switch " On ' and use the electric starter,
when the engine should start at once.

In cold weather the radiator should be filled with
hot water, and five minutes or so allowed for the heat
to flow through the cylinder walls, so as to thin the
lubricant, when the engine can be started as described.
With *magneto ignition* the engine must be started with
the ignition lever well *advanced*, since the magneto
gives its best spark in this position. Most engines are
now fitted with automatic ignition advance devices so
that the instructions for hand-operated control may
therefore be ignored.

The *carburettor throttle* should not be opened very
wide for starting—it may temporarily be opened during
cranking, but not for starting. If the carburettor is
provided with a jet control or air strangler this should
be set in the *"rich mixture"* position for starting.

As soon as *the engine starts* the ignition should be
advanced in the case of battery ignition, but partly
retarded with magnetic ignition. The engine is run at
a slow speed at first to allow the lubricating oil and the
cylinder metal to warm up; it is *a mistake to race* the
engine soon after starting up. When the engine commences to warm up, the choke control of the main air
supply to the carburettor should be returned to its
inoperative position, corresponding to the full-open air
supply. If a variable jet control is fitted—as in the
S.U. carburettor—this should be released to its normal

position. In the case of carburettors fitted with automatically-controlled chokes these instructions do not, of course, apply.

A *new engine should not be raced* for the first 20 to 30 hours of running (i.e., for the first 400 to 500 miles the road speed should not exceed 20 to 25 m.p.h.).

In the case of a *stubborn engine*, which will not start up in the usual way, or which is stiff, due to gumminess of the oil (more particularly in air-cooled engines), a little petrol may be injected into each combustion head prior to switching on. If the engine still *will not start*, examine the contact breaker for a sticky bush in the elbow lever, and for pitted contacts; take out the plugs and examine the points for oil or petrol bridging the gap, and for gap clearance. The throttle may be open too wide, or the switch not working.

Failure to start may also be due to a *choked filter or carburettor jet*, due to dirt or water; the remedy is to remove and clean the filter or jet. The slow running and main jets should both be examined.

If an engine will not start after a few continuous crankings, there is evidently something more serious at fault, and it is useless to continue cranking it over; stop, therefore, and examine the different items referred to.

Sparking Plug.—The *gap clearance* should be as given on p. 162. The plugs should have their points flush with or just below the combustion chamber walls.

Lubrication.—Read carefully the maker's instructions on lubrication. Each engine has its own system, so that no hard and fast rules can be given, although a careful perusal of Chapter VII on Lubrication Systems in this manual will be found helpful. Check the oil level in the sump frequently, and replenish only with fresh, good quality oil of the grade recommended by the makers of the engine. The oil should be *strained* before putting it in the sump; usually, however, the oil filler orifice contains a gauze filter.

In the *case of a new engine*, empty the sump, and wash out with hot lubricating oil after the first 500 miles, whilst engine is hot, for metal particles, such as machining burrs may dislodge in the crankcase of a new engine. It is a good rule to replenish the oil in the sump every 2,000 miles, as it becomes "spent," diluted with fuel, holds carbon and other particles in suspension, and generally loses its lubricating value. When the oil filter is of the edge filter felt or fabric pattern, however, no solid matter is allowed to remain in the oil and its useful life is thus extended. If the engine has been dismantled or decarbonized, smear each cylinder well with lubricating oil by hand, and work the pistons up and down. When the engine is running, frequently check the oil gauge readings; if it does not indicate satisfactorily *stop at once*—many engines are ruined through lubrication failure—and test the system. The oil filter may be choked, the oil pipes stopped up, or there may be an oil leakage; the oil level in the sump should be checked. Finally the gauge or indicator itself may be at fault—this can only be ascertained by disconnecting it, and noting whether the oil flows when the engine is running. The oil pressure relief valve may be at fault; it should be examined, and if necessary the spring pressure adjusted. Excessive oil consumption is caused by worn piston rings, leaky crank-case and pipe joints.

Cooling.—With air-cooled engines it is necessary to keep the cylinder fins clean and unobstructed; if fans are employed, check the belt tension for slipping.

In the case of water cooling, the radiator must be kept clean—it should be washed out regularly, and replenished with rain water. If "hard" or chalky water has been used the fine interstices in the tubes may become clogged with solid deposit; a boiling hot *solution of washing-soda* left in for an hour or two will usually cure this trouble. It is important to remember, however, that if the cylinder unit has an aluminium alloy head or cylinder block, washing soda or other alkaline solution *must not be used*, on account of its chemical action on the alloy. All water should be *filtered* before pouring into the radiator.

The fan belt should be kept properly tensioned. The glands of the water-circulating pump must be inspected for leakage, and adjusted when necessary. See that both the fan and pump bearings are lubricated periodically.

Carburation.*—Every carburettor has its own special characteristics, and the manufacturers' instructions should therefore be followed in each case. Generally speaking, however, attention is necessary, at regular intervals, to ensure that: (1) The petrol unions and plugs do not leak. (2) The petrol is properly filtered, and the filter kept clean. (3) The level of the petrol in the jet is correct—most float needles are provided with petrol-level adjustment means. (4) There is no wear in any of the working parts, such as the float needle mechanism, or the mixture regulation device. (5) There are no air leaks into the induction pipe.

A *weak mixture* generally reveals itself by (*a*) Popping or firing back into the carburettor; (*b*) Sluggishness, poor acceleration and loss of power; (*c*) Firing in the silencer when the car is running downhill with the throttle nearly closed.

A *rich mixture* is indicated by (*a*) Excessive petrol consumption; (*b*) Formation of a powdery type of soot in the combustion chamber and on the valve heads and sparking plugs; also by black smoke at the exhaust outlet; (*c*) Overheating of the engine.

The carburettor should be adjusted so as to give good acceleration, maximum power on the level and on hills, minimum fuel consumption. The following are the principal carburation faults experienced with petrol engines :

Poor Acceleration.—If the car fails to accelerate quickly, or to pick up after slowing down, it is usually a sign that the main jet is too small, or the choke-tube too large. The air supply may be excessive if an extra air valve is fitted. Try a larger main, or compensator jet, or a smaller choke tube.

Irregular Firing.—This is usually caused by a partly choked jet, dirty petrol filter, or water in the carburettor. The petrol supply may not be sufficient due to the fuel feed system not functioning.

Loss of Power.—Too weak or too rich a mixture will cause power loss. A trial should be made with a larger jet or smaller choke, firstly to see whether the

"* A full account is given in Vol. II of this series, entitled *Carburettors and Fuel Systems.*"

former is the case, or with a smaller jet, larger choke or more "extra air" in the latter. The jet and compensator sizes recommended by the makers should be tested. Loss of power is also caused by leaky cylinder head joint due to defective copper asbestos washer or slack nuts.

Petrol Drip from the Carburettor.—Dirt on the float needle seating, a punctured float, or a leaky petrol union will cause this. With some makes, if the car stands on a steep incline the petrol is liable to drip from the jet.

Inability to Run Slowly.—The slow-running device may require adjustment for (a) *Mixture Strength*; (b) *Mixture Quantity*. The former adjustment is usually provided for by the makers, and the latter by means of a small screw stop which limits the throttle closing position. Air leaks at the valve stems and inlet pipe joints will also prevent slow running. It is advisable, with worn engines, to fit one or other of the valve lubricators and leakage prevention fitments such as the Flexigas and S.U.

The carburettor should be thoroughly washed out with petrol and dissembled periodically, and the valve-seating float needle re-ground if required.

Ignition.—Very little trouble is experienced usually with the ignition if care is given to the following points:

(a) Keep the tungsten or platinum alloy contacts on the contact breaker clean and true—clean with a little petrol on a rag, and true with the special fine magneto file provided.

(b) Maintain the correct contact opening gap, namely about $\frac{1}{64}$ in. (or $\frac{1}{2}$ mm.). An adjustment is provided on one of the contacts. and a special gap gauge is obtainable (usually on the magneto file supplied).

(c) Keep the carbon brushes or contacts in the high-tension system clean. The greasy-like polish on brush contact systems can be scraped off with a knife and the brass and ebonite segments cleaned with a petrol rag.

(d) See that the switch and high-tension cables make good connections with their terminals, and that the insulation is not damaged so as to cause a short circuit.

(e) Check the firing order of the cylinders, and see that the high-tension cables are correctly coupled up.

(f) Check the sparking plugs for insulation failure and for correct spark gaps.

Causes of Ignition Failure.—The majority of failures are due not to the magneto or coil-ignition system mechanism, but to secondary causes; it is very seldom that the magneto or coil-ignition system gives trouble.

In damp weather with certain makes of magneto the fibre bush in which the contact-breaker rocker arm works is apt to swell, and to cause the spring-held elbow lever to remain open. The latter should be removed and the hole rubbed with a piece of wood, the thread of the central contact securing screw, or a piece of rolled emery paper of fine grade.

The low-tension cable from the contact breaker side of the magneto to the switch may become shorted on the engine frame; this will stop the engine most effectively. Check by disconnecting this cable at the magneto, when the engine should fire. A faulty ignition switch will also cause ignition failure.

A broken high tension or distributor brush will cause partial or complete ignition failure. With the modern jump spark distributors, there is a tendency for the metal sparking plate on the distributor rotor to erode, so that the spark gap is increased. This point should, therefore, be watched and if necessary after long periods of service a new rotor arm, fitted.

With battery and coil ignition a partially discharged or faulty battery causes ignition troubles. Other possible causes are: (1) The contact breaker. (2) The coil itself—which may be short-circuited internally, or its condenser may have broken down, as shown by excessive pitting of the contact breaker points. (3) The distributor may be very dirty, due to grease or dust.

Engine Timing.—This should be checked after any adjustment. With the piston of the firing cylinder on top centre of its compression stroke, and the ignition lever, if fitted, fully (or almost fully) retarded, the contact breaker points should just commence to open; the same method applies to automatic ignition advance systems. Each cylinder should be checked in turn in cases of doubt. The compression stroke can be ascertained by watching the inlet valve close, then giving another half-revolution, both valves being closed during the period.

General Notes.—So far we have considered the lubrication, cooling, carburation, and ignition maintenance; it now remains to add a few notes concerning the engine itself. Modern engines seldom require attention to their mechanical components, except after long periods of usage, say, after 10,000 to 15,000 miles of road work. Periodical attention to the items mentioned will greatly prolong the life of an engine, and also maintain its efficient working.

Valves.—Occasional attention should be given to the *valve stem and tappet, or rocker clearances;* newly ground valves tend to bed down a little and occasionally in older engines the stems stretch slightly under the influences of the high temperatures and strong spring and impact action. It is not possible to lay down an exact rule for these clearances as they vary with the engine, but it can be stated that the average inlet and exhaust clearances, for *side-valve* engines are $\frac{3}{1000}$ to $\frac{5}{1000}$ in. respectively. The adjustments should be made with the engine hot. In the case of *overhead camshaft* engines, the valve clearances can be kept relatively small, namely from $\frac{3}{1000}$ and $\frac{4}{1000}$ in., and these clearance do not tend to vary to the extent of the two other types previously mentioned; even after relatively long running periods.

The *valve springs*, particularly the exhaust ones, may in time lose their strength; an indication of this is the closing up of the end coils on the valve head side. The old springs should be tested by placing end to end with a new spring, and compressing, when each should deflect equally. After 6,000 to 10,000 miles running, the valves should be ground in on their seatings again—this should be done when the cylinders are decarbonized. A fine abrasive such as Carborundum

FFF, or one of the valve grinding pastes sold for the purpose, will be found satisfactory. See that the stems and valve guides are free from burnt oil.

Engine Faults.—Apart from carburation and ignition faults *an engine may run imperfectly* owing to (*a*) Poor compression.   (*b*) Valves.—Springs too weak. guides and stems worn, wrong clearances, pitted seatings, stretched exhaust valve, incorrect valve timing, valve gummed up with burnt oil, and remaining partly open, or not following the tappet. (*c*) Silencer may be choked with soot (carbon). (*d*) Excessive carbon deposit; this is indicated by overheating, knocking of the engine under load, such as on hills, engine will not take full ignition advance. (*e*) Inadequate lubrication; this causes overheating, partial seizure of pistons, excessive tightness of engine when cranked over.   (*f*) compression leakage at piston rings, sparking plugs or cylinder head joint.

The *engine will stop entirely* for any of the following reasons: (*a*) No petrol. (*b*) Ignition failure. (*c*) Broken valve, the valve remaining on the seat. (*d*) Broken valve spring.   (*e*) Ignition drive, or coupling slipping so as to throw timing out.   Broken piston or piston rings, gummed rings, valve sticking in guide; these will cause a complete or partial loss of compression.   (*g*) Broken connecting-rod or crank-shaft—evident at once by excessive mechanical noises in engine.   (*h*) Throttle or ignition controls not working—probably not connected up.   (*i*) Carburation faults previously mentioned.   (*j*) Plugs sooted or oiled up, or insulation broken inside plug.

A *periodical knock in the engine* may be due to a worn big-end or small-end bearing. If the engine crank-shaft is rocked each way by hand with the piston on top, or bottom centre, the sparking plugs being removed, and the ear situated near to the engine, a knock due to this cause can usually be detected ; the remedy is to re-line the big-end with white-metal, or otherwise to renew the bearing according to its design, or renew the small-end brush, or gudgeon pin.

Excessive valve clearances and *too strong springs* will also cause hammering, but of much higher frequency with multi-cylinder engines than in the previous case.

Backlash in the timing gearing, a slack timing chain and a loose coupling between the secondary driving shaft, and the water pump, or generator will also give rise to knocking. In these cases the *knocking is most noticeable* when the engine is running under a light load, or when the car is coasting down a hill with the engine switched on, and foot off the accelerator; in this case the engine sometimes drives the wheels, and the latter sometimes drives, or over-runs, the former.

A loose timing drive chain under these conditions will knock on the side of the casing, unless an automatic spring tensioner is fitted.

Another *frequent cause of periodical knocking* is the metal fastener on round or flat leather belt-drives to the cooling fan, in the case of old type engines; the fan blades may be out of alignment, and occasionally touch either on the radiator or engine casing; when travelling against a strong head-wind the latter may occur.

Engine knock, due to detonation caused by the use of poor grade (or low octane) fuels in modern high compression engines, pre-ignition, over-heating, over-advanced ignition, weak mixture, or excessive engine load, is readily distinguishable from mechanical knocks by its high pitch and its rapid frequency; a little experimenting with the ignition timing or air supply will indicate the type of knock due to this cause.

Engine knocks may be distinguished from transmission ones by noting whether they cease when the clutch is depressed and engine stopped, or when engine is running in neutral.

No attempt has been made to enter into details of carburettors and carburation, and ignition, or of engine overhaul; these are dealt with in the separate manuals of this series.

CHAPTER XI

Improvements in Fuels.—Since the previous edition, there have been some notable developments in petrol engine fuels, chiefly as a result of the constant demand for increased performance from aircraft engines. The fact that the power output of a given size cylinder can be increased progressively by employing fuels of increasing octane numbers—using higher compression ratios with such fuels—led to the introduction of new fuels containing special blending agents, e.g. iso-octane, iso-pentane, neo-hexane, together with tetra-ethyl lead in small proportions, that gave octane ratings of over 100. Another high octane fuel is that known as Triptane; when used in an aircraft engine of suitable design it enables about 70 per cent. more power to be developed, for the same cylinder capacity, than when 100-octane fuel is used.

Most of these high-octane fuels* are, however, relatively expensive and, further, in order to obtain the best performances, the compression ratios must be raised to the value that will just avoid detonation.

During, and for some time subsequent to, the 1939 War, pool petrol of one quality has been used on British motor vehicles; in 1948 this fuel had an octane rating of about 72—a fact which definitely limited petrol engine development. Thus, if fuel of 80-octane rating were available and the compression ratio and combustion chamber altered to suit this fuel, about 12 per cent. more power could be obtained from the same capacity engine.

Other alternative fuels that were developed and used during the 1939 War period and afterwards were

* Fuller information on these modern fuels and also the other alternative non-petrol fuels is given in "Carburettors and Fuel Systems," 5th Edition (Chapman and Hall, Ltd.).

methane, butane, propane and acetylene. The two former fuels were stored in the liquified form, being stored in steel cylinders and then vaporized under pressure or heat conditions and employed in petrol-type engines in a somewhat similar manner to coal gas. *Modified acetylene*, using methyl or ethyl alcohol as damping, or anti-knock agents has given satisfactory results on vehicles in Switzerland. Fuller details of these fuels are given in the footnote reference.

Octane Rating and Compression Ratio.—It has been shown, on page 21, how, as the octane value of the fuel is increased the compression ratio, and, therefore, the power output from a given size of cylinder, can be increased without incurring detonation effects. It has also been found, as a result of a good deal of experimenting with combustion chambers of various shapes and with different arrangements of the valves and sparking plugs, that each particular shape of chamber and disposition of the inlet and exhaust valve and placing of the sparking plug, corresponds to a given octane value of fuel, such that when using this fuel detonation will just be avoided.

The values of octane numbers and equivalent highest useful compression ratios given on page 21 must, therefore, be regarded as applying to one particular shape of combustion chamber and valve and plug disposition and not as being applicable, generally, to other shapes.

It has been shown that with the plain L-headed chamber shown at F in Fig. 33, on page 82, the highest compression ratio that can be used with petrol of 80-octane value is about 5·8 : 1, whereas with other more efficient designs, compression ratios up to 7·8 : 1 to 8·0 : 1 can be used without knocking or any other ill-effect for the same fuel.

The so-called pool petrol used during, and for some years after, the 1939 War had an octane rating of 70 to 74, but with the efficient combustion chambers used in post-war engines, it was possible to employ compression ratios up to about 6·5 : 1 to 7·2 : 1, and to obtain brake mean pressures of 118 lb. to 125 lb. per sq. in.

Modern Automobile Engine Particulars and Performances.—The trend of development in automobile engines has been one of constant power increase for a given capacity of engine. To achieve this, the compression ratios have progressively increased from their earlier values of 4·5 : 1 to 5·0 : 1, up to 6·0 : 1 to 7 : 5 in recent designs. This has been possible on account of improved fuels and combustion chamber design. Further, engine speeds have increased considerably in order to obtain higher outputs.

The more recent engines fitted to British cars employ *compression ratios* of from 6·3 : 1 to 7·3 : 1, although special sports car engines often use higher values, namely, 7·5 : 1 to 9·5 : 1. The average compresson ratio of 1949 engines was about 6·7 : 1.

The *engine speeds*, for maximum power of various British engines, range from about 3,800 to 4,600 r.p.m. for production models, but in the case of certain sports car engines the speeds were from 4,700 to 5,500 r.p.m.

The more popular of the small to medium sizes of four-cylinder engines were $1\frac{1}{4}$ to 2 litres; the latter correspond to engines of about 16 H.P. Treasury rating.

An analysis of the *power outputs* of various recent car engines of all types showed that these ranged from 29 to 35 B.H.P. per litre of cylinder capacity; in general the smaller sizes of engine, namely, up to about $1\frac{1}{2}$ litres, gave higher outputs per litre than larger ones; this is probably due to the fact that owing to heat exchange considerations, it is possible to employ higher compressions and to use higher maximum speeds in smaller engines. It is of interest to note, despite the claims of adherents to the cubical capacity rating of engines, that the adoption of this method would result in "square" engines of more compact shape, modern engines still employ appreciably longer strokes than their cylinder bores. The *stroke-bore ratios* for the 1948-9 engines ranged from 1·3 to 1·6 being about 1·4 on the average. It is noted that the stroke-bore ratios for higher speed engines is less than for slower speed ones, since the mean piston speeds are about the same in all cases. In regard to *American engines* of similar period to the British ones, previously dealt with, these were invariably of much larger cylinder capacity, namely, from about 3 to $5\frac{1}{2}$ litres, the more favoured size being $3\frac{1}{2}$ to 4 litres in the 6- and 8-cylinder types.

The compression ratios employed ranged from 6·5 to 7·3. The fuel quality restrictions do not apply in the U.S.A., so that with the usual combustion chamber designs higher ratios can be used with these higher-octane fuels. The speeds of various makes of engine were from 3,200 to 4,000 r.p.m., the average value being 3,600 r.p.m.

The outputs of American car engines are appreciably lower per litre of cylinder capacity than for British engines. For production car engines these values ranged from about 26 to 28 B.H.P. per litre.

The stroke-bore ratios were lower than for British engines, their values lying between about 1·1 and 1·35; the average was about 1·23.

It may be mentioned that in order to obtain greater outputs per litre, British engines, which are of much smaller capacity, have adopted, increasingly, the *overhead valve* type of engine, whereas the American designers favour the simpler and cheaper side-valve engine. Thus, about 75 per cent. of 1949 British production car engines have overhead valves, whereas only about 10 per cent. of the American engines employed this type. Further, all sports car engines in this country used overhead valves.

Horse Power and Car Weight.—A convenient and trustworthy method of assessing the possible car performance is that of expressing the relation between the B.H.P. of the engine and the weight of the empty car. If the latter is given in tons, it will be found that for modern small mass-produced cars, in the 8 to 10 H.P. (rating) class, the ratio will be about 35 to 40 B.H.P. per ton. For the 12 to 16 H.P. class, four-cylinder overhead valve engine modern cars, it varies from about 40 to 50 B.H.P. per ton. Cars in the 20 to 30 H.P. category usually give 50 to 55 B.H.P. per ton, whilst sporting models, as distinct from racing cars, have values of about 60 to 90 B.H.P. per ton.

Expressed *in terms of cylinder capacity*, cars in the 8 to 10 H.P. class weigh, on the average, 1·75 lb. per c.c.; those in the medium, or 14 to 18 H.P. class, weigh about 1·2 to 1·3 lb. per c.c., whilst in the larger output and heavier cars the weight is from 1·3 to 1·5 lb. per c.c.

High Output Engines.—It is well known that by supercharging an engine, using high octane fuels and designing the working parts for high speeds of operation, the output from a given size of cylinder can be increased very appreciably above that of normal production car engines. An example of this development is that of the M.G. Midget 750 c.c. four-cylinder engine used by Lieut.-Colonel Gardner to break the flying kilometre and mile record, by exceeding 150 m.p.h. (maximum speed, 164 m.p.h.) with a supercharge pressure of 30 lb. per sq. in. The engine developed 150 B.H.P. at 9,000 r.p.m., giving an equivalent output of 200 *H.P. per litre.* Another engine of 1,100 c.c., when supercharged, gave 198 B.H.P. on the dynamometer at 7,500 r.p.m. and an equivalent output of 180 B.H.P. per litre.

A creditable performance for an unsupercharged engine is that of the four-cylinder Jaguar used by Lieut.-Colonel Gardner to break the flying mile, kilometre and 5 kilometre speed records. This engine had double overhead camshafts and developed 142 B.H.P. at 6,100 r.p.m. for its 1,970 c.c., thus giving about 72 B.H.P. per litre.

Another unsupercharged engine, namely, the Lea-Francis 1½ litre four-cylinder special type designed for American speedway racing, gave 118 B.H.P. at 6,200 r.p.m. The maximum B.M.E.P. was 189 lb. per sq. in. at 4,600 r.p.m. The engine thus had an output of about 79 H.P. per litre. A compression ratio of 15 : 1 was used for the special fuel, which consisted of 90 per cent. methanol, 5 per cent. benzole and 2 per cent. castor oil. Four carburettors were employed. The Lea-Francis 90° overhead valves of this engine were operated by twin side-camshafts, short push-rods and rockers. The combustion chambers were of hemi-spherical form.

Other examples of German, French and Italian high-speed competition engines of the supercharged class develop from about 150 to 180 B.H.P. per litre; in one or two instances higher figures have been claimed.

It may be of interest to note that aircraft engines of the supercharged type usually develop from 45 to 65 H.P. per litre in the larger sizes, e.g. the 2,000 H.P. to 3,000 H.P. class.

Greatest Possible Output of Any Engine.—It is of interest to designers and others to know what is the

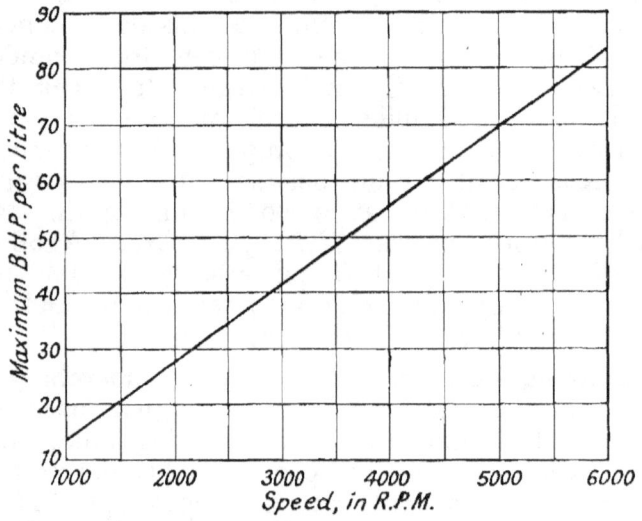

Fig. 291.—Showing Greatest Possible Outputs of any Unsupercharged Engine at Various Speeds.

greatest possible power output that can be obtained from any unsupercharged engine. It is not a difficult matter to ascertain this from considerations of the quantity of air passing through the engine at any given engine speed, per minute, since the amount of oxygen used determines the power output. If this is done it can be shown that the maximum output, in B.H.P. per litre, will vary with engine speed in the manner indicated in Fig. 291,* namely, from about 14 at 1,000 r.p.m. to

* "Modern Petrol Engines." A. W. Judge (Chapman and Hall, Ltd.).

83 at 6,000 r.p.m., and in direct proportion at higher
speeds. It should, however, be made clear that these
values represent the limiting outputs theoretically attain-
able using certain assumptions concerning volumetric,
mechanical and thermal efficiencies, etc.

Typical Engine Performance Curves.—To illustrate
the improvement that has occurred in the performances

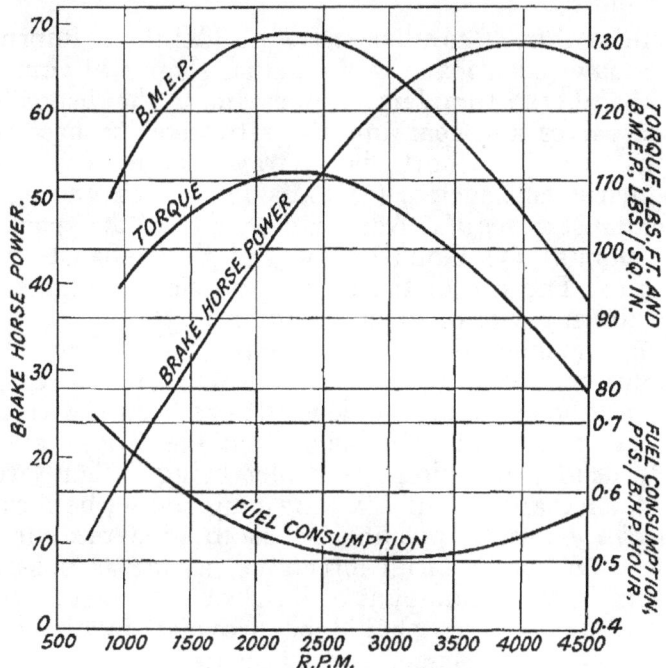

Fig. 292.—Typical Performance Curves of a Modern
Four-cylinder Engine.

of modern automobile engines, the curves shown in
Fig. 292 for a typical four-cylinder mass-production
engine, rated at 18 H.P. The bore and stroke are 85
mm. and 92 mm., and cylinder capacity 2,088 c.c. (127·6
cu. in.) The compression ratio is 6·8 : 1.

Referring to the brake horse power curve, the scale
for which is given on the left, it will be seen that the
maximum B.H.P. is about 68, at 4,000 r.p.m. The
maximum B.M.E.P. occurs at about 2,300 r.p.m. and
is 131 lb. per sq. in., whilst the maximum torque, which
also occurs at the same engine speed is about 111 lb. ft.

The fuel consumption curve indicates a minimum fuel consumption of about 0·5 pints per B.H.P. per hour at 3,000 r.p.m. At lower engine speeds, as is typical of modern petrol engines, the fuel consumption increases fairly quickly. Thus at 1,000 r.p.m. the figure is 0·67 pints. Similarly, at full power output the fuel consumption increases, but at a lower rate, being about 0·575 pints at 4,500.

Combustion Chamber Design.—Marked improvements have been made in the actual shapes of combustion chambers of modern engines and in the disposition of the valves and sparking plugs to give the best conditions for (1) Short flame travel path; (2) Good volumetric efficiencies; (3) Satisfactory scavenging of the exhaust gases; (4) Adequate cooling of the sparking plug points; (5) Good cooling of the exhaust valve head; (6) The use of the highest possible compression ratio for any fuel of given octane rating.

The principles discussed on page 83, in regard to combustion zones have been followed and developed with the above objects in view. It has been shown that the best results are obtained when the valves are of ample head proportions, the inlet being rather larger than the exhaust. The sparking plug should be located in a fairly central position so as to give the shorter flame travel paths in all directions, as far as possible. Further, a certain amount of swirl or turbulence of the compressed mixture should be provided by the combustion chamber shape, in relation to the piston at the top of its compression stroke, so that the sparking plug will be cooled sufficiently by the mixture movement.

To obtain such results it is necessary to depart, somewhat, from the combustion head and valve arrangements shown in Fig. 33 on page 82, and to use inclined valves or one vertical and one inclined valve. There is a relatively large number of alternative arrangements to those shown in Fig. 33, but if these are examined in the light of the principles described on page 83, the number of practicable designs will be found to be very limited. Here, it may be added that the L-head and T-head combustion chambers are now obsolete, but the Ricardo and Whatmough modifications to the former type have increased its efficiency and enabled good performance results to be obtained.

Fig. 293* shows, diagrammatically, some alternative combustion head arrangements that have been developed with the object of fulfilling some or all of the conditions enumerated previously; it will be observed that in each case the valve-actuating mechanism is more complicated than for the side-valve engine.

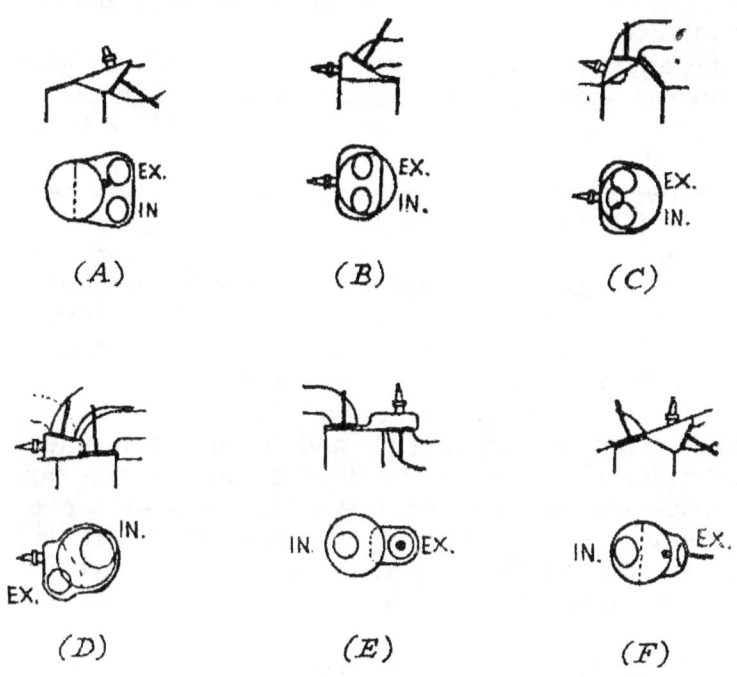

Fig. 293.—Alternative Arrangements of Valves, Sparking Plugs and Combustion Chambers.

Referring to (A) Fig. 293, this represents a development from the side-valve engine to give a more compact combustion head and approximately central sparking plug position, but with a bias towards the exhaust valve. Owing to the valves being in an offset pocket, the full amount of mixture may not be drawn into the cylinder; further, the sparking plug and exhaust valve would not be fully cooled by the mixture.

* "Poppet Valve Cylinder Heads, etc." J. Swaine. Proc. Inst. Mech. Engrs., 1948

The arrangement shown at (B), namely, with two inclined overhead valves, with parallel stems, is an improvement on the vertical overhead method as it gives a wedge-type combustion zone with central plug position. The mixture during the final compression stage is squeezed out from the extreme right side and thus given some degree of turbulence which will provide valve and plug cooling. The initial flame front near the plug is of wide dimensions, which is a distinct advantage.

The alternative arrangement shown at (C), to give somewhat similar results, necessitates a dished piston head, but enables vertical-type overhead valves to be employed.

The overhead valve disposition and combustion head shape in (D) known, generally, as the F-head, appears to fulfil most of the enumerated requirements; the combustion space is compact and there is directional turbulence. The plug is located correctly, to give anti-knock combustion conditions, and it has a certain amount of mixture swirl cooling effect.

The arrangement of vertical overhead inlet and side exhaust valve shown at (E), and the partly-turbulent combustion chamber on the lines of the Ricardo one for side-valve engines, gives suitable combustion conditions and tends to simplify the valve gear. It will be observed that the plug is located over the exhaust valve and that the flame front is a wide one in this region.

As a final example of the many possible alternative arrangements, that shown at (F) corresponds to the modern Rover car engine combustion head, shown in sectional view in Fig. 294. It approximates to the hemispherical shape, with the plug at the centre of the flat side, giving minimum flame travel path, but permitting large valve diameters for induction and exhaust clearance. The piston at the end of the compression stroke forces out the mixture past the overhead inlet valve and across the plug points, thereby tending to cool them. This gives a high efficiency combustion chamber, such that it enables compression ratios of 7·75 : 1 to 8·0 : 1 to be employed with 80-octane fuel, and about 7·2 : 1 to 7·5 : 1 for 72-octane fuel. In the former case, brake m.e.p.'s up to 130 lb. per sq. in. are obtainable. Further, it has been shown that weaker mixtures down to about 20 parts air to 1 part petrol can be employed without misfiring and with very low fuel consumption.

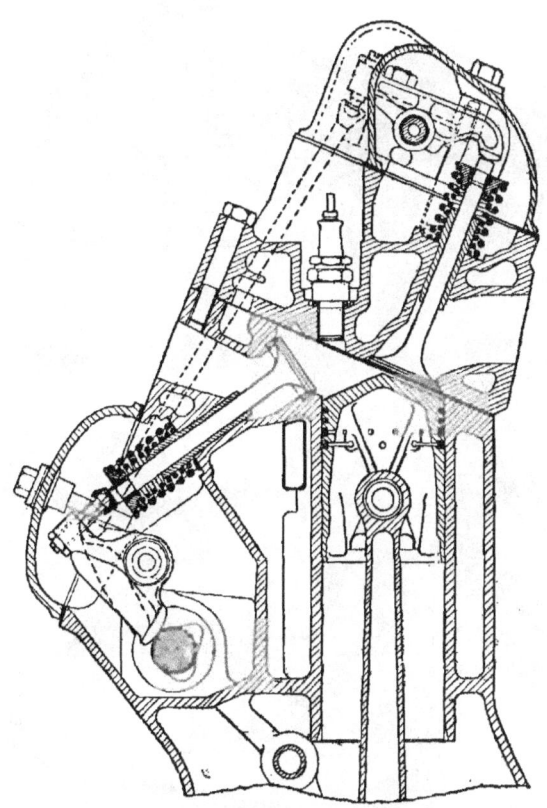

Fig. 294.—Sectional View of the Rover Car Engine Cylinder Head.

Typical Automobile Engine.—A typical engine, illustrating the trend of development in mass-production models, is shown in Fig. 295. It represents the four-cylinder overhead class which covers the 10 to 16 h.p. rated power range, the actual model shown being the Austin A.40 of 10·6 h.p. rating. The engine has a bore and stroke of 65·4 mm. and 89 mm., respectively, giving a capacity of 1,200 c.c. It has a compression ratio of 7·2:1 and develops about 40 B.H.P. at 4,300 r.p.m., with a maximum torque of 59 lb. ft. at 3,000 r.p.m.

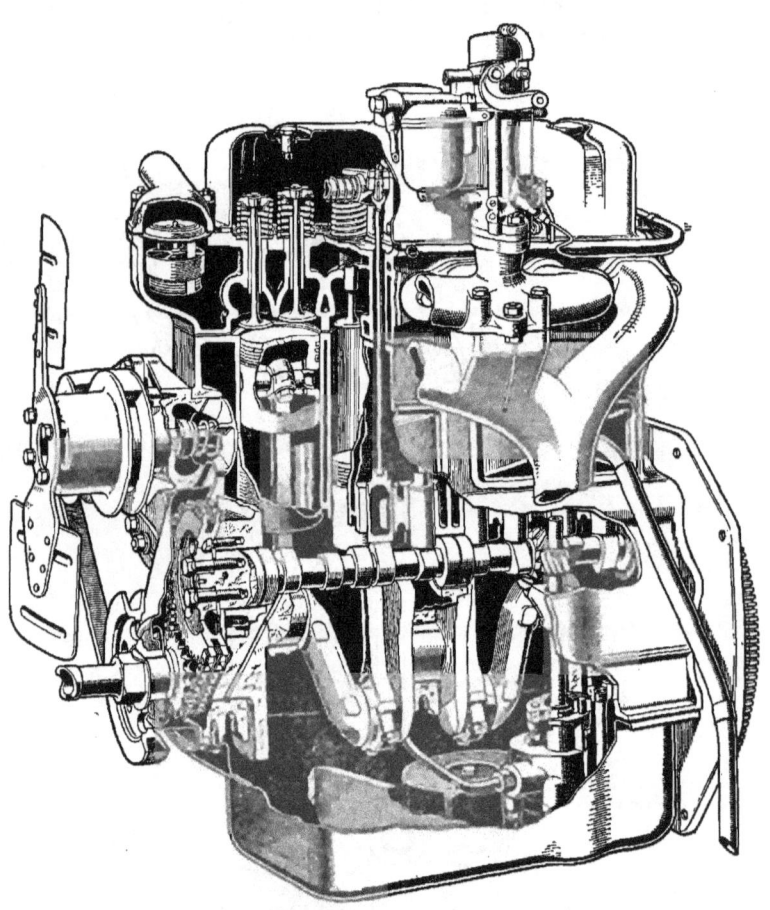

Fig. 295.—The Austin A40 Engine.

Relatively large inlet valves are employed and the combustion chamber has been designed to give a certain degree of turbulence to the mixture. The proper vaporization of the fuel is assisted by the provision of an exhaust-heated hot spot at the bottom of the inlet manifold, so that any fuel deposited from the mixture is rapidly vaporized. The torque curve of the engine is relatively high at the lower engine speeds so that the car pulls well on top gear at speeds as low as 6 to 8 m.p.h. The smooth running of the engine is assisted by the substantial dimensions of the flywheel, the automatic timing chain tensioner* and rubber-cushioned mountings.

Special attention has been given to the engine lubrication details and the results of wartime experience with their engines influenced the makers in this matter. The system employed involves the use of a large output spur-type gear pump which feeds oil under pressure to four feed points in each camshaft main bearing, thus ensuring even oil distribution over the bearing surfaces, whilst drillings in the crankshaft feed oil to the big-end bearings uniformly. An external by-pass oil filter is fitted. Special oil feeds are provided to the camshaft and overhead valve rocker shaft bearings and to the tappets and timing chains.

The salient features of this engine are shown in Fig. 295, and of these special mention may be made of the following items, namely, (1) Silent and automatic timing chain sprocket; (2) Thermostat in upper left water outlet; (3) Simple overhead valve operating mechanism with large bearing area tappets operated directly from the cams; (4) Shrouded valve guides; (5) Belt-driven two-bladed fan, the shaft of which also drives the water impeller; (6) Gudgeon pins clamped to the small ends of the connecting rods; (7) Vertical shaft for oil pump drive, operated through a pair of spiral gears from the camshaft; (8) Common joint of exhaust and inlet manifolds, for heating the stainless steel hot-spot inside; (9) Down-draught carburettor. This is fitted with an air cleaner and silencer, but is not illustrated in Fig. 295.

Four-cylinder Opposed Engine.—The Jowett four-cylinder engine is a good example of a modern high performance unit having several novel design features. It consists of two pairs of opposed cylinders, each cylinder of an opposing pair being offset in relation to the opposite one, thus enabling a separate connect-

* Vide page 439.

ing rod and crank journal to be used for each cylinder. The cranks of Nos. 1 and 2 cylinders and also those of Nos. 3 and 4 are at 180° apart, but the cranks of Nos. 2 and 3 are parallel, thus giving the usual four-cylinder vertical engine crankshaft arrangement. The balance of the rotating and reciprocating members is theoretically perfect since the rocking couples due to Nos. 1 and 2 cylinders being offset are balanced by those of Nos. 3 and 4.

The bore and stroke are 72·5 mm. and 90 mm., respectively, giving 1,485 c.c. (90·9 cu. in.) capacity

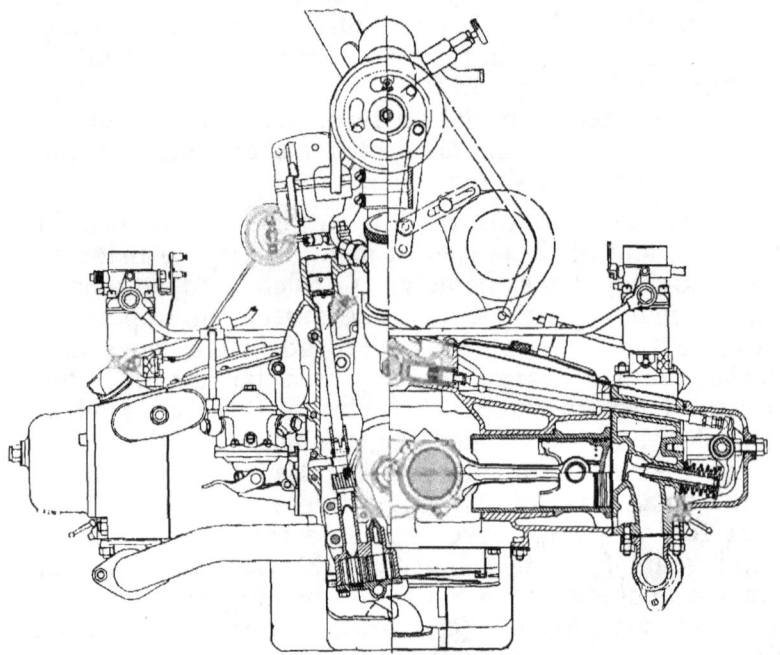

Fig. 296.—The Jowett Four-cylinder Opposed Engine, in Part Sectional View.

and a stroke-bore ratio of 1·24. The compression ratio is 7·2 to 1. The engine develops a maximum output of 52·5 B.H.P. at 4,500 r.p.m. and a maximum torque of 76 lb. ft. at 2,600 r.p.m. The maximum B.M.E.P. is 126 lb. per sq. in. at 2,600 r.p.m. The output is equivalent to 35·2 B.H.P. per litre, or 2·05 B.H.P. per sq. in. of piston area. The lowest fuel consumption is 0·555 pints per B.H.P. hour. The piston speed at maximum B.H.P. is 2,660 ft. per min. Overhead valves, with the push-rods actuated by a central camshaft, above the crankshaft, are used for both inlet and exhaust valves. The tubular push-rods are ball-ended and the valve clearance is automatically maintained by Zero-Lash hydraulic tappets. The connecting rods

Fig. 297.—The Jowett Four-cylinder Opposed Engine.

(Fig. 296)* are split, diagonally at 45° to the cylinder axis, to allow the piston and rod to pass through the cylinder barrel. Die-cast tin-plated pistons having two compression and one oil-control ring are used. Two Zenith down-draught carburettors are fitted, namely,

* Courtesy, " The Automobile Engineer."

O

one to each pair of cylinders; this enables a short and efficient induction system to be employed. The water circulation pump and four-bladed fan are driven by vee-belt and pulley from the crankshaft; the same belt also drives the dynamo. Wet-type cylinder liners of Vacrit cast iron are employed, but, as will be seen from Fig. 296, only the upper parts are water-cooled. The gear-wheel type of oil pump is driven by a shaft, inclined slightly to the vertical, by means of skew gearing from the crankshaft front end. This shaft is continued upwards and operates the ignition distributor unit. The pump delivers high pressure oil to a full-flow oil filter mounted accessibly above the engine, whence the oil is delivered to horizontal oil galleries running parallel to the camshaft; from here oil is led through drillings to the main crankshaft bearings. Tappings in the side of each gallery supply oil through pipes to the overhead valve gear. Small oil jets are provided in the system for lubricating the timing chain and crankshaft skew gearing. Provision is made, by spring-loading the filter element against a seating in the housing, to by-pass the filter element in the event of its becoming choked.

In the Jowett engine cooling system the radiator is mounted above and behind the engine. Separate pipes connect the lower tank to the rear of each cylinder block and after circulating through the heads it passes out through ports high up in the front of each block and finally to a single vertical outlet on which the water pump is mounted by a short hose connection. There is a bellows-type thermostat in the pump outlet pipe.

An Efficient Valve Gear.—In high efficiency engines, operating at medium to high engine speeds it is necessary to reduce the weight of the reciprocating members of the valve unit to a minimum, to minimise operating power and vibration effects. Further, the valves and their seatings should be water-cooled as far as practicable. The $2\frac{1}{2}$-litre Lagonda engine valve gear, shown in Fig. 298, conforms, largely, with these requirements. The valves are operated directly from overhead camshafts which are arranged so that the valve axes are at 62° to one another; this enables large diameter valves and a kind of conical combustion chamber with centrally disposed sparking plug to be used. The valve cams operate large diameter plungers or tappets enclosing the compound-type valve springs. The spaces around the valve guide and the seatings are water-jacketed so that the valves and seatings are cooled to some extent. It will be observed that insert valve seatings of hard wear-resisting metal are used. The engine in question develops a maximum output of 105 B.H.P. at 5,000 r.p.m., which is equivalent to 42 B.H.P. per litre.

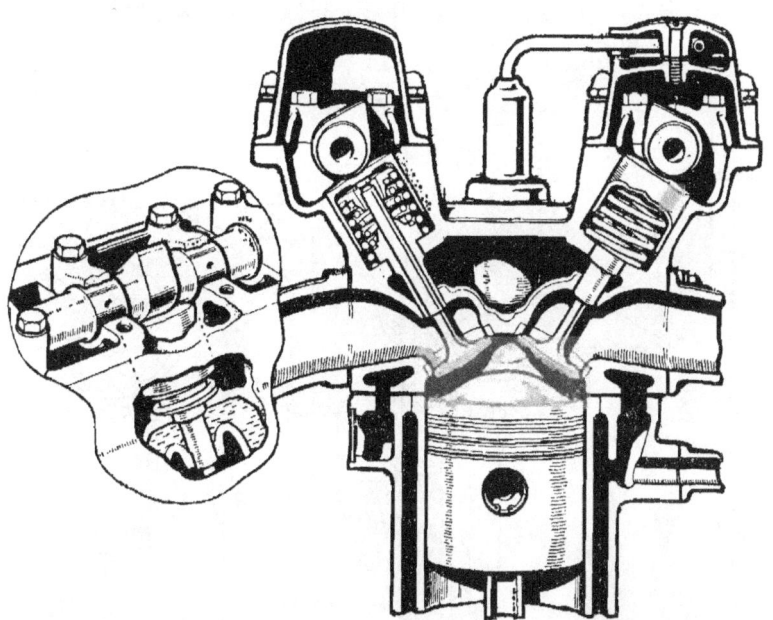

Fig. 298.—Valve Gear of Lagonda 2½-litre Engine.

High Camshaft O.H.V. Engine.—The employment of camshafts located relatively high up in respect to the crankshaft axis enables a good deal of weight to be saved in the push-rods of overhead valve engines thus reducing inertia effects and, to some extent, simplifies cylinder block design. Fig 299 shows the Lea Francis engine in two part-sectional views, namely one on either side of the cylinder axis. The high camshafts and short tappet push-rod unit are clearly shown. It will be observed that the push-rods are inclined inwardly towards the top, thus making for a compact valve gear

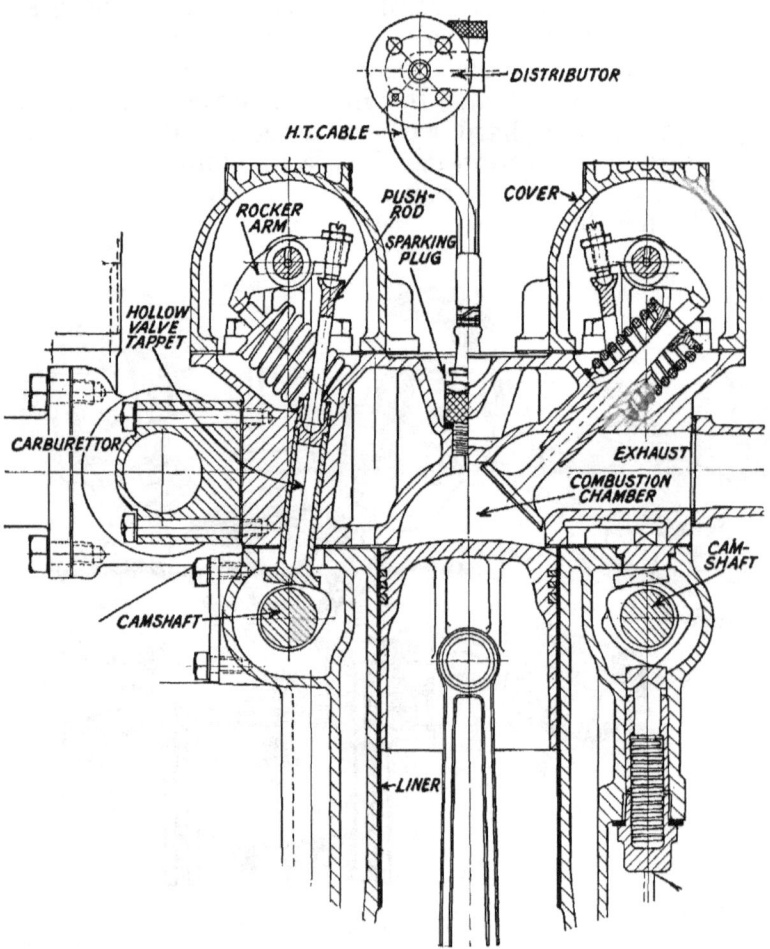

Fig. 299.—Lea-Francis High Camshaft Overhead Valve Engine.

above the valves. The combustion chamber is hemispherical in this engine, with slightly domed piston and centrally-located sparking plug.

The valve tappets are made hollow, for lightness, and both ends of the push-rods have ball-and-socket connections with the tappet heads and rocker arms respectively. The engine illustrated has dry cylinder liners.

Progress in Automobile Engine Design.—A notable amount of progress has been made in motor vehicle engines since the last edition of this book, chiefly in detail design features and improved performances. There have, however, been certain mitigating circumstances that, for a time, retarded full development, namely, the enforced use of relatively low octane Pool petrol in the first few post-war years and the necessity of producing relatively cheap motor cars for export purposes; the latter condition has had an appreciable influence upon engine design from the mass-production viewpoint.

The Four-Cylinder Engine.—One striking feature of more recent engines has been the more general adoption of the four- in place of six-cylinder engine for sizes up to 16 h.p. (treasury rating) and six-cylinder ones in the 16 to 25 h.p. class. For larger output engines the straight-eight engine has been almost predominant, with only one or two production-type vee-eights. The principal reasons for the adoption of the four-cylinder engine have been cheapness of manufacture in comparison with the six-cylinder type and the possibility of obtaining smooth performance and ample power for the size of car to which it is fitted. With improvement in engine balancing means, and in rubber mountings for the engine and gear box unit, the four-cylinder engine is fully satisfactory for its purpose.

Valve Arrangements.—The greater output per given cylinder capacity has been obtained, despite the relatively poor Pool petrol, by much improved combustion chamber design,* the adoption of *overhead valves* in many instances or, where the combustion chamber shape has dictated it, combination overhead and side valve engines. There is a further merit in the use of

* See page 410.

overhead valves, namely, that it is easier to adjust the
valve stem clearances than with side valves; this is
more especially the case with modern engine bonnet
design, whereby the engine is often located in a kind
of inaccessible " well." *Overhead camshafts* have
come into favour again, in some engines, notably those
designed for much higher outputs per litre than hitherto.
In one or two instances the combustion chamber shape
and valve location has necessitated the overhead camshaft or twin camshafts. It is notable that the more

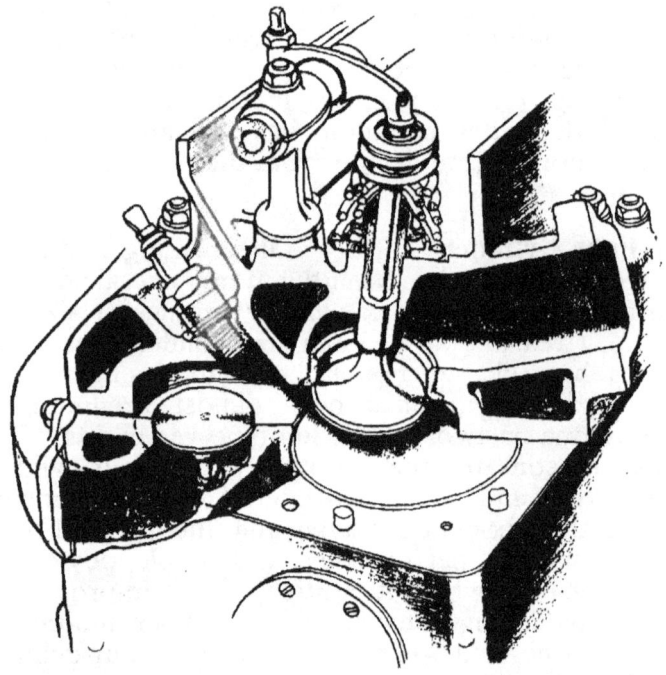

Fig. 300.—Rolls Royce Engine Cylinder Head.

recent Rolls Royce F-head engines (Fig. 300)* employ
overhead inlet valves, operated by push-rods and
rockers and side-type exhaust valves. As in certain
other modern designs the *inlet valve is of rather larger
diameter* than the exhaust valve, on account of the better
volumetric efficiency that is obtainable.

One objection to the use of overhead valves, namely,

*Courtesy *The Autocar*.

that of *noisier operation* than side valves, has to some extent been overcome by careful design. In the case of the Austin 16 h.p. engine oil-cushioned push-rods are used and there is also a felt-insulated double cover for the rocker gear. The use of the hydraulic automatic valve stem clearance adjusters and noise silencers has increased in British engines.

A novel method of holding the valve spring collar without the aid of split cones, cotters or other devices, is used on the Standard Vanguard engine overhead valves. The valve collar (Fig. 301) has two holes, the outer one H being large enough to allow the stem of the valve to pass through; the

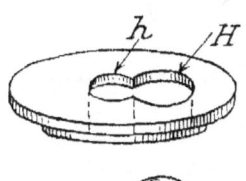

other *h* is of smaller diameter. The valve stem is recessed, as shown, and after the collar is placed over the stem, through the larger hole, it is moved sideways to the central position, the spring being compressed. Upon releasing the spring pressure the smaller hole part of the collar is forced against the tapered portion T of the valve stem, where it is held centrally and securely.

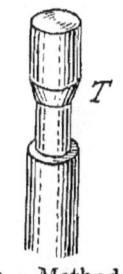

Fig. 301.—Method of Holding Valve Collar without the use of Cotters on Vanguard Engine.

Cylinders.—The principal changes in cylinder block and head design have been those connected with better cooling of the valve ports and sparking plug bosses. In some cases positive delivery of cooling water to these parts is provided by directed flow of the water from the pump through carefully designed ports or channels. The exhaust valve—which is the hottest part of the combustion chamber and is, usually, the limiting item from the viewpoint of the highest compression ratio that can be employed without the occurrence of detonation—has received particular attention in the matter of valve seat cooling. The usual arrangement is now to cool the regions mentioned by pump-directed flow, but to employ thermo-syphon cooling for the cylinder jackets. In some instances rather longer water jackets around the cylinder barrels have been provided for improved cooling of the metal and the lubricating oil. In general, there is now better positioning and dimensioning of the water passages in the cylinder block and head, allowing a freer flow of cooling water; further, the thickness of the head metal tends to be more uniform in order to avoid the *distortion* that occasionally occurred in previous models.

The use of *aluminium alloy cylinder heads* has increased; this is on account of their better heat conduction and, therefore, the possibility of using higher

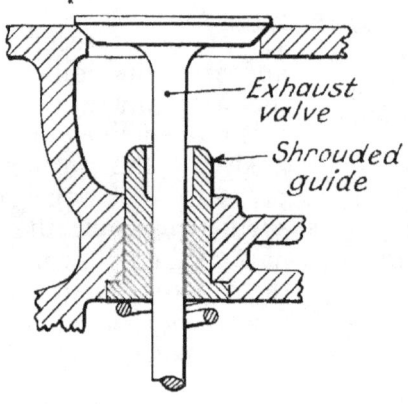

Fig. 302.—Austin Shrouded Valve Guide.

compressions than for cast iron heads. One notable disadvantage of such heads is that they are apt to corrode around the cylinder studs and become difficult to remove for decarbonizing and valve grinding purposes. We advocate the *use of graphite* on the cylinder studs before the head is placed in position to obviate this trouble.

In regard to the *valve guides* in the cylinder block, when *leaded fuels* have been used over an appreciable period, deposits of lead salts on the exhaust valve stems have often been the cause of *sticking valves*. To avoid this the Austin engine valve guides are shrouded in the manner indicated in Fig. 302, so that the valve stem is protected from the lead-salt laden exhaust gases.

Cylinder blocks are now made of a better *quality alloy cast iron* than in previous engines in order to increase their wear resistance. Thus, the use of *chromidium iron alloys* has resulted in extending the useful life of the cylinders before re-boring becomes necessary by at least twice that for the plain cast irons hitherto used for cylinder blocks.

The use of *wet-type cylinder liners*, made from alloy cast irons or nitrided ones, is increasing. The heat-treated centrifugally-cast sleeves in loded cast iron-nickel-chromium and other alloy irons, are much harder than ordinary cylinder block irons and give at least 50 per cent. greater wear resistance. Moreover, when the cylinders are eventually reconditioned, standard replacement liners of similar dimensions are employed, so that a standard size of replacement piston can be used; this is obviously an advantage over the method of reboring worn cylinders and using oversize pistons of various step-up dimensions. Wet-type cylinder liners used in car engines are, relatively, thicker than dry liners; a typical wet liner would be from $\frac{3}{16}$ in. to $\frac{1}{4}$ in. thick and a dry liner from $\frac{1}{16}$ in. to $\frac{1}{8}$ in.

In order to prevent the appreciable wear in the upper parts of cylinders due to a combination of corrosion and frictional influences, this portion or the whole of the bore is sometimes electroplated with *porous chromium*, an extremely hard metal that contains fine surface cavities which act as oil reservoirs. Cylinders thus coated will operate for periods of 90,000 miles and upwards before reconditioning is needed. The Van der Horst process of chrome-hardened cylinder bores that is much used on automobile petrol and Diesel engines is a typical example of this method.

In the Rolls Royce unit engine a short hardened alloy cast iron sleeve is used. This is arranged at the top of the cylinder barrel, where the maximum wear occurs.

Pistons.—The use of aluminium alloy pistons is now universal, although during the second world war, owing to restricted supplies of aluminium in the U.S.A., steel and also cast iron pistons of light design were used on car engines.

o*

Fig. 303 illustrates an alloy *cast iron piston* employed on Studebaker engines of the war period. Although cast iron is about three times as heavy as aluminium, the 3⅜ in. diameter piston shown weighed only about 10 per cent. more than the aluminium one that it replaced, weighing only 1·23 lb. The crown tapered from 0·12 in. to 0·13 in., while immediately below the gudgeon pin bosses the skirt tapered from 0·06 in. to 0·04 in. The making of such pistons necessitated efficient cast-

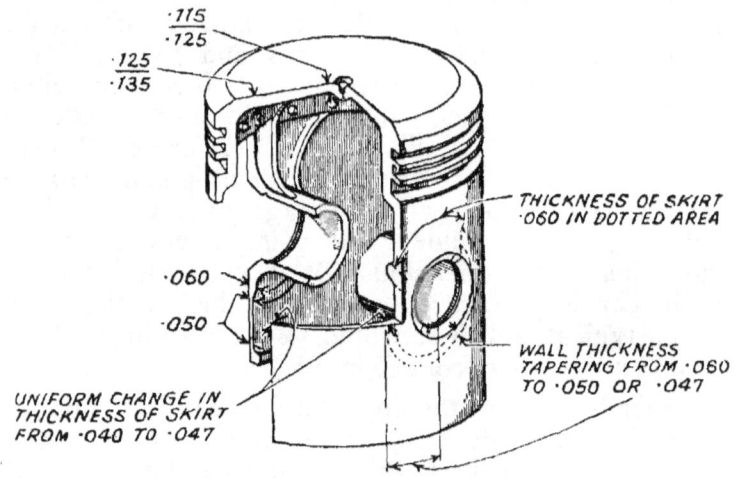

Fig. 303.—Modern Light Cast Iron Piston.

ing methods, involving thin sections cast to fine toler-ances. The pistons were stated to compare most favourably with aluminium alloy ones in regard to oil consumption, friction, freedom from scuffing and fit.

The *aluminium alloy pistons* now favoured in recent engines are of a similar kind to those described on pages 94 to 97, but certain improvements have been introduced and new designs developed. The modern piston is of the cam-ground oval section type, having its minimum diameter coincident with the gudgeon pin axis and major diameter to the thrust face diameter. Thus, as the engine warms up and heat is conducted downwards to the gudgeon pin boss metal first, thus causing the minor axis of the oval section to expand, with the result that the section in question becomes circular.

With this type of piston the thrust faces are separated to some extent from the piston ring lands, by means of a slot; this gives thermal insulation and directs the heat from the piston crown to the boss metal. There is no piston slap with this type when cold. A typical example of a piston designed along these lines is the Specialloid "Easy Start" one shown in Fig. 304. The bracing struts for the piston crown also act as heat conductors to the bosses, whilst the slots shown on the thrust faces act as heat insulators.

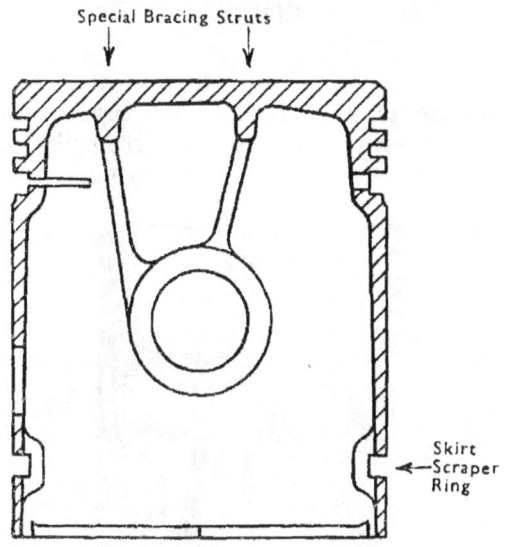

Fig. 304.—The Specialloid "Easy Start"' Aluminium Alloy Piston.

The Wellworthy "Welflex" piston (Fig. 305) is also designed on a somewhat similar principle, automatic compensation being provided for heat expansion effects. It has a split skirt and this is supported at its upper ends on most of the trailing side, thus making it much stronger that the usual split skirt design of piston. For this reason there is less risk of "piston rock" owing to the skirt closing; neither do the rings become convex-faced. It is significant that with this type of piston the cold cylinder clearance can be reduced to 0·00075 in. per inch diameter, so that for a 3 in. piston the total clearance is only 0·00225 in.

A recent design of piston, known as the Wellworthy "O.T." (oval-turned) one is similar in principle to the Welflex one but, in addition, is provided with a channel-sectioned scraper ring about half-way down the skirt. This type has been developed, in particular, for oily engines and is recommended as a replacement piston for rebored engines.

In other types of compensating piston, including the cam-ground oval section and the bimetallic one, having cast-in steel compensating wires or rods, the cold clearances are smaller than for the plain cylindrical types.

A marked improvement in piston manufacture has been that of the *pressed light alloy type*. The results of tests have shown that the tensile strengths of pressed aluminium alloy pistons are about 50 per cent. greater

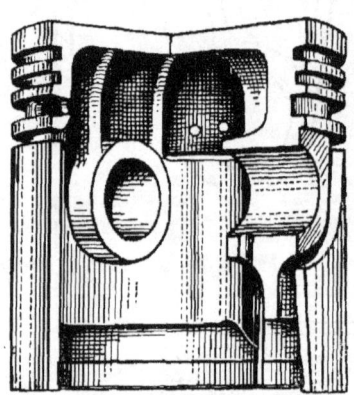

Fig. 305.—The Wellworthy 'Welflex' Aluminium Alloy Piston.

than for sand-cast and 25 to 30 per cent. more than for die-cast pistons of the same metal; moreover the ductility is much better so that the pistons are considerably less brittle than the two other types mentioned. Another advantage is that owing to the precision with which the pistons can be pressed the final dimensions are such that less machining is required than for sand-cast and forged pistons.

A more recent innovation in aluminium alloy pistons is that of *knurling the piston skirt*, by means of closely-pitched fine serrations—similar to those that might be made with a screw-cutting tool. These serrations (Fig. 306) of from 0·01 in. to 0·015 in. deep are made by a combined rolling and cutting operation, known as knurling. The breaking up of the otherwise plain smooth skirt into a very large number of tiny rhomboidal areas, separated by fine grooves, greatly improves skirt lubrication and reduces the temperature of the metal, owing to the better oil circulation

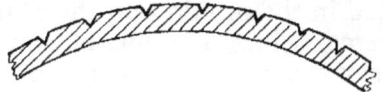

Fig 306.—Section of Knurled Piston Skirt.

Further, in the case of tin-coated and other types of surface treated pistons the coatings, being better lubricated, last much longer.

The *tin-coating* of pistons, when new, mentioned on pages 98 and 99, is more widely used as it has proved an effective preventative of scuffing and seizure during

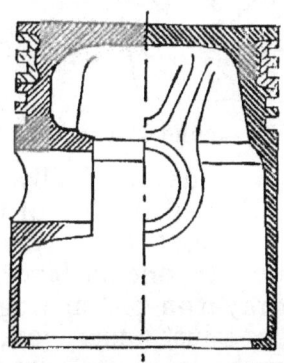

Fig. 307.—Two-ring Nickel-iron Piston Insert.

the running-in stages. The tin coating required need only be very thin; thus, deposits of 0·0002 in. thickness have given excellent results. Other surface treatments, including anodising and Parco-lubricizing are also used for pistons to prevent initial running-in troubles and to give subsequent low friction surfaces. In some cases the deposited layer is porous in order to act as an oil absorbent and thus prevent seizure during the early stages of the piston's life.

The wear that occurs in the two upper piston ring grooves of aluminium alloy pistons is due to the high frequency hammering effects of the harder piston rings on the relatively softer piston metal. In order to reduce such wear to a practically negligible extent, hard nickel-iron ring inserts are employed (Fig. 307). These are located in the mould and the aluminium alloy cast around them. Since the nickel-iron ring carrier has about the same expansion coefficient as the piston metal there is no tendency to loosen during service.

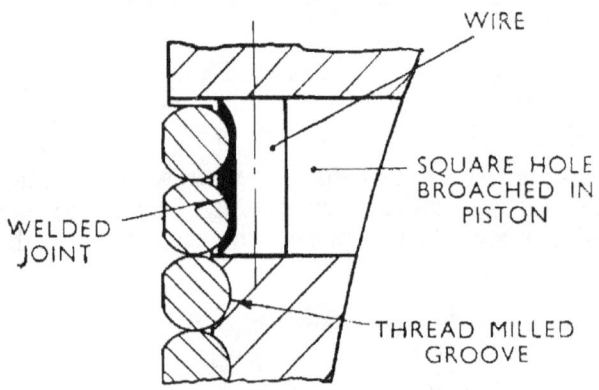

Fig. 308.—Wire-wound Piston Skirt Method.

The results of tests made originally in Germany in 1931-33, showed clearly that the life of the insert piston was increased by at least twice that of the plain aluminium alloy one. In one instance, the measured wear between the gray-iron piston rings and the cast-iron insert grooves was just under 0·004 in. after 64,000 miles. The pistons with new rings were then replaced in the cylinders but without any machining work having been done on them and they gave a further long period of efficient service. Wear of the cylinder was so small that there was no measurable increase in the oil consumption and new liners were fitted only after 120,000 miles. The piston ring insert is of particular value in high-speed Diesel engines.

Wire-wound Pistons.—The split-skirt piston, of oval-ground form, is fitted so that, when cold, the skirt portion is tight enough on the thrust axis to prevent piston slap. When, however, the piston heats up, the radial expansion of the skirt portion increases the pressure against the cylinder walls and thus gives rise to additional friction. To overcome this drawback the wire-wound piston was introduced by The Automotive Engineering Company, Twickenham. The wire spiral band is arranged around the aluminium alloy skirt to resist the expansion, and since this band expands no more than the bore it does not increase pressure on the cylinder walls as the engine warms up.

The band, which is shown in sectional detail in Fig. 308, is arranged both above and below the gudgeon pin, the piston being ground after the wire is in place. The wire in the original pistons was held in place by spot-welding the last two turns to two pieces of similar wire inserted in a hole broached in the skirt. In later pistons the wire is fixed with two small steel pegs; further, in the smaller sizes of piston, there is only one wire winding and not two as shown in Fig. 309.

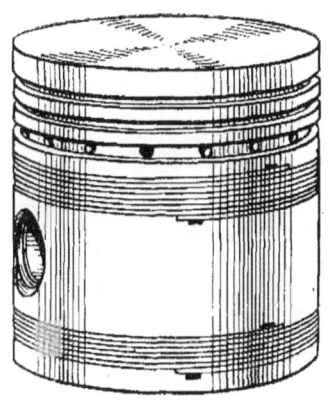

Fig. 309.—External View of Wire-Wound Aluminium Alloy Piston.

The control of expansion is so effective that *practically no clearance* is required for pistons up to 3½ in. diameter; for this reason the wire-wound piston is referred to as the *constant clearance one*, at all working temperatures.

Piston Rings.—Modern piston rings are of the narrow axial thickness type, of good quality alloy cast iron having elasticity properties, or flexibility, so that they can readily be sprung over the piston without breakage or permanent set. The rings are now always supplied by specialist piston ring firms and are of a high standard. The type preferred is the diagonal-cut one with uniform or graduated radial pressure produced by peening followed by peripheral grinding with the ends clamped together. It has been shown that if the correct ring gap measurement is given there is no more tendency for gas leakage past this kind of joint than any other one. The gap clearance is usually taken as being $\frac{1}{400}$ of the cylinder diameter; so that a 3 in. ring would have $\frac{3}{400}$ = ·008 in.; a 4 in. one, 0·010 in., and so on. The principal reason for *using narrow axial thickness rings* is that in the event of gas leakage into the back space of the slot, there is less radial pressure on the cylinder wall than for wider rings. Other advantages include lighter weight, greater flexibility and less wear in the piston grooves owing to their lower inertia force. Rings should not, however, be too narrow or they will wear more quickly. The clearance of the ring in its groove is from 0·002 in. to 0·0025 in. in car engine types.

The eccentric type of ring is now obsolete; only the concentric ones are used.

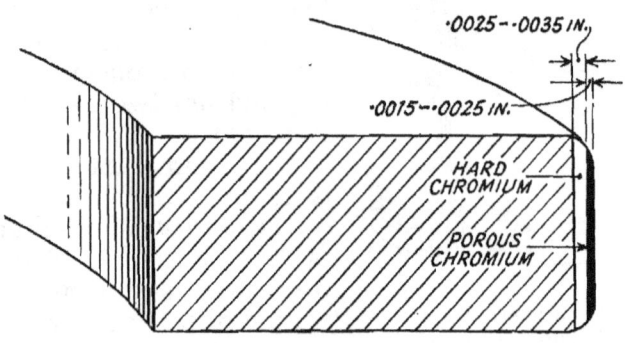

Fig. 310.—The Muskegon Porous Chromium-Plated Piston Ring.

In regard to the radial thickness, this is usually made to $\frac{1}{30}$ of the diameter with a tolerance of plus $\frac{8}{1000}$ in. and a free gap equal to 3 to 4 times the thickness; these dimensions correspond to a minimum wall pressure with alloy iron rings of 9 lb. per sq. in. for sand-cast and 10·5 lb. per sq. in. for centrifugally-cast rings.

A new innovation in piston rings of the compression type is to *chromium plate* the area that bears on the cylinder wall. In the Muskegon ring there is an inner layer of 0·0025 in. to 0·0035 in. of extremely hard chromium and on this another layer of 0·0015 in. to 0·0025 in. of *porous* chromium, as depicted in Fig. 310; the total thickness is 0·004 in. to 0·006 in. It will be observed that the bearing portion of the ring is radiused. This type of ring gives a much longer useful cylinder and also ring life, reduced friction and gives freedom from scuffing action; in some instances the chromium-plated ring gives from 3 to 5 times the wear resistance of the uncoated ring. Chromium-plated rings should *not be used in chromium-plated cylinders*, however.

It is necessary to use only one such ring, namely, in the top piston ring groove in order to extend the ring life and very appreciably *reduce cylinder wear*.

Apart from the chromium-plated ring, cast iron rings are sometimes *surface treated* with films of *iron oxide, iron sulphide* or *metallic phosphates*, to facilitate bedding in, reduce surface friction and prevent scuffing action. In most cases the film wears off during bedding in and leaves a brightly polished metal surface.

Some notable advances have been made in the design of *oil-control piston* rings in recent years, with the object of obtaining increased flexibility, both in an axial and radial sense. This type of ring, however, is intended for replacing the usual oil control ring when the cylinder shows signs of uneven wear such that, although it affects oil consumption and, to some extent, the power output, it does not actually necessitate re-boring of the barrels. The principle of this type of replacement ring, of which typical examples are the Simplex, Duaflex, Cords, Trancoseal, Muskegon Type XSS, Perfect-Circle Type 86, etc., is to use a very flexible alloy cast iron ring of the slotted oil control type and to employ an internal spring steel expander—usually of polygonal form, although helical springs are sometimes used—which forces the former ring to conform with the section of the cylinder at all times. In most of the rings mentioned it is also arranged that there shall be flexibility of the ring in the piston slot, in order to prevent oil leakage behind the ring.

The multiple flexible steel type piston ring has been popular in the U.S.A. as a replacement one for worn cylinders. In this connection it may be noted that there are two all steel piston rings available in this country. One is the Wellworthy expander type, comprising a number of thin steel rings, about five being used in a typical application to fill the piston ring slot. These rings have no special side flexibility properties but use a steel polygonal expander ring as in the Simplex model. They act mainly as compression rings but with

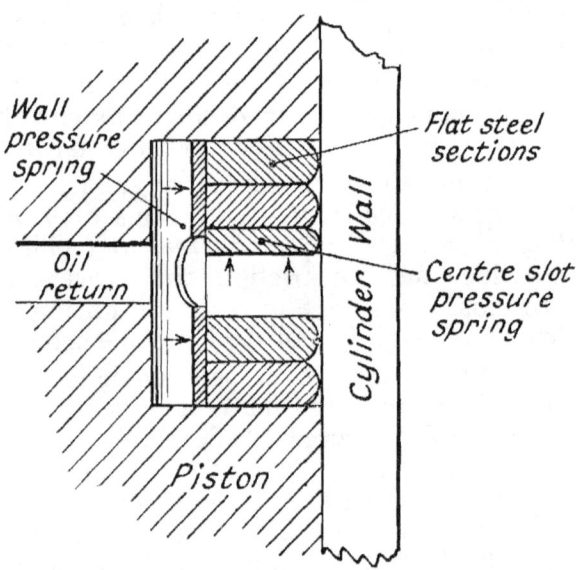

Fig. 311.—The Wellworthy Duaflex Piston Ring.

a certain amount of oil control action. The other type has an identical set of thin steel rings possessing radial flexibility but without the use of an expander. They are used for piston slots that are not deep enough for the expander type or cannot conveniently be machined to size.

The Simplex and Cords rings are described on pages 106 and 107. Fig. 311 illustrates the Wellworthy Duaflex, a more recent example of flexible oil-control ring, having advantages over the two types mentioned.

It consists of a number of narrow flat section alloy steel rings with their cylinder wall edges rounded off. In a typical example there would be four of these rings. Between each set of two rings there is arranged a crimped expander ring of spring steel which forces the rings to maintain contact with the sides of the slot. In order to ensure cylinder wall contact all round, the four rings are pressed against the wall by means of a steel expander of octagonal shape—similar to the type used in the Simplex ring; since the thin unit rings are very flexible, good contact is made against the cylinder wall. The rounded edges of the rings ensure immediate bedding-in so that the walls are not injured. This type of ring tends to wipe the surplus oil from the walls, instead of scraping it, as in the case of the ordinary oil control ring. Tests made with these rings show that a marked reduction in oil consumption is obtainable.

The Perfect Circle Type 86 oil-control ring has a long coil spring of circular periphery, with its ends together and placed behind the flexible alloy cast iron oil-control ring; this gives uniform wall pressure.

The Hepolite delayed action oil-control ring is somewhat similar in general design to the ring shown at C, Fig. 55, page 105, but, instead of having a truly cylindrical periphery, it is of polygonal form, i.e. has a large number of small flats all around its circumference, which allow free passage of oil to the upper rings. During the engine life the "corners" gradually wear away until, finally, the periphery becomes truly cylindrical and the ring then exerts its full scraping efficiency. This type of ring is *additional* to the ordinary oil-control ring or rings, so that an extra piston slot must be provided; it should not be used in place of the usual oil-control ring. The Simplex and Duaflex replacement rings can, however, be used instead of oil-control rings and are intended to replace these when the engine begins to show signs of excessive oil consumption; they must not be used in place of the compression rings, as they are not designed primarily for gas-tight purposes.

A later type of piston ring, known as the *Wedge ring*, has a tapered or wedge section of about 10° angle. It can be shown that with this shape of ring there is less friction on the piston slot faces and the ring maintains better contact with the cylinder wall than for the usual rectangular-section rings. It has also been found to be much less prone to *piston-ring flutter* and *blow-by*. To ensure that there is no increased groove wear, the wedge faces of the ring must be hone- or lap-finished. This type is particularly suitable for high gas pressure engines, such as Diesel ones. Referring to Fig. 312, the angle $\theta = 5°$; the recommended side clearance C=0·001 in. to 0·002 in. and distance D of the ring below piston land, 0·002 in. to 0·006 in.

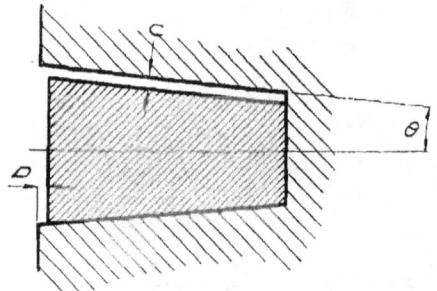

Fig. 312.—The Wedge-Type Piston Ring, for Diesel Engines.

Gudgeon Pins.—There have been no changes in the methods of fixing and preventing end movement of gudgeon pins in pistons. The usual method now is to use the *thermal fit* in which the piston is heated by immersion in boiling water, in order to expand the piston boss holes sufficiently to enable the cold gudgeon pin to be pressed into place by hand. The reverse method is used for removal of the pin, the piston being heated whilst the pin is kept cool by means of cold water. The end movement is prevented by the circlips at either end. Alternatively, the gudgeon pin is clamped to the small end of the connecting rod, as shown in Fig. 38 on page 88. In some cases the wearing surface of the pin is chromium plated to increase its useful life.

Big-end and Main Bearings.—The white-metal lined connecting-rod big-end and main bearings have now been superseded by the shell-type of bearing, consisting of two semi-cylindrical steel shells lined with white-metal. The half-shells are held together by the big-end cap and bolts or, in the case of the main bearings by the bearing caps, studs and nuts. When the crank journals wear it is necessary to have them re-ground and to fit under-size replacement shell bearings. It is not satisfactory to attempt to take up bearing wear by the methods employed for white-metal lined bearings, so that new bearings only are used. The half-shells are provided with oil grooves and oil feed holes; small projections on the steel back fitting into recesses in the connecting-rod or main bearing holes ensure that the shells do not rotate. In some cases the steel shell is first copper-plated before the white-metal is applied. Another method, used in America, consists in applying to the steel shell an intermediate layer or matrix of copper-nickel alloy of a porous nature to which is bonded a thin high lead Babbitt metal overlay.

Apart from the use of white-metal for automobile engine bearings other alloys are employed for high output and Diesel engines. In addition to those mentioned on page 145, there are aluminium alloys such as Hiduminium R.R.56, Chromet (consisting of 90 per cent. aluminium and 10 per cent. silicon) and an aluminium-tin alloy A.C.9 that has been used on Bentley engines with good results. Cadmium alloys with copper and cadmium-silver bearing alloys have proved satisfactory for high bearing pressures and running speeds. Indium, when alloyed with copper and lead, has given excellent wear and corrosion resistance in Diesel engine bearings.

More recently lead-bronze shell-type bearings with a coating of indium have been used in Leyland engines.

Camshafts.—There have been few changes in camshafts other than the more general adoption of the large valve clearance (0·015 in. to 0·020 in. for side valves) type of cam. In general, with the adoption of the new designs of combustion head, with side and overhead or both overhead valves, the high-position camshaft is often employed since it gives a lighter valve operating gear, with reduced inertia effects. A typical example is that of the Lea Francis engine, shown in Fig. 299.

In most engines the camshaft runs in cast iron bearings machined in the cylinder-crankcase casting, but often the front bearing is of phosphor bronze or gunmetal with an outer flanged face. Experience has shown that with camshafts running at one-half engine speed there is very little wear in this type of bearing. Usually, for a four-cylinder engine there are four bearings, although in some cases only three are employed. In the Morris commercial four-cylinder engine there are three bearings, the front one being of white-metal and the others cast iron; as these are of ample length and diameter the camshaft is well supported.

A small amount of end-movement or "float," usually 0·002 in. to 0·003 in., is allowed for the camshaft in its bearings, shims being provided for adjusting this clearance.

Timing Gear Drives.—The roller chain and sprocket method of driving the camshaft is now practically universal for side valve and overhead valve engines of the push-rod and rocker arm type. The duplex pattern roller chain with some form of automatic tensioner is favoured. The tensioner is generally of the spring blade kind and gives satisfactory results.

A method of tensioning and silencing the timing chain drive to the camshaft used in the more recent Austin engines employs an oil-resistant synthetic rubber ring between the two sets of teeth of the larger (camshaft) sprocket wheel (Fig. 313). The timing chain makes contact with the rubber which projects above the bottom of the sprocket teeth, so that the chain action is quietened and also tensioned, on account of the resiliency of the rubber ring. It may be mentioned, also, that the overflow oil from the front camshaft bearing is collected in a hollow steel ring on either side of the sprocket wheel and is fed from this ring by centrifugal action on to the teeth of the sprocket and also the side links of the chain.

In a few instances the jockey-sprocket method of tensioning the timing chain, as used on commercial vehicle engines, is employed on car engines. With this system there is a spring tensioner and eccentric pin to give the take-up to the driving chain.

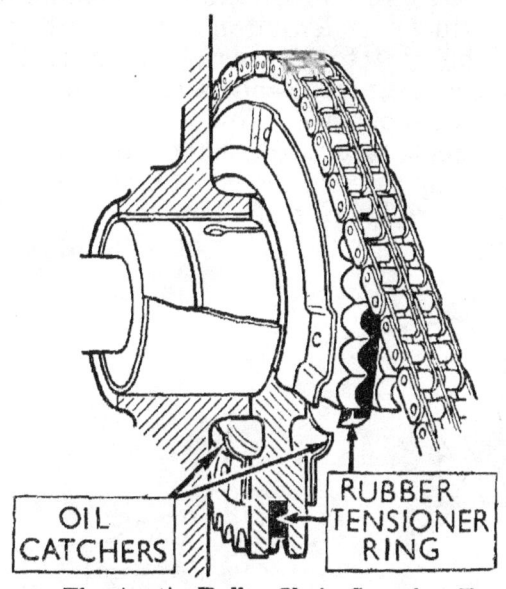

Fig. 313.—The Austin Roller Chain Sprocket Tensioning and Silencing Device.

The Rover engine employs a hydraulically-operated jockey sprocket chain tensioner with a ratchet and spring-loaded pawl device; engine oil-pressure is used for this purpose.

The overhead camshaft drive of more recent Wolseley engines is by non-backlash gears and a vertical shaft. Spiral gears are used and with the object of obtaining silent operation the spiral gear on the camshaft is made in two unequal parts, with a spiral spring device between.

Induction Manifolds.—There have been certain improvements in the design of inlet and exhaust manifold systems, in connection with the down draught type carburettor with the objects of giving (1) Better mixture distribution to individual cylinders; (2) Higher volumetric efficiencies; (3) Better exhaust scavenging or clearance and (4) Freedom from deposition of solid fuel in the inlet manifold.

There is still a lack of uniformity in manifold design. but the general result aimed for is to provide a water- or exhaust-heated region of the inlet manifold at the place where fuel condensation or deposition would occur, with the object of vaporizing this fuel. Whilst in American engines thermostatically controlled air chokes and hot-spots are provided for cold starting and arrangements are also made for fast-idling in order to warm up the engine as soon as possible, there was, in 1949, only one British car, namely, the Vauxhall, provided with a thermostatically controlled hot-spot.

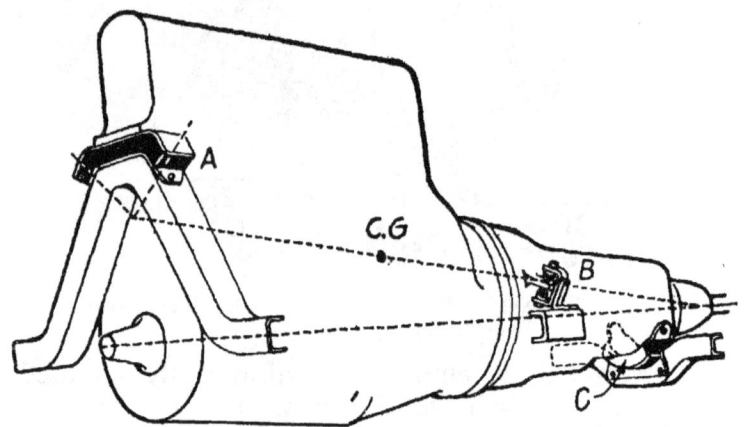

Fig. 314.—The Floating Power Method of Flexible Engine Mounting.

As the subjects of manifold design and modern carburettors is more fully dealt with in Volume II of this Motor Manual series,* it is not proposed to discuss these further.

The Floating Power Method of Engine Mounting.— Introduced by the Chrysler Company, of U.S.A., the

* "Carburettors and Fuel Systems," 5th Edition, 1949 (Chapman and Hall, Ltd.).

" floating power" method consists in mounting the engine and gearbox unit on rubber blocks positioned in respect to the centre-of-gravity of the unit in the manner depicted in Fig. 314. There is a relatively large rubber mounting at the front end A of the engine and another at C near the rear end. In addition there is a pair of rubber compression members B, one pair on either side of the engine axis (only one pair being shown in Fig. 314). These members limit the rotational movements about the axis passing through the C.G. as indicated by the upper dotted line, which is the normal rocking axis. The supports are located and

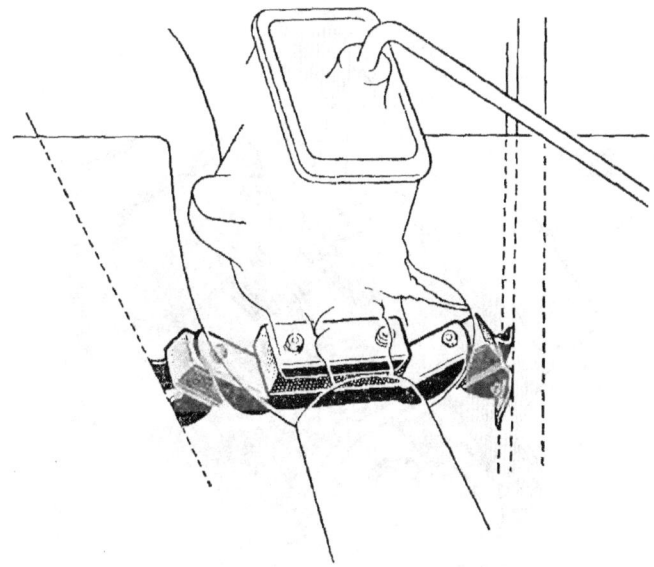

Fig. 315.—Austin Engine-Gearbox Rubber Mounting Method.

their elastic properties selected so that the unbalanced engine vibrations, weight and torque are taken account of and the general result is that of an elastic support at the C.G. of the engine.

Rubber Mountings.—Rubber can be used in tension, shear or compression in flexible mountings for the engine-gearbox unit, but the former method is not used since in the event of small surface injuries, e.g. cuts, the rubber would rapidly lose its strength properties. In engine mountings rubber bonded to metal is employed in shear or compression and the bond is such that it gives a direct tensile strength of 800 lb. to 1,400 lb. per sq. in. Whilst in compression loadings up to 10,000 lb. per sq. in. can be employed. There is a wide range of alternative rubber-bonded metal mountings for automobile engines of the shear, compression and combined shear-compression type, the usual method being to ensure that the engine metal side does not move sufficiently to make contact with the chassis

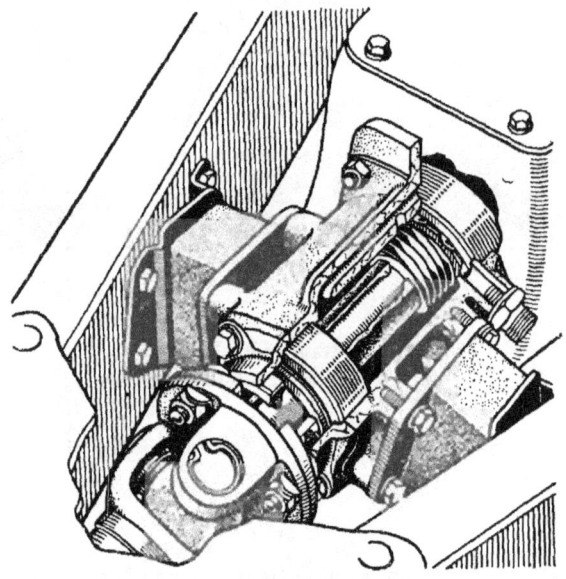

Fig. 316.—Austin 16 h.p. Car Rear End Gearbox Flexible Mounting.

mounting side; in some cases engine movement is limited by the use of stop units on the chassis-fixed part of the mounting. Synthetic rubbers, on account of their marked resistance to the effects of heat, sunlight, oil, petrol and air influences, are preferred to natural rubbers.

The usual arrangement is the three- or four-mounting

one, whereby, in the former case there is a single rubber mounting at the front and two at the back of the engine-gearbox unit. The latter method employs two front and two rear mountings. In some instances the rear mounting consists of one continuous metal-rubber unit bolted to the rear end of the gearbox at its central portion and, on either side, to side- or sub-frame members of the chassis, inside the main side frames, as shown in Fig. 315 for the Austin engine. An alternative method, used in recent Austin 16 h.p. cars is shown in Fig. 316; the separate rubber mountings, in this case, are arranged on the side of the

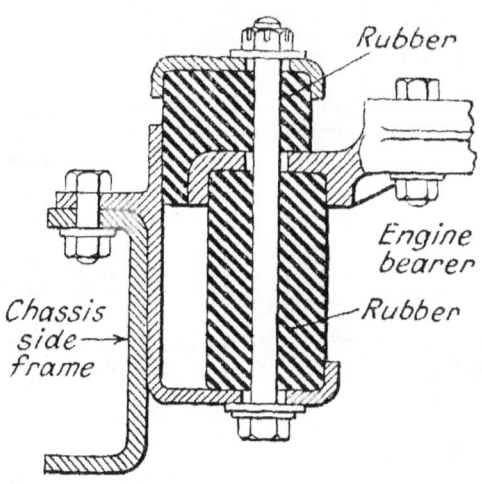

Fig. 317.—Lea-Francis Engine Mounting Method.

gearbox mainshaft extension. Fig. 317 shows a rubber mounting, used on the Lea Francis engines of the direct compression type, for the weight and with shear-resistance for side movements; this provides for ample damping resistance in all directions.

Simplified Torsional Vibration Damper.—The modern vibration damper is much simpler than that shown in Fig. 133 on page 188. It consists of a metal flywheel attached by a rubber mounting ring to the fan-drive pulley on the front end of the crankshaft.

The rubber insulating ring provides the vibration damping medium and, in practice, is most effective for automobile engines, on account of its high energy absorption per unit weight. Fig. 318* illustrates the Metalastik vibration damper which consists of a steel flywheel F to which is attached, by an efficient rubber-to-metal bonding process, a rubber ring of angle-section; this is also bonded to a steel pressing D, attached to the fan-drive pulley P, the latter being keyed to the end of the crankshaft C. The damping action takes place between the plate D —which may be regarded as being in a state of axial vibration—and the flywheel F, through the medium of the rubber unit E. The inertia effect of the rubber-insulated flywheel is an important factor in the final vibration damping out result.

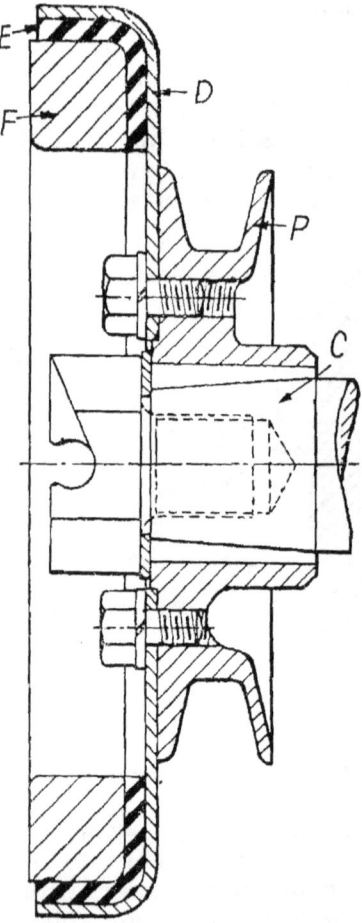

Fig. 318.—Rubber Ring Vibration Damper.

The Lubrication System. —The high pressure system is now universal for modern car engines, with normal working pressures of 40 lb. to 60 lb. per sq. in. Whilst the gear type of oil pump is widely used, the lobed-rotor pattern has been fitted to certain more recent engines. Usually this is a four-lobe rotor which rotates within an inner rotor of corresponding shape mounted

* "The Mechanics of Petrol and Diesel Engines." A. W. Judge (Sir I. Pitman Ltd., London).

eccentrically; the two rotors are of cast iron. A typical car engine pump of this kind will deliver 4 gallons of oil per min. at 50 lb. per sq. in. when operating at 2,000 r.p.m.

The floating pattern oil intake (Fig. 243) is used on several makes of engine and a recent development consists of a shaft with an external indicator fixed to it, for showing the level of the oil in the sump on the outside of the crankcase; this is an improvement on the still widely used *dipstick method* with its tendency to become inaccessible in modern deep-bonnet engines.

Oil filters have been much improved and it is now usual to provide external filters placed in accessible positions. These belong to one or other of two types,

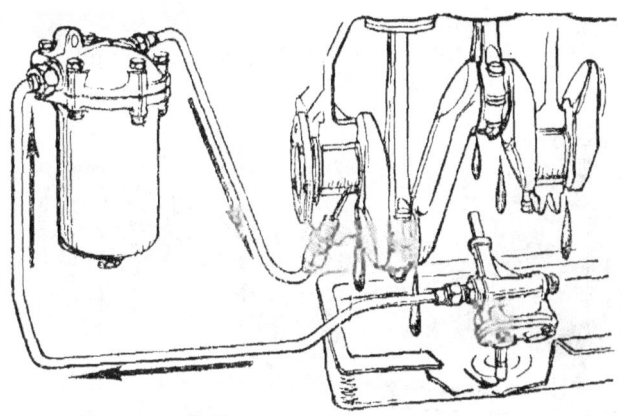

Fig. 319.—The Full Flow Type Oil Filter.

namely, the *Full-Flow* or *By-Pass* patterns. In the former model (Fig. 319)* the full oil flow passes through the filter. This is satisfactory so long as the oil is hot and clean, but when dirty and cold it may be too viscous to pass through the filtering element and the oil flow in the system would stop, were it not for the provision of a by-pass valve (Fig. 320) which opens and allows the oil to by-pass the filter, by providing a direct connection between the inlet and outlet orifices

* Courtesy, British Filters Ltd., Market Harborough.

of the filter. In Fig. 320 the thick black arrows show the thick or dirty oil by-passing the filter. Another disadvantage of the full-flow type is that the large volume of oil and the speed of flow are not conducive to the most efficient filtering of the oil, unless large capacity filters are provided.

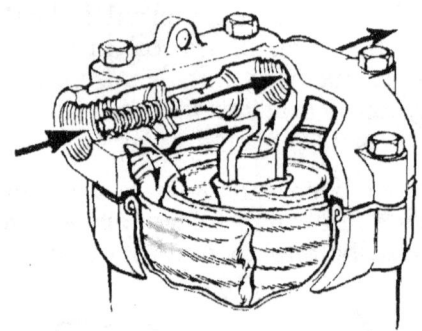

Fig. 320.—Showing By-Pass Valve in Full Flow Oil Filter.

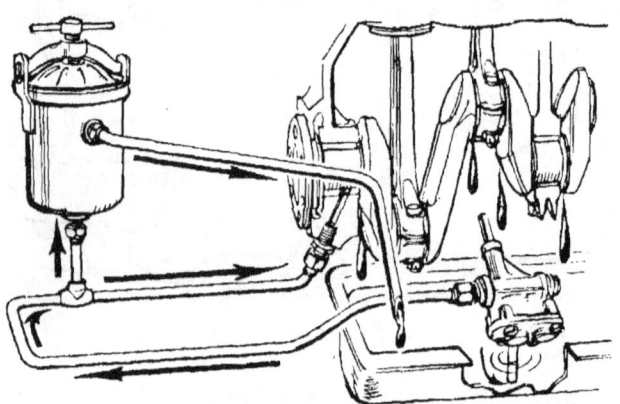

Fig. 321.—The By-Pass Type Oil Filter.

In the by-pass type (Fig. 321) only a fraction, namely, about $\frac{1}{8}$ to $\frac{1}{16}$ of the full delivery is passed through the filter; this oil is cleaned and returned direct to the oil sump. In the course of time all of the oil in the system will have been filtered and the process repeated continuously. The felt or fabric elements are usually discarded after a stated number of miles, usually 8,000 to 10,000 miles, and new ones fitted. Felt-type filters are usually cleaned in petrol every 2,000-3,000 miles and replaced at the intervals previously stated.

More attention is now given to the lubrication of the valve tappets, rocker arm bearings, timing gears and, in many instances, the cylinder walls; in the latter case oil jets from the big-end bearings are used to lubricate the cylinder walls. In others, oil is led up the connecting-rod, which is drilled for this purpose, to the small-end bearing, from which it escapes through small drillings as jets to the cylinder walls.

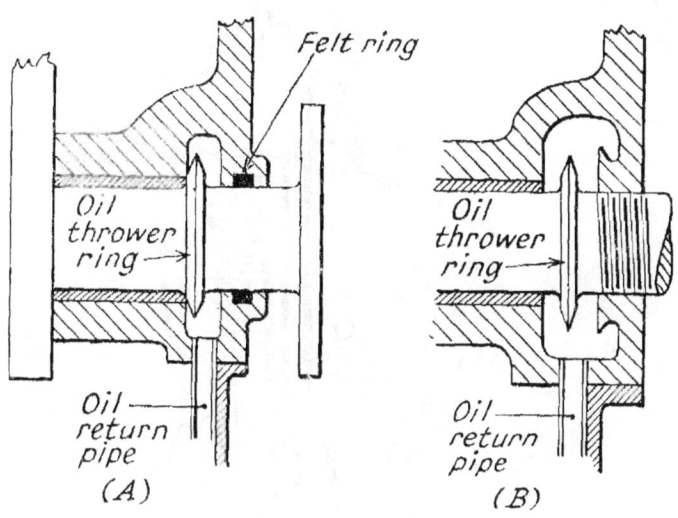

Fig. 322.—Methods of Preventing Oil Leakage from Rear Main Bearing.

The usual method of *preventing escape of oil under pressure* from the *rear main crankshaft bearing* into the clutch housing is to provide an oil-thrower ring between the rear bearing and crankcase wall, so that any oil that escapes past the bearings is caught and flung outwards by this ring, by centrifugal action. The oil flows down into a chamber below the ring and thence by an oil return pipe to the sump, as depicted in Fig. 322 (A). The orifice in the crankcase wall through which the crankcase passes is usually sealed with a suitable oil-seal of the felt type, although in some cases the shaft is screw-threaded so that any oil on it is returned by screw-action back to the oil-thrower ring, as indicated at (B).

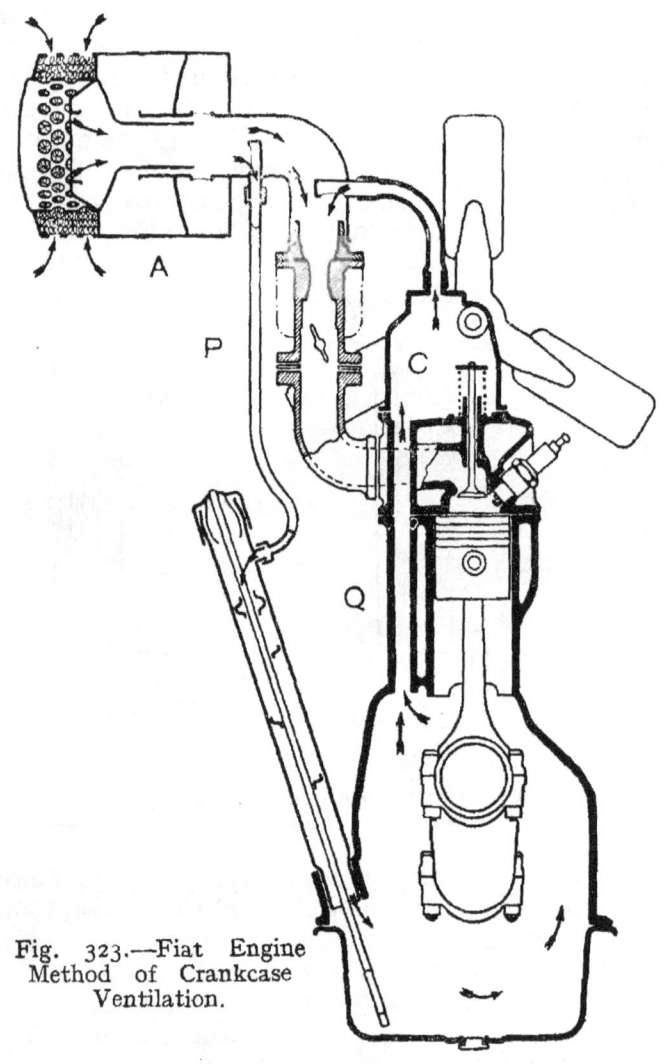

Fig. 323.—Fiat Engine Method of Crankcase Ventilation.

Crankcase Ventilation.—Although the subject of crankcase ventilation is a rather old one and the method was used on certain earlier American car engines it is now receiving much more attention in this country. The principle employed is to take advantage of air pressure differences, as between the crank chamber and inlet manifold or by means of a venturi device, as shown in Fig. 248, to circulate filtered air through the crank case. This not only prevents oil dilution by petrol vapour or gaseous products escaping past the pistons, but tends to keep the bearings and lubricating oil much cooler. Fig. 323 illustrates the method used in Fiat post-war four-cylinder engines. In this example there is an air pipe connection P between the dipstick tube and the main air intake to the carburettor on the latter side of the air cleaner A, so that filtered air only is supplied to the crankcase. This air circulates through the crank chamber, as shown by the arrows, and eventually passes upwards through the passages, as shown in Fig. 323, to the overhead valve rocker air-tight casing C, from which it passes to the main air intake of the carburettor.

In the Standard Vanguard engine an air pipe connects the filtered air side of the air cleaner with the breather connection on the crankcase, whilst another air pipe connects the sealed rocker casing to the induction manifold *on the engine side of the throttle* by means of a restriction valve having a constant orifice area of 0·007 sq. in. In this way a circulation of air through the crank chamber and rocker casing is provided, the blow-by gases being led over the valve gear and thence into the induction manifold.

Lubricating Oils.—Hitherto, many of the engine lubricating oils have shown a tendency to create gumming up of the pistons and ring slots, due to carbon separation under engine temperature conditions. In recent oils, known as *detergent oils*, there is a much reduced tendency to produce carbon deposits in the cylinder components and sludge in the sump oil; this sludge, if allowed to persist, would clog up the holes in the piston oil control ring slots and cause packing or gumming up of the rings in their grooves.

Cooling System Developments.—Mention has been made of the tendency to cool the valve ports and spark-

P

ing plug bosses by direct pump supply, leaving the cooling of the cylinder barrels to thermo-syphon action. In all cases thermostats are fitted in the upper side of the water circulation system, so that the radiator is cut off during the engine warming up period. Of the available types the bellows pattern is now favoured; the by-pass system, shown at A in Fig. 272, page 362, is also employed.

One disadvantage of thermostats in very cold weather is that since the radiator is shut off from the rest of the cooling system during the starting and warming up period there is a risk of the water freezing in the radiator narrow water spaces before the thermostat opens; several cases of "frozen" radiators have been due to this cause.

The pressure-cooling method is finding adherents in both the car and commercial engine classes. In the more recent Vauxhall engines pressure-cooling has been adopted. The principle employed is to seal the cooling system by making the radiator filler cap joint air-tight and providing two valves, one opening to the atmosphere when the pressure within the cooling system exceeds a predetermined amount, and the other opening inwards against the pressure of a spring. The object of this valve is that when the water and vapour in the system cool down or condense and a partial vacuum is thereby produced the valve opens under atmospheric pressure and thus prevents excessive pressures on the cooling system components which, in the case of the radiator might cause its collapse. The advantages of pressure-cooling are that it permits the use of higher water temperatures which give better efficiencies and allows smaller radiators to be employed. since a reduced amount of cooling water is required.

The Cold-Filling Radiator.—When engines are used under very cold weather conditions, unless anti-freeze coolant is used, it is necessary to empty the water from the cooling system when the vehicle has to stand for any appreciable period. In order to prevent the water from freezing in the radiator block whilst filling the system, in the cold-filling radiator method, illustrated for the Austin vehicles in Fig. 324, the majority of the water poured in goes direct to the bottom tank, then across the tank to the outlet pipe where it is circulated by the pump on the closed thermostat system whilst the radiator is still filling. The engine is generally started *before re-filling* with water. The water entering the radiator during the filling process, *via* the direct filling pipe is warmed in contact with the bottom tank before reaching the radiator, so that there is no risk

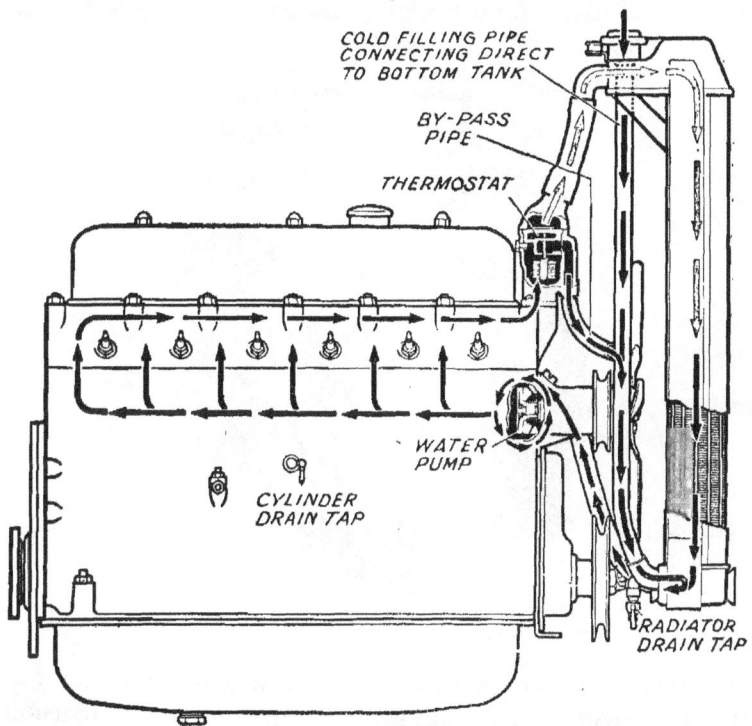

COLD FILLING PIPE
CONNECTING DIRECT
TO BOTTOM TANK

BY-PASS
PIPE

THERMOSTAT

WATER
PUMP

CYLINDER
DRAIN TAP

RADIATOR
DRAIN TAP

Fig. 324.—The Austin Commercial Vehicle Type Cold Filling Radiator.

of freezing in the latter. The cold filling pipe therefore obviates the necessity for any water to be passed down the cooling tubes or water spaces of the radiator. Thus, when the engine is first started this system allows warm water to be by-passed across the bottom tank instead of it going direct from the cylinder head to the water-pump as in previous types; it also fills the radiator with warm

P*

water. This system has no disadvantages when used in ordinary or warm climates.

Referring again to Fig. 324, the black arrows refer to the water circuit before the thermostat opens and the grey arrows to the isolated portion of the system whilst warming up.

Automatic Water Loss Compensator.—It is advantageous to prevent water or anti-freeze solution loss in the cooling system, due to expansion. The method used on Morris commercial vehicles is illustrated in

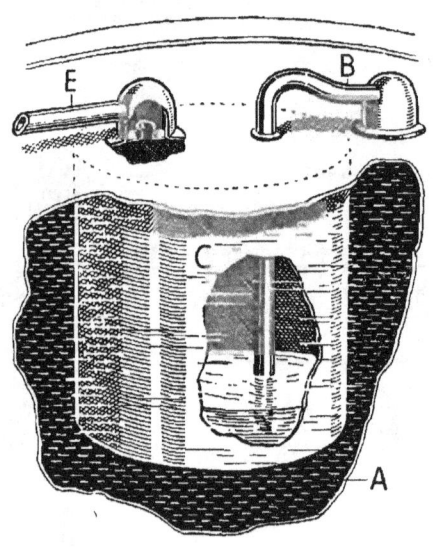

Fig. 325.—Morris Automatic Water Loss Compensator.

Fig. 325, the unit shown being arranged near the top of the upper header tank. C denotes the expansion chamber which is a small tank contained in the header tank A. As the water in the cooling system heats up it expands and the excess passes through the pipe B into the expansion chamber C. The inlet pipe extends to the bottom of C and on the cooling down of the water in the system a partial vacuum is created in the main header tank which causes any water in the expansion chamber to be drawn back through pipe B into the header tank, thus compensating for the loss of water in the main system due to expansion. An outlet pipe E open to the atmosphere prevents any pressure rise in the cooling system. The radiator cap must be screwed down so as to be airtight for this system to operate satisfactorily. The same result is obtained in Austin radiators by another device.

Fan and Pump Drives.—The almost universal method adopted in British mass-produced cars for driving both the cooling fan and also the water pump is by means of an endless rubber belt of vee-section. Usually, the dynamo is driven by the same belt. In many instances the fan shaft drives the impeller or water pump so that this arrangement dispenses with an additional belt pulley. The belt is tensioned by swinging the dynamo holding plates in a similar manner to that depicted in Fig. 263, page 354. Instead of hand-screwed greasers, however, a grease-gun nipple is provided.

In regard to cooling fans these are now mostly of the four-bladed pattern of the pressed steel type or as a single aluminium alloy casting; in the smallest sizes of engines, however, two-bladed fans are generally employed.

Petrol Injection Spark Ignition Engines.*—Development in this class of medium-compression engines in this country has been confined largely to aircraft engines, where maximum power per unit weight and minimum fuel consumption are of primary importance. There is little doubt that the well designed fuel-injection system gives higher volumetric efficiency, better mixture distribution to individual cylinders and lower fuel consumpton. For this reason the single-injection supercharger and the multiple point systems have largely superseded the aircraft carburettor method.

For automobile purposes, however, the injection system has certain drawbacks to offset the advantages previously mentioned. Thus, instead of a relatively

---

* A fuller illustrated account of these engines is given in " Carburettors and Fuel Systems " and " Aircraft Engines," Volume I (Chapman and Hall, Ltd., London).

simple carburettor of a well developed design, there is substituted a complicated and much more costly fuel injection pump made to high precision limits, together with individual fuel injection nozzles and fuel pipe lines to the cylinders ; these are, of course, additional to the ignition system, including the H.T. leads and sparking plugs. Moreover, the small size of petrol-injection pump required for the average car engine would require very careful design and construction to meter the extremely small quantities of petrol required—more particularly at slow running and cruising speeds.

Other disadvantages include the extra weight of the injection equipment and the additional items that would

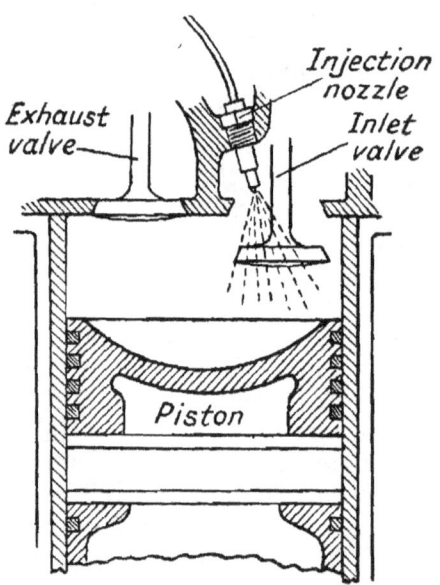

Fig. 326.—The Thornycroft Petrol-Injection Method.

require maintenance and servicing, since the fuel injection nozzles are prone to carbon up if of the direct cylinder injection pattern.

Altogether, the advantages at present are well in favour of the much more simple, light and relatively inexpensive carburettor. The additional power output of the petrol-injection system would be allowed for in the carburettor engine by a slight increase in cylinder dimensions.

It is, however, of interest to note that some promising performance results have been obtained from a converted Thornycroft high speed Diesel engine of the NR6 type. This Diesel engine has six cylinders of 4⅛ in. bore and 6 in. stroke, giving 481 cu. in. (7,880 c.c.) cylinder capacity. The compression ratio is 16:1, giving maximum compression and combustion pressures of 560 lb. and 960 lb. per sq. in. respectively. The output is 100 B.H.P. at 1,800 r.p.m. with a maximum B.M.E.P. of 105 lb. per sq. in., and torque, 337 lb. ft.

In order to operate on the petrol-injection method, the compression ratio was reduced to 6·92:1. The system used was to inject a 30° spray cone of petrol into the inlet valve port (Fig. 326) during the first 80° or so of the induction stroke. A special C.A.V. fuel injection pump was employed. The modified engine developed 150 B.H.P. at 1,900 r.p.m. and gave a maximum B.M.E.P. of 140 lb. per sq. in. and torque of 445 lb. ft. both at 800 r.p.m.

In regard to fuel consumptions the Diesel engine had an equivalent of 8·25 m.p.g. on the 22-tons Thornycroft vehicle, whereas the petrol-injection converted engine, under similar conditions, gave 6·2 m.p.g. As compared with a carburettor engine giving 100 ton miles per gallon the petrol-injection one had a figure of 130 and the Diesel vehicle 180—200.

The advantage of this petrol-injection engine is that it enables vehicles to be used in countries where Diesel fuel is not obtainable and to operate at much higher speeds; in the Thornycroft vehicles mentioned, the petrol-injection vehicle averaged, over a 20-mile course, 25·9 m.p.h. and the Diesel one, 16 m.p.h.

It may be mentioned that a commercial Waukesha engine is made that can readily be converted to run as a petrol engine, with a carburettor; a Diesel engine, or to operate on a hydrocarbon gas, such as methane or butane.

Two-cycle Engines.—There have been few developments in these engines in recent years, except in the small motor cycle and engine-assisted pedal cycle (autocycle) and the cheap French car classes. In the

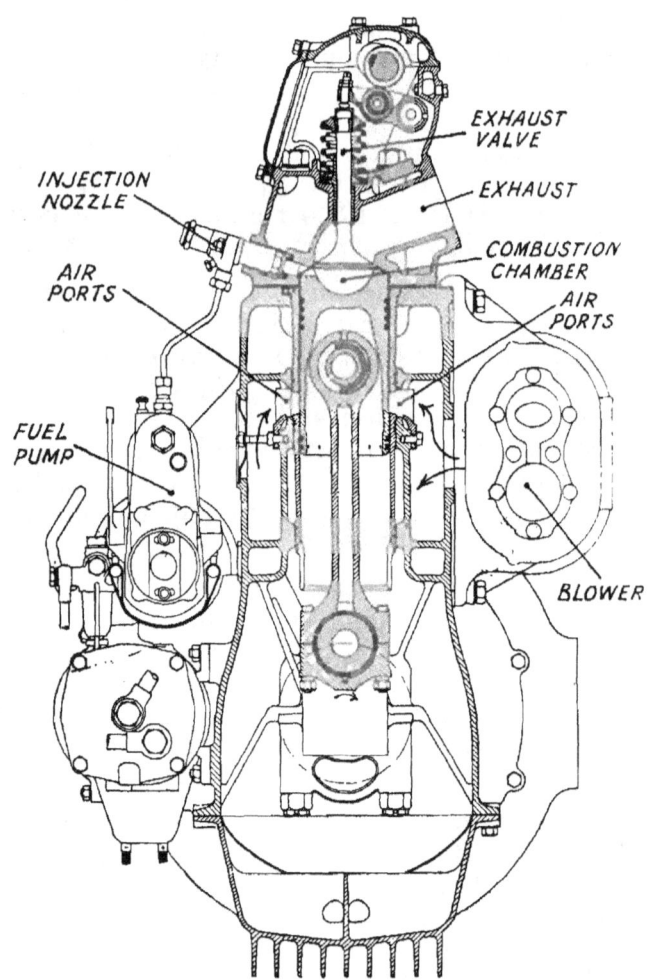

EXHAUST VALVE

INJECTION NOZZLE

EXHAUST

AIR PORTS

COMBUSTION CHAMBER

AIR PORTS

FUEL PUMP

BLOWER

Fig. 327.—The Foden Two-Cycle Air Scavenged Engine.

case of Diesel engines of the commercial vehicle class, notable examples of efficient and powerful two-cycle engines are the Foden six-cylinder two-cycle and the American G.M.C. ones which are made in several models. The Foden engine operates on the Kadency principle (Fig. 327) in which a low-pressure rotary type compressor delivers air to ports arranged at the bottoms of the cylinders, whilst the exhaust gases are ejected through poppet valves in the cylinder heads. The momentum effect of these gases is utilized to create a partial vacuum in the exhaust manifold, thus assisting the scavenging and cylinder charging by air under pressure from the piston uncovered ports at the bottom of the cylinder. It has been proved that for the same output the Kadency type operates at appreciably lower cylinder pressures and, owing to the efficient scavenging of the enables the efficiency to be maintained better at part consumption is lower and the power output, for a given engine size, is greater than for the ordinary two-cycle engine.

The G.M.C. engine made by General Motors Company, of America, operates on a somewhat similar principle, namely, with exhaust valves in the cylinder head and air ports in the cylinder which are uncovered by the descending piston; this method of scavenging and charging the cylinders is termed the *uniflow* one.

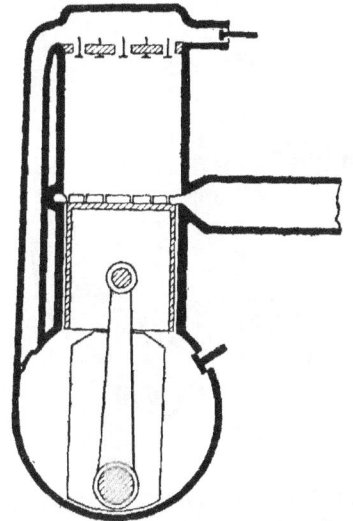

Fig. 328.—The Kylen Air-Scavenged Two-cycle Engine.

The Foden six-cylinder vertical engine, of 85 mm. bore and 120 mm. stroke, develops 126 B.H.P., which is equivalent to 31 B.H.P. per litre. This engine has two mechanically-operated exhaust valves per cylinder. An interesting engine, known as the Kylen, of Swedish origin, is provided with a number of poppet valves in the cylinder head opening inwards automatically (Fig. 328).* The exhaust gases are ejected through ports uncovered by the piston near the end of its stroke and

* Courtesy *The Automobile Engineer.*

the momentum of the escaping gases, owing to the special design of the exhaust system results in a negative pressure within the cylinder. As the piston in descending has compressed the air within the crankcase, and there is a direct connection between the latter and the space above the cylinder head, the effect is to cause the automatic air inlet-valves in the head to open and allow scavenging air to flow into the cylinder and thus clear out most of the remaining exhaust gases. The satisfactory operation of this system depends upon careful design of the exhaust system to give the correct pulsations resulting in a negative pressure when the exhaust ports are uncovered; also upon the correct

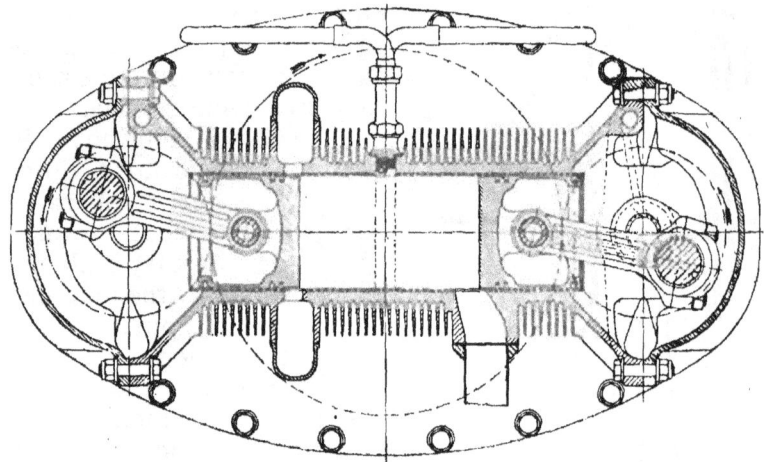

Fig. 329.—The McCulloch Opposed-Piston Two-Cycle Engine.

weights and spring pressures of the air inlet valves. Experimental engines operating on this system show better output and fuel economy than the normal types of three-port engines, with good flexibility. The best results appear to be obtained with petrol-injection into the cylinder and the use of compression ratios of 7:1 to 8:1.

Another novel design of two-cycle engine, which appears to bear some resemblance to the well-known Junker's opposed-piston two-cycle engines, is the American McCulloch 120 h.p. three-cylinder one, shown in Fig. 329. There are three double cylinders, each of the two opposed-piston pattern, one of which is illustrated. Each piston is provided with its own connecting-rod and, as shown, there are two crankshafts which rotate in the same direction and are geared to a common larger diameter gear wheel (shown dotted) the shaft of which is the power take-off one. The mixture ports are on the left, in Fig. 328, and the exhaust on the right. The right piston has an angular lead on the left one, so that it uncovers the exhaust ports before the inlet ports on the left are opened, thus minimising the escape of scavenging air or mixture into the exhaust when the inlet ports are uncovered by the piston. This type of engine can be made of compact form, its external shape being of oval character.

Sleeve Valve Engines.—Although eminently satisfactory in aircraft engines, giving better results than for poppet valves, the sleeve valve engine has made very little progress as a possible alternative to its poppet-valve rival in automobile engines. Typical aircraft engines using single sleeve valves include the Bristol 14-cylinder radial and 18-cylinder radial and Napier 24-cylinder opposed engines.

The fact that it is possible to obtain higher power outputs per litre with single sleeve valve engines, the absence of the multitudinous poppet valve gear components, the markedly silent operation and the much reduced maintenance attention with these engines, makes it somewhat difficult to understand why this type has not been adopted for the automobile engine in post-war years. The principal reason is, no doubt, the appreciably greater cost and also the difficulty of mass-production manufacture, which place it at an economic disadvantage to the present type, in the cheaper car category.

Certain detail improvements have been made in regard to the operation of the sliding sleeve and the port arrangements. Thus, in the more recent Schmid engine a thin sleeve (1 mm. thick) split along its entire length is employed. The sleeve has only an up-and-down motion and the ports are all above the highest piston position and, therefore, are not passed by the piston rings; further, no junk rings are needed for the cylinder head. The sleeve has a gap of 0·03 mm. only, so that leakage is negligible; it is important that the materials of the sleeve and cylinder should have the same expansion coefficients to preserve this gap. The sleeve is quite flexible and the engine design is such that speeds up to 10,000 r.p.m. are practicable.

Gas Turbines.—In its own sphere of application, namely, for aircraft engines and land and marine installations, the gas turbine has already made considerable progress. Thus, for aircraft use, where the backward momentum of the exhausted gases is utilized for propulsion purposes, or in the combined airscrew-jet propulsion application, the gas turbine* has clearly established itself as a rival to the more expensive and complicated high output (2,000—3,000 h.p.) reciprocating petrol engine. In these applications the utilization of the ejected gases for propulsion purposes is the principal factor and one which would be absent in land propelled vehicles.

For large locomotives, power stations and certain marine applications the gas turbine has proved a serious competitor to the steam turbine, but in order to achieve its best efficiencies it is necessary to depart from the simple type of gas turbine, i.e. compressor-combustion chamber-turbine unit, and to introduce multiple turbines, intercoolers, heat-exchangers, etc. These complicate and increase the weight and bulk of the gas turbine installation. As it stands the simple gas turbine (Fig. 330) is much less efficient than the petrol and Diesel engines and, in the smaller sizes that would be necessary for automobile engines, would be still less efficient, since there would be appreciable losses due to the scaling-down effect and little room for such refinements as heat-exchangers. Moreover, to maintain the efficient turbine and compressor blade speeds the smaller units would require to operate at considerably higher

* Vide "Modern Gas Turbines." A. W. Judge (Chapman and Hall, Ltd., London).

rotational speeds, namely, from 40,000 to 60,000 r.p.m. This would bring in complications in regard to reduction gears and controls. In order to minimise fuel costs, due to the lower efficiency of the gas turbine, the aircraft types operate on paraffin fuels. If these were used for road vehicles the exhaust odours might be objectionable.

On the other hand, the gas turbine undoubtedly possesses some marked advantages over the reciprocating petrol engine, which may be very briefly enumerated as follows: (1) Smoother running, due to absence of reciprocating parts; (2) Absence of vibration; (3) Much higher mechanical efficiency; (4) More compact shape; (5) Silent operation due to absence of reciprocating members and to the continuous exhaust; (6) Low internal pressures (these seldom exceed about 75 lb. per sq. in.); (7) Reduced maintenance costs due to simplicity of design; (8) Very much lower oil consumption, namely, about $\frac{1}{50}$ that of equivalent petrol engine; (9) Much lower weight, namely about one-third that of the equivalent petrol engine; (10) Absence of clutch or gearbox, except for reversing and emergency low gear. Whilst it is quite feasible to produce small gas turbines of output suitable for larger cars and commercial vehicles, and experimental engines have already been built here and in the U.S.A., the possibility of this type of prime mover replacing the petrol engine will depend upon a number of factors, such as the comparative fuel costs, output per unit weight, solution of the mechanical problems of transmission, road speed and idling speed controls, response to fuel regulation, etc.

Operating on cheaper fuels the lower thermal efficiency drawback may be overcome from the fuel cost viewpoint and with suitable combustion arrangements a free exhaust could be obtained, although the regulation of combustion over the wide range of operational speeds would require special consideration. The starting of a small gas turbine with its purely rotational units is a relatively simple matter compared with that of the petrol or Diesel engine but a more powerful starting motor and larger battery would be needed. Moreover, if the gas turbine could be run at a few thousand r.p.m. with little power output it would be unnecessary to use a clutch for vehicle starting purposes, since the power developed would probably be less than the transmission resistance losses; otherwise a simple design of transmission brake might be effective.

Referring to Fig. 330, this shows the working principle of the most successful type of gas turbine, known as the continuous pressure or constant combustion one. It consists of three main units, namely, the

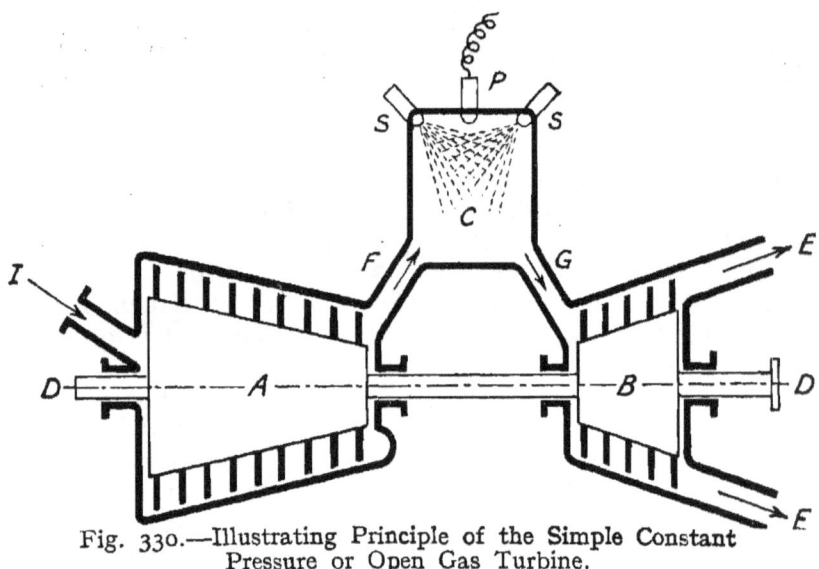

Fig. 330.—Illustrating Principle of the Simple Constant Pressure or Open Gas Turbine.

rotary (axial or centrifugal) type of air compressor A, combustion chamber C and turbine or power unit B. The compressor is driven by the turbine rotor, both being fitted to a common shaft DD, with power output flange at D on the right. The axial compressor shown draws air into its casing at I and compresses this in a series of stages, in each of which the volume is reduced and the pressure increased, until the final stage, shown at the right of A is reached, when the compressed air is delivered through the duct F into the combustion chamber C, where the liquid fuel is sprayed through fuel injection nozzles S. When starting up from rest the air-fuel mixture is ignited by the electric resistance plug P.

In aircraft gas turbines there are several small combustion chambers and their spraying nozzles, but the former are all connected to a common outlet leading to the turbine unit; this is adopted in order to give a compact unit. Referring to Fig. 330, the burnt gases under about the same pressure as at F are forced along the outlet duct G into the turbine casing and there impinge at very high velocity upon a series of rows of turbine blades of increasing diameter until, finally, the gases—having lost a considerable amount of their energy to the rotor B—pass out into the exhaust E.

The energy given to the turbine unit by the combustion gases is greater than that required to compress the air in A, so that the difference is available for useful power output purposes at A. In applying the gas turbine to road vehicles, instead of the arrangement shown in Fig. 330, the exhaust gases from the turbine B would be supplied to a second independent turbine which would provide the power output for propulsion purposes. Thus, the turbine B would be employed, only, to drive the compressor A. The independent turbine arrangement enables the efficiency to be maintained better at part loads and provides an easier method of control. For fuller particulars the reader should consult footnote reference given on page 460.

It should, however, be pointed out that of the power developed by the turbine unit B, only a small proportion —usually about one-fifth—is available at the output coupling D. Thus, in the case of a Swiss gas turbine locomotive the turbine unit developed about 12,000 B.H.P., of which about 10,000 B.H.P. was required to drive the compressor, leaving the difference, 2,000 B.H.P., as useful output. It will be thus evident that the turbine-compressor shaft must be designed so as to be strong enough to transmit considerably more power than the useful output.

INDEX